Ionic Soft Matter: Modern Trends in Theory and Applications

NATO Science Series

A Series presenting the results of scientific meetings supported under the NATO Science Programme.

The Series is published by IOS Press, Amsterdam, and Springer (formerly Kluwer Academic Publishers) in conjunction with the NATO Public Diplomacy Division.

Sub-Series

I. **Life and Behavioural Sciences**	IOS Press
II. **Mathematics, Physics and Chemistry**	Springer (formerly Kluwer Academic Publishers)
III. **Computer and Systems Science**	IOS Press
IV. **Earth and Environmental Sciences**	Springer (formerly Kluwer Academic Publishers)

The NATO Science Series continues the series of books published formerly as the NATO ASI Series.

The NATO Science Programme offers support for collaboration in civil science between scientists of countries of the Euro-Atlantic Partnership Council. The types of scientific meeting generally supported are "Advanced Study Institutes" and "Advanced Research Workshops", and the NATO Science Series collects together the results of these meetings. The meetings are co-organized by scientists from NATO countries and scientists from NATO's Partner countries — countries of the CIS and Central and Eastern Europe.

Advanced Study Institutes are high-level tutorial courses offering in-depth study of latest advances in a field.
Advanced Research Workshops are expert meetings aimed at critical assessment of a field, and identification of directions for future action.

As a consequence of the restructuring of the NATO Science Programme in 1999, the NATO Science Series was re-organized to the four sub-series noted above. Please consult the following web sites for information on previous volumes published in the Series.

http://www.nato.int/science
http://www.springeronline.com
http://www.iospress.nl

Series II: Mathematics, Physics and Chemistry –Vol. 206

Ionic Soft Matter: Modern Trends in Theory and Applications

edited by

Douglas Henderson
Brigham Young University,
Provo, U.S.A.

Myroslav Holovko
Institute for Condensed Matter Physics,
National Academy of Sciences of Ukraine,
Lviv, Ukraine

and

Andrij Trokhymchuk
Institute for Condensed Matter Physics,
National Academy of Sciences of Ukraine,
Lviv, Ukraine and
Brigham Young University,
Provo, U.S.A.

Published in cooperation with NATO Public Diplomacy Division

Proceedings of the NATO Advanced Research Workshop on
Ionic Soft Matter: Modern Trends in Theory and Applications
Lviv, Ukraine
14–17 April 2004

A C.I.P. Catalogue record for this book is available from the Library of Congress.

ISBN-10 1-4020-3663-9 (PB)
ISBN-13 978-1-4020-3663-7 (PB)
ISBN-10 1-4020-3662-0 (HB)
ISBN-13 978-1-4020-3662-0 (HB)
ISBN-10 1-4020-3659-0 (e-book)
ISBN-13 978-1-4020-3659-0 (e-book)

Published by Springer,
P.O. Box 17, 3300 AA Dordrecht, The Netherlands.

www.springeronline.com

Printed on acid-free paper

Printed in the Netherlands.

This book is dedicated to the people of the Ukraine whose devotion to freedom is undiminished in spite of a state organized famine, Nazi occupation and terror, Soviet totalitarianism, and an attempted fraudulent election.

Contents

Preface

Recently there have been profound developments in the understanding and interpretation of liquids and soft matter centered on constituents with short-range interactions. Ionic soft matter is a class of conventional condensed soft matter with prevailing contribution from electrostatics and, therefore, can be subject to possible long-range correlations among the components of the material and in many cases crucially affecting its physical properties. Among the most popular representatives of such a class of materials are natural and synthetic saline environments, like aqueous and non-aqueous electrolyte solutions and molten salts as well as variety of polyelectrolytes and colloidal suspensions. Equally well known are biological systems of proteins. All these systems are examples of soft matter strongly influenced, if not dominated, by long-range forces.

For more than half of century the classical theories by Debye and Hückel as well as by Derjaguin, Landau, Verwey and Owerbeek (DLVO) have been at the basis of theoretical physical chemistry and chemical engineering. The substantial progress in material science during last few decades as well as the advent of new instrumentation and computational techniques made it apparent that in many cases the classical theories break down. New types of interactions (e.g. hydrodynamic, entropic) have been discovered and a number of questions have arisen from theoretical and experimental studies. Many of these questions still do not have definite answers.

The Advanced Research Workshop (ARW) on ionic soft matter that has been held in Lviv (Ukraine) during April 14-17, 2004 under the NATO Physical and Engineering Science & Technology Programme, was dedicated to discuss some actual theoretical problems of ionic soft matter physics as well as some selected model applications of ionic media in the technological, environmental and biological processes that are governed by electrostatic forces. Our approach to gain these goals was to combine the recent advances in the understanding of ionic systems by covering various aspects from applied research to basic science. To succeed in this, the ARW brought together an international group of experts from different research sectors that includes mathematicians, physicists, chemists, biologists and engineers.

The choice of Lviv as a host site of the Workshop was not by an accident. The traditions of physical and mathematical science in this historic Western Ukrainian city are linked to the names of Marian Smoluchowski and Stefan Banach who lived and worked in Lviv in the early and middle part of the last century. The Lviv physical school in its present form was founded by Ihor Yukhnovskii and is known as the Institute for Condensed Matter Physics.

The lectures given during the ARW have provided an overview of the current state of the art on the interlaced research frontiers of ionic soft matter. Special attention has been focused on the fundamentals of ionic substances that include such important phenomena as association, adsorption, solvation, double layer, criticality, screening, thin film stability, collective dynamics and transport through biological and polymer membranes. Selected applied aspects of ionic soft matter have been discussed in details as well. These include: the behavior of ions at the ice surface and ice/water and lipid/water interfaces that is a fundamental process encountered in a wide range of biological systems and has relevance to problems in atmospheric chemistry; the proton transport properties in single pores polymer electrolyte membrane materials in order to understand those chemical and physical parameters which lead to an optimum performance in a low temperature fuel cell; the conformational, structural and dynamical features of the highly-charged macroions, polyelectrolyte chains and ionic surfactant complexes that are important for a wide range of applications ranging from drug formulations to coating materials; the charge and size selectivity of biological membranes that control a wide range biological functions; the sensitivity, specificity and the dynamic range of DNA array devices. Finally, some aspects related with the Chornobyl disaster have been included into the Workshop agenda.

We would like to thank all those who contributed to this book. We also thank all the participants who attended the ARW and contributed to the success of both the meeting and this volume. We thanks the local organizers for their hospitality and profound contribution in promoting an exciting and rewarding meeting. A special thanks to Olexandre Ivankiv for his initiative at the early beginning as well as Olesya Mryglod who helped us at the final stages to prepare this book. Finally, we gratefully acknowledge the financial support of NATO under the PST.ARW.979653 grant that made the Workshop and this volume possible.

DOUGLAS HENDERSON

MYROSLAV HOLOVKO

ANDRIJ TROKHYMCHUK

FIELD THEORETICAL APPROACH FOR IONIC SYSTEMS

An introduction of a classical field theory

D. di Caprio[1], J. Stafiej[2], J.P. Badiali [1]

[1] *Laboratoire d'Electrochimie et de Chimie Analytique,*
ENSCP et Université P. et M. Curie,
UMR 7575, 4. Place Jussieu, 75005 Paris, France

[2] *Institute of Physical Chemistry, Polish Academy of Sciences,*
ul. Kasprzaka 44/52, 01–224 Warsaw, Poland

Abstract   We introduce a classical field theory starting from the idea developed in quantum field theory. A partition function is written as a functional integral depending on real valued fields and the usual interaction potential. A hamiltonian is introduced in which we have a classical contribution and a term reminiscent of quantum physics; this term is treated as a local functional. Strictly speaking we have no proof that this hamiltonian is exact but when we use it with a simple renormalization procedure we recover the results of the usual liquid state theory. However our main goal in developing a field theory is to introduce a new point of view on the structure and thermodynamics of dense phases. Here we focus on the following ingredients: the local density fluctuations, the role of ideal entropy, the existence of Dyson-like equations. At a microscopic level we investigate the density fluctuations of a homogeneous Yukawa fluid and the structure of an ionic solution in contact with a wall. A field theory can also be used at a mesoscopic scale, the coupling constants represent the physics at a scale that we do not want to explore. As an example we analyze the differential capacitance of a metal/solution interface and we investigate a very general phenomena observed for many solvents and ionic solutions. Inspired by the methods of quantum field theory the theoretical results are described using a Lie group analysis.

Keywords: Field theory, ionic systems, local density fluctuations, Dyson-like equation, Yukawa fluids, Lie group analysis

1.   Introduction

Starting during the 19th century with the gas-kinetic theory, statistical mechanics has been very successfully developed during the 20th century. Today we have at our disposal efficient theoretical tools (diagrammatic expansion,

1

D. Henderson et al. (eds.), Ionic Soft Matter: Modern Trends in Theory and Applications, 1–17.
© 2005 *Springer. Printed in the Netherlands.*

integral equations, density functional theory, numerical simulations, etc.) to describe both thermodynamics and structure of a large class of systems. Based on a description at the level of the phase space and using the Gibbs ensemble approach, numerical methods are in principle exact provided the interaction potentials are known; they allow to predict macroscopic properties starting from an initial description at a microscopic level. The analytic approaches like integral equations are based on the same ingredients but introduce some approximations needed to solve the n-body problem. In practice, for instance in the case of ionic solutions in contact with a charged metal, we have to deal with many interaction potentials that are not very well known and the results can be extremely sensitive to the choice of these potentials. Therefore, for complicated systems we have to elaborate new methods or even new problematic as it will be illustrated below on an example.

In contrast with the case mentioned above, near a second order critical point the interaction potentials are irrelevant at least for the determination of critical exponents. In this domain new technics inspired by the methods of the renormalization group theory, developed first in quantum field theory (QFT), have been introduced. In QFT, the variables are primarily fields, not particles, we do not work in the phase space but in a functional space having an infinite degrees of freedom. This leads to the following question: is it possible to use in a wider context the methods inspired by QFT to describe dense phases in general? This has been done first for the so-called critical objects like polymers or liquid crystals for which there are long range correlations [1, 2]. Today, this kind of approach is widely used to describe the world at a mesoscopic scale, the main goal is to discover some general trends observed on complicated systems [3]. In this domain we use a Landau type Hamiltonian which is expected to be correct at the mesoscopic scale. This field of investigation is at the origin of what we call today the soft matter science.

In parallel there exist some attempts trying to introduce a field theory (FT) starting from the standard description in terms of phase space [4–6]. Of course, the best way to derive a FT for classical systems should consist in taking the classical limit of a QFT in the same way as the so called classical statistical mechanics is in fact the classical limit of a quantum approach. This limit is not so trivial and the Planck constant as well as the symmetry of wave functions survive in the classical domain (see for instance [7]). Here, we adopt a more pragmatic approach, assuming the existence of a FT we work in the spirit of QFT.

At this level we may question ourselves as to why to try to introduce a new approach of the liquid state? We do not expect this approach to be more efficient in order to describe one specific system, what we hope is that a new approach may introduce a new point of view and may bring new insights concerning the properties of dense phases. For instance, a FT forces us to focus on the local density fluctuations as the natural variables instead of focusing on the particles. In what follows we will show how the ideal entropy that is practically absent in the phase space description plays an important role here, it enters in a direct competition with the contribution from the interactions. Moreover, in the spirit of QFT some results can be deduced from symmetry arguments independently from the specific form of the interaction potentials. If we decide to work at a mesoscopic scale, QFT suggests we should introduce coupling constants that represent an underlying physics that we do not wish to take into account explicitly. The existence of a Lie group is of crucial importance in QFT, as we shall see below such groups may lead to a new analysis of very traditional problems in the domain of charged interfaces.

In the next section we first introduce the main ingredients of a FT by mimicking as far as possible what is done in QFT. To be illustrative, in Sec. 3 we show how it is possible to derive the virial expression for the pressure in a FT. In Sec. 4 we show that we have at our disposal some relations between field-field correlation functions that are obtained from symmetry arguments or from the fact that fields are dummy variables. In Sec. 5 on two examples, we derive some results starting from Dyson-like equations. Finally, in Sec. 6, leaving the purely microscopic level we introduce a model showing the existence of a demixion transition at a metal-solution interface using a mean field approximation. Brief conclusions are presented in Sec. 7.

2. General aspects of field theory

In QFT, fermions are primarily described by field operators $\hat{\rho}$, these operators can be expanded in terms of annihilation and creation operators that account for how we may generate or destroy particles. A given QFT is based on a lagrangian density that is the measure used to weight the fields contribution in a generic functional. In general the lagrangian is chosen in such a way that the mean field theory is meaningful by itself, *i.e.* it allows to recover well known or well accepted equations. The general results are obtained by performing a functional integration in which we give to the fields all possible shapes and magnitudes.

How to translate such a theory to describe a classical system? In order to do that we have to perform several steps. First, we have to choose the relevant fields and then to introduce the Hamiltonian that we will use for weighting these fields. This Hamiltonian must include both classical and quantum ingredients. Finally, we have to explain how to give a meaning to the functional integrals. In this section, to save notations we assume the existence of only one field.

2.1   *Hamiltonian*

Instead of field operators we consider a function $\rho(\mathbf{r})$ and, in the absence of annihilation/creation operators, we introduce particles by demanding that the integral of $\rho(\mathbf{r})$ over the volume of the sample, V, represents the number of particles associated with this field. Thus N is a functional of $\rho(\mathbf{r})$ defined as,

$$N = \int_V \rho(\mathbf{r})d\mathbf{r}. \tag{1}$$

We define a partition function, Θ, according to

$$\Theta = \int D\rho(\mathbf{r}) \, \exp\left\{-H[\rho(\mathbf{r})]\right\}, \tag{2}$$

in which the symbol $D\rho(\mathbf{r})$ means that we have to perform a functional integration over the fields and $H[\rho(\mathbf{r})]$ is a dimensionless functional that we use in order to weight the contribution of the field $\rho(\mathbf{r})$ to Θ. In principle, we may imagine many such functionals, however in order to reproduce the thermodynamics we use as a guideline the grand-canonical ensemble although we are not in a Gibbs ensemble description. Thus, $H[\rho(\mathbf{r})]$ will contain a term describing the field-field interaction that we write,

$$\frac{1}{2}\int_V \rho(\mathbf{r}_1)\rho(\mathbf{r}_2)v(\mathbf{r}_1,\mathbf{r}_2)d\mathbf{r}_1 d\mathbf{r}_2, \tag{3}$$

where $v(\mathbf{r}_1,\mathbf{r}_2)$ is the interaction potential in units $k_B T = 1/\beta$, where T is the temperature. Moreover we introduce in $H[\rho(\mathbf{r})]$ a term,

$$\mu \int_V \rho(\mathbf{r})dr, \tag{4}$$

were μ is the chemical potential in $k_B T$ units. The sum of (3) and (4) represents the classical ingredients that enter into $H[\rho(\mathbf{r})]$.

From the standard approach [7] we know that we have to introduce in the partition function two quantum ingredients. To a given field $\rho(\mathbf{r})$ we associate N entities via (1) and to each of them corresponds a given elementary cell in space. The volume of this cell results from the uncertainty relations; in the thermal regime this volume is Λ^3 where Λ is the thermal de Broglie wavelength [7]. In addition, we have to introduce the symmetry of the wave function. First, let us consider the case of an ideal gas. In a volume V the number of cells is V/Λ^3 and to each of the N particles corresponds a given cell. Quantum physics states that any permutation in the particle's positions over the whole volume cannot change the physics. Taking into account that we have to distribute N particles amongst V/Λ^3 cells, a simple combinatorics gives the number of states N_s corresponding to a field $\rho(\mathbf{r})$. Assuming that N and V/Λ^3 are very large we may use the Stirling formula to write $1/N_s$ as,

$$\frac{1}{N_s} = \exp -N \left[\ln \frac{N}{V} \Lambda^3 - 1 \right]. \tag{5}$$

This formula shows that the arguments of the exponential cannot be written $\int_V \rho(\mathbf{r})\phi(\mathbf{r})d\mathbf{r}$ *i.e.* as a local functional, this is clearly related to the fact that the N particles can be distributed on the overall volume. However, Eq. (5) has to be included in a functional integral that requires a given discretization and, as explained below in (2.2), there is a given scale a for which we may replace Eq. (5) by its local form that we write as,

$$\frac{1}{N_s} = \exp - \left[\int_V \rho(\mathbf{r}) \left[\ln \rho(\mathbf{r})\Lambda^3 - 1 \right] d\mathbf{r} \right]. \tag{6}$$

Since $1/N_s$ must weight the contribution of a field, in general, the final form of the Hamiltonian $H[\rho(\mathbf{r})]$ is,

$$H[\rho(\mathbf{r})] = \int_V \rho(\mathbf{r}) \left[\ln \rho(\mathbf{r})\Lambda^3 - 1 \right] d\mathbf{r} + \frac{1}{2} \int_V \rho(\mathbf{r}_1)\rho(\mathbf{r}_2)v(\mathbf{r}_1, \mathbf{r}_2)d\mathbf{r}_1 d\mathbf{r}_2$$
$$- \mu \int_V \rho(\mathbf{r})d\mathbf{r}. \tag{7}$$

In the spirit of QFT we have to check whether the mean field theory resulting from Eq. (2) and (7) is meaningful. The mean field approximation, $\bar{\rho}(\mathbf{r})$ of $\rho(\mathbf{r})$ is the solution of the equation $\delta H[\rho(\mathbf{r})]/\delta\rho(\mathbf{r}) = 0$. From Eq. (7) we derive,

$$\mu = \ln \bar{\rho}(\mathbf{r}_1)\Lambda^3 + \int_V \bar{\rho}(\mathbf{r}_2)v(\mathbf{r}_1, \mathbf{r}_2)d\mathbf{r}_2, \tag{8}$$

which is a well accepted approximate form for the chemical potential. For an homogeneous system and a radial potential $v(\mathbf{r}_1, \mathbf{r}_2) = v(\mathbf{r}_{12})$, $\bar{\rho}(\mathbf{r})$ can be written $\bar{\rho}$ and (8) is reduced to

$$\mu = \ln \bar{\rho}\Lambda^3 + \bar{\rho}v, \tag{9}$$

in which $v = \int_V v(\mathbf{r})d\mathbf{r}$.

2.2 Functional integral and renormalization

Written in terms of continuous functions, a functional integral represents a formal writing. The calculation of such an integral requires to perform a discretization of the space to reduce the integral to Riemann sums. We may imagine that the sampling points form the sites of a square lattice of lattice spacing a. In the volume V the number of such points is V/a^3 and the meaning of $D\rho(\mathbf{r})$ is $D\rho(\mathbf{r}) = \prod_{i=1,V/a^3} d\rho(\mathbf{r}_i)a^3$, the volume a^3 has been introduced in such a way that $D\rho(\mathbf{r})$ is dimensionless.

As an example of a functional integral calculation, we show in which conditions the local form introduced in Eq. (6) may reproduce the exact value of $\ln \Theta$ in the case of an ideal system. Starting from Eq. (9) we may relate μ to an intermediate density $\bar\rho$ via $\mu = \ln \bar\rho\Lambda^3$, when this definition of $\bar\rho$ is inserted in $H[\rho(\mathbf{r}]$ we have,

$$H[\rho(\mathbf{r}] = \int_V \rho(\mathbf{r}) \left[\ln \frac{\rho(\mathbf{r})}{\bar\rho} - 1 \right] d\mathbf{r} . \tag{10}$$

Due to the local form of $H[\rho(\mathbf{r}]$ and after simple mathematical transformations we may write $\ln \Theta$ according to

$$\ln \Theta = \frac{V}{a^3} \ln \left[\int_0^\infty \exp \left\{ x \left[\ln \frac{x}{\bar\rho a^3} - 1 \right] \right\} dx \right] . \tag{11}$$

It is easy to verify [8] that $\ln \Theta = V\bar\rho + O\left(\ln \bar\rho a^3 / \bar\rho a^3\right)$ and if $\bar\rho a^3$ is such that $\ln \bar\rho a^3 \ll \bar\rho a^3$, we have $\ln \Theta = V\bar\rho$. It is traditional to introduce the pressure, p, as the conjugate of the volume via $\ln \Theta = \beta p V$, after the functional integration we have now $\beta p V = \bar\rho V$ leading to the equation of state for the ideal system $\beta p = \bar\rho$ provided we identify $\bar\rho$ with the real density. Note that this identification also leads to the exact expression for the chemical potential since we started from $\mu = \ln \bar\rho\Lambda^3$. Thus the results of the ideal system are reobtained provided the discretization is not too fine ($\bar\rho a^3 \gg 1$).

If $v(\mathbf{r}_{12}) \neq 0$, we would like to prolongate the functional integration even for very small values $\bar\rho a^3$ in order to explore the potential for all values of the distance. Thus it seems difficult to treat on the same footing the quantum effects that determine the properties of the ideal gas and the interaction potential. The existence of such a kind of problem is not specific to FT. When taking the classical limit of the quantum statistical mechanics the results are exact whatever Λ in the case of ideal gas while in presence of a potential we have to take the limit $\Lambda \to 0$ and not to perform an averaging of the potential on the size of elementary cells. Here we have a similar problem and we decide to prolongate the ideal part whatever the value of a. Of course this may originate spurious contributions and even diverging terms. In what follows, to give a meaning to functional integrals we subtract all the diverging terms in a systematic way. When this renormalization procedure is associated with $H[\rho(\mathbf{r})]$ we have verified on a large number of examples that the results are equivalent to those obtained in the liquid state [9]. Of course, this is not a strict proof that the hamiltonian $H[\rho(\mathbf{r})]$ is the right one but this seems to us a highly likely conjecture.

In the next section we will show that p introduced in the case of the ideal system verifies the usual virial expression in presence of a potential.

3. Virial expression for the pressure

As in the previous section we introduce the pressure via $\ln \Theta = \beta p V$ and we consider the discrete form of Θ that we write according to

$$
\Theta = \int \prod_{i=1,V/a^3} d\rho(\mathbf{r}_i)a^3 \exp\left[-\frac{1}{2} \sum_{i,j} v(\mathbf{r}_{ij})\rho(\mathbf{r}_i)a^3\rho(\mathbf{r}_j)a^3 \right.
$$

$$
\left. + \sum_i \rho(\mathbf{r}_i)a^3 \left[\ln \frac{\rho(\mathbf{r}_i)a^3}{\exp\mu} \left(\frac{\Lambda}{a}\right)^3 - 1 \right] \right]. \quad (12)
$$

To change the volume in this expression we change a into λa, but keeping constant the number of points in the discretization. This leads to change the volume V to $V' = \lambda^3 V$ and the relation between Θ and p becomes $\ln \Theta[\lambda] = \lambda^3 \beta p V$. If we take $\lambda = 1 + \epsilon$ and $\epsilon \to 0$ we may expand the previous relation in term of ϵ leading to

$$
\beta p V = \frac{1}{3} \left[\frac{d\ln\Theta(\lambda)}{d\lambda} \right]_{\lambda=1}. \quad (13)
$$

In Eq. (12) each $\rho(\mathbf{r}_i)a^3$ becomes $\rho(\mathbf{r}_i)\lambda^3 a^3$, in the potential that we have assumed to be radial we have a dilation of the distance *i.e.* $v(r_{ij})$ is changed into $v(\lambda r_{ij})$ and an extra dependence in λ appears in the logarithm term. Now we may shift the field into $\rho(\mathbf{r}_i)\lambda^3$ that we may rename $\rho(\mathbf{r}_i)$, as it is a dummy variable. Since the number of points in the discretization is fixed the new system has the same number of particles as before dilation. The scaling parameter λ appears only in the pair potential and in the logarithm term. The derivative $d\ln\Theta(\lambda)/d\lambda$ can be easily performed and after returning to the continuous expression we obtain,

$$
\beta p = \frac{\langle N \rangle}{V} - \frac{1}{6} \int_V \langle \rho(\mathbf{r}_i)\rho(\mathbf{r}_j) \rangle r_{ij} \frac{dv(r_{ij})}{dr_{ij}} d\mathbf{r}_{ij}. \quad (14)
$$

This result is formally identical to the standard one. However it is obtained with a different Hamiltonian and the logarithm term has a crucial importance in the derivation of (14).

4.   Correlation functions

In Eq. (14) the brackets mean a statistical average and $\langle \rho(\mathbf{r}_i)\rho(\mathbf{r}_j)\rangle$ is a field-field or a density-density correlation function. In this section we first derive some properties of this correlation function.

4.1   *Ward-Takahashi identities*

In a previous paper [10], we have shown that the Wertheim, Lovett, Mou and Buff [11, 12] relations can be derived as consequences of the translational invariance of the interaction potentials. These relations are analog to the Ward-Takahashi identities obtained in field theory, which represent relations between the correlation functions and are consequences of invariance property of the Hamiltonian. Here we briefly give the derivation of these identities in the case of translational invariance for our Hamiltonian $H[\rho(\mathbf{r})]$. We assume the system is subject to an external potential $V_e(\mathbf{r})$, there is then a supplementary term in the Hamiltonian $\int V_e(\mathbf{r})\rho(\mathbf{r})d\mathbf{r}$. The external potential limits the system in space. The fields extend further but vanish at large distances, their precise behavior is not relevant as the average values of the fields will be vanishing in these regions as a consequence of the external potential. In this context, we can consider that the spatial invariance of the interaction potential leads to the invariance of the Hamiltonian when the fields are modified such that the new field $\rho'(\mathbf{r}) = \rho(\mathbf{r}+\varepsilon)$, this corresponds to a translation of constant infinitesimal vector ε. Following [10], using the translation invariance of the Hamiltonian in Θ leads to

$$\int V_e(\mathbf{r})\nabla\rho(\mathbf{r})d\mathbf{r} = 0. \tag{15}$$

We introduce $F[\rho]$ the Legendre transform of the grand-potential $\beta F[\rho] + \beta pV = \mu \int \rho(\mathbf{r})d\mathbf{r}$ and we have $\delta F[\rho]/\delta\rho(\mathbf{r}) = V_e(\mathbf{r})$. By functional differentiation we define the set of correlation functions,

$$C^{(n)}(\mathbf{r}_1,\ldots,\mathbf{r}_n) = \frac{\delta^n F[\rho]}{\delta\rho(\mathbf{r}_1)\ldots\delta\rho(\mathbf{r}_n)}, \tag{16}$$

which are the equivalent of the so called one particle irreducible correlations. We then have the following hierarchy of relations for $C^{(n)}$,

$$\sum_{i=1,n} \nabla_i C^{(n)}(\mathbf{r}_1,\ldots,\mathbf{r}_n) = \int d\mathbf{r}_{n+1}\,\nabla_{n+1}\rho(\mathbf{r}_{n+1})\,C^{(n)}(\mathbf{r}_1,\ldots,\mathbf{r}_{n+1}). \tag{17}$$

For $n = 1$, we have,

$$\nabla \mathcal{C}(\mathbf{r}_1) = \int d\mathbf{r}_2 \nabla_2 \rho(\mathbf{r}_2) \mathcal{C}^{(2)}(\mathbf{r}_1, \mathbf{r}_2) , \tag{18}$$

as for the derivation of the Baxter [13], Lebowitz [14] relations presented in [10], we can introduce and infinitesimally slowly varying external field and in this limiting case it is justified to use a local approximation, which writes,

$$\frac{d\mathcal{C}[\rho]}{d\rho} = \int d2 \, \mathcal{C}^{(2)}(1, 2) , \tag{19}$$

where $\mathcal{C}^{(2)}(1, 2)$ is the function for the homogeneous system. This is formally equivalent to the relations in [13, 14], except that $\mathcal{C}^{(2)}$ includes a local correlation term. These relations are a typical example of a hierarchy of relations for the correlation functions which are obtained in the spirit of the field theory, using symmetry properties of the system.

4.2   *Dyson-like equations*

In addition to the relations established above that are formally identical to those used in liquid state theory we may derive a set of relations based on the fact that the field is a dummy variable [15]. The physical meaning of this derivation is the following. Let us consider two fields $\rho_0(\mathbf{r})$ and $\rho_1(\mathbf{r}) = \rho_0(\mathbf{r}) + \delta\rho(\mathbf{r})$ in which $\delta\rho(\mathbf{r})$ is a given function. Independently of the field from which we start the calculation of the functional integral, the result must be the same. The physical reason of this is clear; since we must give to $\rho_0(\mathbf{r})$ all possible shapes and magnitudes we are certain that a transformation of $\rho_0(\mathbf{r})$ during the integration will be identical to $\rho_1(\mathbf{r})$ thus starting from $\rho_0(\mathbf{r})$ or $\rho_1(\mathbf{r})$ we must have the same result. To give a mathematical content to this we consider an arbitrary functional $A[\rho(\mathbf{r})]$. We must have $\langle A[\rho_0(\mathbf{r})] \rangle \equiv \langle A[\rho_1(\mathbf{r})] \rangle$. If we take $\rho_1(\mathbf{r}) = \rho_0(\mathbf{r}) + \delta\rho(\mathbf{r})$ and assume that $\delta\rho(\mathbf{r})$ is vanishingly small we may expand $\langle A[\rho_1(\mathbf{r})] \rangle$ in terms of $\delta\rho(\mathbf{r})$, the result is $\langle A[\rho_1(\mathbf{r})] \rangle = \langle A[\rho_0(\mathbf{r})] \rangle + \langle \delta A[\rho_1(\mathbf{r})] \rangle$ and we must have $\langle \delta A[\rho_1(\mathbf{r})] \rangle = 0$ whatever $\delta\rho(\mathbf{r})$. If we apply this result to the partition function we have immediately,

$$\left\langle \frac{\delta H[\rho(\mathbf{r})]}{\delta\rho(\mathbf{r})} \right\rangle = 0 . \tag{20}$$

This result illustrates quite well the difference between FT and the density functional theory. In both cases, we start from a given hamiltonian; it is a functional of the field in one case and a functional of the density in the other case. In a FT we have Eq. (20) while in a density functional approach the functional derivative of the functional free energy gives us directly the mean density. The two approaches coincide only if we consider a mean field approximation.

With relation (20) and the Hamiltonian given in Eq. (7) we have the following result,

$$\mu = \langle \ln \rho(\mathbf{r}_1)\Lambda^3 \rangle + \int_V \langle \rho(\mathbf{r}_2) \rangle \, v(|\mathbf{r}_1 - \mathbf{r}_2|) d\mathbf{r}_2 . \tag{21}$$

The meaning of this equation is quite clear. If we define the ideal entropy functional S_{ideal} according to

$$S_{\text{ideal}} = \int_V \rho(\mathbf{r}) \left[\ln \rho(\mathbf{r})\Lambda^3 - 1 \right] d\mathbf{r} . \tag{22}$$

The first term in (21) is, on average, the change of ideal entropy when we try to change the field at one point, i.e. $\delta S_{\text{ideal}}/\delta\rho(\mathbf{r})$. The chemical potential appears simply as the result of this change and the corresponding change in the energy. This illustrates the importance of the ideal entropy in a FT.

In the case of a homogeneous system for which the mean density is ρ we have,

$$\mu = \langle \ln \rho(\mathbf{r})\Lambda^3 \rangle + \rho v . \tag{23}$$

The exact relations (21) and (23) are quite simple but different from those corresponding to the mean field approach given in (8) and (9).

More generally for a functional $A[\rho(\mathbf{r})]$ we have,

$$\left\langle \frac{\delta A[\rho(\mathbf{r})]}{\delta\rho(\mathbf{r})} \right\rangle = \left\langle A[\rho(\mathbf{r})] \frac{\delta H[\rho(\mathbf{r})]}{\delta\rho(\mathbf{r})} \right\rangle . \tag{24}$$

5.   Example of the results

To illustrate a FT, we use the Dyson equation that is characteristic of fields and we focus on the local density fluctuations that is the most natural variable in this approach. Two systems are considered: a homogeneous Yukawa fluid and an ionic solution near a hard wall.

5.1   *Density fluctuations in a homogeneous Yukawa fluid*

From the intermediate density, $\bar{\rho}$, defined in Eq. (9) we write the field as $\rho(\mathbf{r}) = \bar{\rho} + \delta\rho(\mathbf{r})$. The exact density, ρ, is given by $\rho = \bar{\rho} + \langle \delta\rho(\mathbf{r}) \rangle$ in which $\langle \delta\rho(\mathbf{r}) \rangle$ results from the fluctuations around the mean field regime. By performing an expansion of $\ln \rho(\mathbf{r})$ in terms of $\delta\rho(\mathbf{r})$ in (21) we have the following result,

$$\frac{\langle \delta\rho(\mathbf{r}) \rangle}{\bar{\rho}} = \frac{1}{1 + \bar{\rho}\bar{v}} \left[\frac{1}{2} \frac{\langle [\delta\rho(\mathbf{r})]^2 \rangle}{\bar{\rho}^2} + \sum_{n>2} \frac{(-1)^n}{n} \frac{\langle [\delta\rho(\mathbf{r})]^n \rangle}{\bar{\rho}^n} \right] . \tag{25}$$

This equation is remarkable since it relates different powers of the density fluctuations, $\delta\rho(\mathbf{r})$, at the same point. To evaluate the r.h.s. of (25) we expand the hamiltonian (7) in terms of $\delta\rho(\mathbf{r})$, we have,

$$H[\rho(\mathbf{r})] = H_0[\bar{\rho}] + H_2[\delta\rho(\mathbf{r})] + \delta H[\delta\rho(\mathbf{r})] , \tag{26}$$

in which $H_2[\delta\rho(\mathbf{r}]$ is the part of $H[\rho(\mathbf{r})]$ that is quadratic in terms of $\delta\rho(\mathbf{r})$ and $\delta H[\delta\rho(\mathbf{r})]$ contains higher powers in $\delta\rho(\mathbf{r})$.

As an example for which analytical results are possible we consider the case of a repulsive Yukawa potential, $v(r) = \delta_Y \exp(-\alpha r)/r$, where δ_Y and α are respectively the amplitude and the inverse range of the potential. In the quadratic approximation we have the following result,

$$\langle \delta\rho \rangle = \frac{\delta_Y}{2} \frac{\bar{\rho}\bar{\alpha}}{1 + \bar{\rho}v} , \tag{27}$$

with $\bar{\alpha}^2 = \alpha^2(1 + \bar{\rho}v)$. The same result can be obtained using different routes [16]. From the partition function Θ we may calculate the pressure p and from standard thermodynamics we have $\langle \rho \rangle = [\partial p/\partial \mu]_{T,V}$ or, since p is calculated in terms of $\bar{\rho}$, we have also $\langle \rho \rangle = [\partial p/\partial \bar{\rho}]_{T,V} \, d\bar{\rho}/d\mu$. Another route consists in calculating directly $\langle \delta\rho(r) \rangle$ by expanding the hamiltonian. However, to obtain the same result the calculations require different orders of expansion of the term related to the ideal entropy. This illustrates the importance and the role of the ideal entropy in our approach and indicates that it is the precise form of this term that allows self consistent calculations.

5.2   *Inhomogeneous ionic solutions*

We consider the case of a 1:1 electrolyte near a hard wall, the solvent is a dielectric continuum that fills all the space and therefore there is no-image potential. Here we have to introduce two fields $\rho_i(\mathbf{r})$ with $i = +$ or $-$ or a combination of them such as the density $s(\mathbf{r}) = \rho_+(\mathbf{r}) + \rho_-(\mathbf{r})$ and the charge density $q(\mathbf{r}) = \rho_+(\mathbf{r}) - \rho_-(\mathbf{r})$. It is easy to write the Dyson-like equations [17] for this system. If the interaction potential is only the Coulomb potential, we have,

$$\left\langle \ln \left[\frac{\rho_i(\mathbf{r})}{\rho_\infty} \right] \right\rangle + \beta(i) \langle V(\mathbf{r}) - V_\infty \rangle = 0 , \tag{28}$$

where the bulk phase is characterized by the density ρ_∞ and by V_∞ the mean value of the electric potential $\langle V(\mathbf{r}) \rangle$ far from the wall. In terms of $s(\mathbf{r})$ and $q(\mathbf{r})$ we have,

$$2 \left\langle \ln \left[\frac{s(\mathbf{r}) + q(\mathbf{r})}{s(\mathbf{r}) - q(\mathbf{r})} \right] \right\rangle + \beta \langle V(\mathbf{r}) - V_\infty \rangle = 0, \qquad (29)$$

and

$$\left\langle \ln \left[\frac{(s(\mathbf{r}) + q(\mathbf{r}))(s(\mathbf{r}) - q(\mathbf{r}))}{\rho_\infty^2} \right] \right\rangle = 0. \qquad (30)$$

For a symmetric electrolyte near a neutral hard wall, there is no net charge and consequently no potential across the interface. In this case the mean field version of (28) that corresponds to the traditional Gouy-Chapman (GC) theory gives $\rho_i(\mathbf{r}) = \rho_\infty$ and thus $s(\mathbf{r}) = 2\rho_\infty \equiv s_\infty$. Of course, for symmetrical ions we must have $q(\mathbf{r}) = 0$ but there is a density profile. If we write $s(\mathbf{r}) = s_\infty + \delta s(\mathbf{r})$ and assume that $\delta s(\mathbf{r})$ and $q(\mathbf{r})$ are small we may expand relation (29) in terms of these quantities and we obtain,

$$\langle \delta s(\mathbf{r}) \rangle = \frac{\langle q(\mathbf{r})^2 \rangle - \langle q(\mathbf{r})^2 \rangle_\infty}{4\rho_\infty^2}. \qquad (31)$$

In [17] we have calculated $\langle \delta s(\mathbf{r}) \rangle$ and shown that the value of $\langle s(\mathbf{r}) \rangle$ on the wall verifies the contact theorem. The integral of $\langle s(\mathbf{r}) \rangle$ gives the adsorption Γ. From the calculation of the free energy we have obtained the surface tension and shown that the Gibbs isotherm leads to the same Γ as the one calculated via the density profile. Thus, as in the case of the homogeneous Yukawa fluid we have a totally self consistent calculation.

6.   Field theory at a mesoscopic scale

From an experimental point of view a metal/ionic solution interface is characterized by its differential capacitance C_d. We have at our disposal a set of experiments showing how C_d depends on electric variables (charge σ and/or electrode potential V), on the nature of the solvent, on the nature and the concentration of electrolyte, on the nature of the metal, etc. The variation of C_d with V exhibits extrema, in general, and a lot of papers have been devoted to explain their origin without clear success or convincing explanation. The difficulty of this task is related to the fact that we have to analyze the fourth derivative of the surface tension versus V.

Here we develop a problematic in which we do not focus on the explanation of a specific interface. We observe that for a lot of solvents and electrolytes there exists a part of the curve $C_d(\sigma)$ for which the specific nature of the salt disappears [18]. Depending on the nature of the solvent this phenomena appears more or less near the potential of zero charge and its extension in term of V is more or less pronounced, but the existence of this fact takes place whatever the shape of $C_d(\sigma)$ observed [18]. Thus we are in presence of a quite general process and we may expect that it must have a quite general interpretation not related, for example, to the specific nature of the solvent. Moreover since we do not want to reproduce the details of $C_d(\sigma)$ we can work at a scale for which we may forget some details of the microscopic scale. The phenomena that we want to investigate is clearly related to the charge distribution across the interface but it is obvious that the GC theory cannot explain the observed fact. Thus, our first task is to introduce a new hamiltonian in comparison of the one used above.

6.1 Hamiltonian

Starting from the ingredients already contained in Eq. (7) *i.e.* ideal entropy and Coulom interaction we may introduce extra local and/or non-local terms and possibly a new kind of interaction with the surface. The simplest approach consists in introducing a term related to the square gradients of the fields like in the Van der Waals theory and commonly used in the Landau-Ginzburg hamiltonian. In a mean field approach the functional derivative of this term leads to a Dirac distribution at the wall. This means that near the wall we have to introduce a more sophisticated description, we do it by assuming that near the wall this effect is in competition with an adsorbing potential localized on the wall and that we represent by a Dirac distribution. Thus, in addition to the elements in relation (7) we introduce an extra hamiltonian $\delta H[\rho_i(\mathbf{r})]$ containing a square gradient term and a non-electrostatic coupling with the wall, this non-electrostatic effect has an extremely short spatial extension compared to the Coulomb interaction and we reduce it to a Dirac distribution. Thus we have,

$$\delta H[\rho_i(\mathbf{r})] = \frac{1}{2}b \int \left[(\nabla s(\mathbf{r}))^2 \right] d\mathbf{r} + h \int s(\mathbf{r})\delta(z)d\mathbf{r}, \qquad (32)$$

where b is a coupling constant depending on the nature of the solution while h is connected to both the solution and the wall placed at $z = 0$.

6.2 Mean field description and Lie group

In the mean field description of a system defined by $H[\rho_i(\mathbf{r})] + \delta H[\rho_i(\mathbf{r})]$ the densities $\rho_i(\mathbf{r})$ verify two coupled non-linear differential equations. These profiles $\rho_i(\mathbf{r})$ depend on the boundary constraints of charge σ and adsorption h at the wall. The situation is equivalent to the GC theory except that in this case there is a single external parameter σ. However, we may present the well known GC theory under a different viewpoint. In the GC theory if we consider two arbitrary values of the charge, σ_1 and σ_2 the two solutions for the profiles are not independent. The profiles for the two different charges correspond to different sections of a one and only given charge profile, solution of the differential equations. From infinity (bulk) where the charge is zero, following this unique profile, the profile for each system reaches a point such that the integral of the charge profile corresponds to the adequate value of the charge. Changing the charge is then equivalent to changing the position of the wall on this unique charge profile. The consequence of this description is that different possible walls, i.e. charges can be parameterized by their corresponding position on the unique profile: $\sigma(z)$. The same kind of description can be generalized here, the description is more elaborate as two external parameters define the system.

Let us consider the plane of coordinates (h, σ), each point represents a given system. In this situation, there is not a unique profile representing all possible systems. However, as for the GC theory, amongst these systems there are infinite sets of them which are related to one another: they correspond to different possible truncations of the same profile. This way of classifying the systems using the profiles is intimately related to the structure of the differential equations. We show that these equations can be rewritten as four first order differential equations which have a one-parameter Lie Group structure.

A representation of the different profiles in the (h, σ) plane is given in Fig. 1a. The profiles start near the origin $\sigma = 0$, $h = 0$ which represents the neutral bulk system and end on a limiting curve. On each trajectory the charge and adsorption parameter can be written as functions of the corresponding position parameter on the profile: $\sigma(z)$, $h(z)$. A supplementary parameter is then sufficient to characterize the different trajectories.

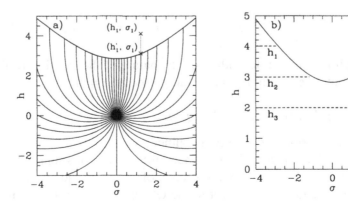

Figure 1.   Profiles in the h, σ plane (a). Three model systems, with different values of the adsorption parameter (b).

The phase space of *a priori* independent (h, σ) systems reveals interesting features. A direct consequence of the Lie group analysis is the existence of three invariants. One of them has a clear and familiar physical meaning, it corresponds to the pressure which is obviously constant throughout the system therefore along the profile. Another unexpected feature is the existence of the limiting curve parameterized by another invariant and where the Lie group symmetry breaks down. The upper part of the (h, σ) plane beyond this curve does not correspond to any physical solution for the profile. The profile for parameters (h_1, σ_1) in this region corresponds to systems with the same charge σ_1 but for a value h'_1 on the limiting curve. The density profiles for both ions $\rho_i(\mathbf{r})$ vanish when reaching this curve. The interest of the analysis above is that the topology of the phase space induced by this interpretation of the differential equations, has measurable implications.

Let us consider a real system, where the charge of the system is varied thus spanning the horizontal axis of the phase space. It is natural to assume that a given physical wall is characterized by a given value of the adsorption parameter h_i (Fig. 1b). We present three different systems. For h_1 and h_2, the two systems with the strongest repulsion at the wall, by varying σ the systems cross the limiting curve. In the solution, the values of h can only correspond to physically acceptable profiles, therefore we have to follow the limiting curve. The analysis then predicts that the behavior of the walls characterized by h_1 and h_2, will be the same when they follow the common limiting curve. Exploring a region of the charge on the wall, is typically what is done when differential capacitance is measured. We have simulated the general shape of differential curves for systems like h_1 and h_2. The behavior of the curves is exactly what is seen on a series of experimental differential capacity curves for various electrolytes [18].

By adding additional parameters we may fit specific experimental data but the main result is that an extremely simple hamiltonian can reproduced qualitatively this particular behavior. The interpretation is also very simple the fact that some parts of the differential capacitance curves are independent from the ions is naturally correlated to the fact that in this region the ion profiles vanish an thus the wall surface is insensitive to their specific nature.

A more sophisticated analysis shows that the crossing of the limiting curve can be considered as a kind of phase transition. It is characterized by features like the non-analytic behavior for the free energy, a Lie group symmetry breaking and, starting from the terminal wall position, the profiles $\rho_i(\mathbf{r})$ exhibit a non-analytic behavior with respect to the distance at the wall. Of course, it should be very interesting to investigate the previous hamiltonian going beyond a mean field description.

7.   Summary

We have elaborated a classical field theory by extending the ingredients of quantum field theory. In this approach the fields are real quantities related to the particles density by a functional. A hamiltonian has been introduced in which a local hamiltonian reproduces the symmetry of the wave function, it can be associated with the ideal entropy. In presence of an interaction potential we extend the validity of this functional to a very small scale producing diverging terms. We renormalize the diverging terms when they corresponds to ill defined mathematical expressions. Functional integral and renormalization procedure are used to derive some results in homogeneous Yukawa fluids or for ionic solutions near a wall. These results derived by using Dyson-like equations depend on the shape of the hamiltonian and their self-consistency tends to validate the form of the hamiltonian. We have extended the previous results to a mesoscopic scale for which the theoretical tools renew the analysis of classical problems at charged interfaces.

The extension of a FT to hard sphere systems has been performed by Caillol [6] recently showing that a FT may be applied in a large domain in chemical physics. Recently a first attempt to use a field theory for associated systems has been presented [19].

References

[1] de Gennes, P.G. (1985). *Scaling Concepts in Polymer Physics*. London, Ithaca: Cornell Univ. Press.

[2] des Cloizeaux, J., and Jannink, G. (1987). *Les Polymères en Solution: Leur Modélisation et Leur Structure*. France: Editeur Les Ulis, Les éditions de physique.

[3] Parisi, G. (1988). *Statistical Field Theory in Frontiers in Physics*. USA, MA: Addison-Wesley, Reading.

[4] Nabutovskii, V.M., Nemov, N.A., and Peisakhovic, Yu.G. *Mol. Phys.*, 1985, **54**, p. 979; *Sov. Phys. JETP*, 1980, **52**, p. 1111.

[5] Netz, R.R. *Eur. Phys. J. E*, 2001, **5**, p. 189.

[6] Caillol, J.M., and Raimbault, J.L. *J. Statist. Phys.*, 2001, **103**, p. 753–776; Raimbault, J.L., and Caillol, J.M. *J. Statist. Phys.*, 2001, **103**, p. 777–799.

[7] Hill, T.L. (1956). *Statistical Mechanics*. New York: Mc Graw Hill.

[8] Dieudonné, J.A. (1980). *Calcul Infinitésimal*, 2nd Edition, ed. Hermann. Paris.

[9] di Caprio, D., Badiali, J.P., and Stafiej, J., to be published

[10] di Caprio, D., Zhang, Q., and Badiali, J.P. *Phys. Rev. E.*, 1996, **53**, p. 2320–2325.

[11] Wertheim, M.S. *J. Chem. Phys*, 1976, **65**, p. 2377–2381.

[12] Lovett, R., Mou, C.Y., and Buff, F.P. *J. Chem. Phys.*,1976, **65**, p. 570.

[13] Baxter, R.J. *J. Chem. Phys.*, 1964, **41**, p. 553.

[14] Lebowitz, J.L., and Percus, J.K. *Phys. Rev.*, 1961, **122**, p. 1675; Lebowitz, J.L., Percus, J.K. *J. Math. Phys.*, 1963, 4, p. 116.

[15] Feynman, R.P., and Hibbs, A.R. (1965). *Quantum Mechanics and Path Integrals.* New York: McGraw Hill.

[16] di Caprio, D., Stafiej, J., and Badiali, J.P. *Mol. Phys.*, 2003, **101**, p. 2545–2558.

[17] di Caprio, D., Stafiej, J., and Badiali, J.P. *Mol. Phys.*, 2003, **101**, p. 3197–3202.

[18] Payne, R. (1970). *Advances in Electrochemistry and Electrochemical Engineering*, eds. P. Delahay and C.W. Tobias, vol. 7, p. 1–76. New York: Interscience Publishers.

[19] di Caprio, D., Holovko, M.F., and Badiali, J.P. *Condens. Matt. Phys.*, 2003, **6**, p. 693–702.

INDUCED CHARGE COMPUTATION METHOD

Application in Monte Carlo simulations of inhomogeneous dielectric systems

D. Boda [1], D. Gillespie[2], B. Eisenberg[2], W. Nonner[3], D. Henderson[4]

[1]*Department of Physical Chemistry,*
University of Veszprém,
H–8201 Veszprém, P.O. Box 158, Hungary

[2]*Department of Molecular Biophysics and Physiology,*
Rush University Medical Center,
Chicago, Illinois 60612, USA

[3]*Department of Physiology and Biophysics,*
University of Miami School of Medicine,
Miami, Florida 33101, USA

[4]*Department of Chemistry and Biochemistry,*
Brigham Young University,
Provo, Utah 84602, USA

Abstract   In a recent publication we have introduced the induced charge computation (ICC) method for the calculation of polarization charges induced on dielectric boundaries. It is based on the minimization of an appropriate functional. The resulting solution produces an integral equation that is transformed into a linear matrix equation after discretization. In this work, we discuss the effect of careful calculation of the matrix element and the potential by treating the polarization charges as constant surface charges over the various surface elements. The correct calculation of these quantities is especially important for curved surfaces where mutual polarization of neighboring surface elements is considerable. We report results for more complex geometries including dielectric spheres and an ion channel geometry with a surface of revolution.

Keywords: Dielectric boundaries, Monte Carlo simulation, polarization charge

1.   Introduction

Understanding the behavior of physical systems containing many degrees of freedom requires considerable computational time unless we treat a certain portion of the degrees of freedom as continuous response functions. These

D. Henderson et al. (eds.), Ionic Soft Matter: Modern Trends in Theory and Applications, 19–43.

response functions vary widely depending on the interactions they are intended to replace (electrostatic, dispersion, repulsion) and on the nonlocal nature of the medium (uniform homogeneous, inhomogeneous, anisotropic response functions). Because electrostatic interactions play a basic role in many fields such as molecular biology, quantum chemistry, electrochemistry, chemical engineering, and colloid chemistry (without any claim of this being a complete list), one of the most important response functions is the dielectric response of fast atomic and molecular motions. This procedure uses constitutive relations and macroscopic conservation laws and reduces to solving the Poisson's equation for source charges (which are the degrees of freedom treated explicitly) in an inhomogeneous dielectric medium characterized by a space dependent dielectric coefficient, $\varepsilon(\mathbf{r})$.

One field where such a procedure is commonly used is the study of solvation of molecules. The solute (which can be treated quantum mechanically) is hosted in a cavity built in a dielectric continuum representing the solvent. This approach is called the polarizable continuum model (PCM) [1–3] which is studied by the apparent surface charge (ASC) method. This approach determines the surface charge induced on the surface of the cavity so that the appropriate boundary conditions are fulfilled at the boundaries. Using Green's functions, the problem can be written in the integral equation formalism (IEF) [4–6], whose numerical solution results in a linear matrix equation. This matrix equation was first developed by Hoshi et al. [7] and named the boundary element method (BEM). Later Cammi and Tomasi [8, 9] adopted the method of Hoshi et al. and the group of Tomasi have developed several numerical procedures for the fast solution of the matrix equation using various iterative methods [10]. The PCM has been extended to cases where the molecule is hosted in anisotropic solvents, ionic solutions, at liquid interfaces and metals [11]. There are a large number of studies using various BEM procedures including those implementing a linear interpolation across each boundary element to improve accuracy [12–18].

Another field where dielectric continuum models are extensively used is the statistical mechanical study of many particle systems. In the past decades, computer simulations have become the most popular statistical mechanical tool. With the increasing power of computers, simulation of full atomistic models became possible. However, creating models of full atomic detail is still problematic from many reasons: (1) computer resources are still unsatisfactory to obtain simulation results for macroscopic quantities that can be related to experiments; (2) unknown microscopic structures; (3) uncertainties in developing intermolecular potentials (many-body correlations, quantum-corrections, potential parameter estimations). Therefore, creating continuum models, which process is sometimes called "coarse graining" in this field, is still necessary.

The most obvious example is the so called restricted primitive model (RPM) of electrolytes where the ions are represented as charges hard spheres, while the solvent is modelled as a dielectric continuum. Examples for inhomogeneous systems include electrochemical interfaces [19], semiconductor junctions [20], and cell membranes [21, 22].

A biologically crucial field where dielectric continuum models have a basic importance is ion channels embedded in the cell membrane. Several works have been published that use various methods to solve Poisson's equation for channel-like geometries. These include interpolation methods using lookup tables to store discretized Green's functions [23–27], BEM procedures [28–30], generalized multipolar basis-set expansion of the Green's function [31], and analytical solutions [30, 32–34]. The statistical mechanical methods also have a wide variety including the Poisson-Nernst-Planck (PNP) equation [32, 35, 36], Brownian dynamics (BD) simulations [23, 25, 26, 29, 31, 35, 36], the mean spherical approximation (MSA) for homogeneous fluids [37, 38], and Monte Carlo (MC) simulations [34, 39–42]. Special attention must be paid to the MC simulation works of Green and Lu [44–46] who developed a method to calculate dielectric boundary forces that practically equivalent to the BEM resulting in a matrix equation that corresponds to that developed by Hoshi et al. [7].

Ion channel studies motivated Allen et al. [47] who have developed an elegant variational formalism to compute polarization charges induced on dielectric interfaces. They solved the variational problem with a steepest descent method and applied their formulation in molecular dynamics (MD) simulations of water permeation through nanopores in a polarizable membrane [48–50]. Note that the functional chosen by Allen et al. [47] is not the only formalism that can be used. Polarization free energy functionals [51–53] are more appropriate for dynamical problems, such as macromolecule conformational changes and solvation [54–57].

In our two previous papers [58, 59] the induced charge computation (ICC) method for the calculation of polarization charges in an inhomogeneous dielectric system has been introduced. The method is based on the variational formulation of Allen et al. [47]. We have developed a different solution [58] for the minimization of their functional. The discretization of the resulting integral equation leads to a matrix equation that can be readily applied in a computer simulation. Further simplification of the method for the special cases where the source charges are point charges and the dielectric interfaces are sharp boundaries leads to the matrix equation of Hoshi et al. [7] and Green and Lu [44]. We have implemented the method and proved its usefulness in MC simulations of hard sphere ions in simple inhomogeneous dielectric geometries, where the dielectric boundaries are planar [58]. We have presented results for a system for which an analytic solution is available (one flat sharp boundary), and we have shown that our ICC method provides results in excellent agreement with the simulations using the exact solution (on the basis of electrostatic image charges). Furthermore, we have reported results for the more general case of two parallel flat sharp boundaries (slab geometry) where the matrix is not diagonal.

The generalization of ICC method allows to impose arbitrary boundary conditions on various boundaries in the system [59]. Furthermore, a numerical approach to calculate the surface integrals appearing in the matrix elements has also been introduced [59]. The correct calculation of these integrals is especially important in the case of curved surfaces; therefore, it is sometimes called "curvature correction". In this work, we present the method of [58] supplemented by the "curvature corrections" introduced in [59] and we report results for more complex geometries than those considered in [58]. We show that "curvature corrections" are important not only for curved surfaces but also for the slab geometry if the slab is thin. We study the potential of a charge in a dielectric sphere and show results for the effective interaction between two charged dielectric spheres. We also show some preliminary results for a calcium channel that have a rotational geometry.

2.   Method

2.1   *Variational formulation*

Let us consider a discrete or continuous charge distribution $\rho(\mathbf{r})$ confined to a domain $\mathcal{D}$ of volume V with a boundary S. For the geometries considered in this work, the potential can be chosen to be vanishingly small on S. This makes the elimination of some surface terms possible. This does not mean that we impose zero potential on the boundary of the system. Instead, we use infinite systems (simply assuming that the system is infinite, or, in the case of simulation, applying periodic boundary condition), or, in the case of a finite simulation cell, we assume that the cell is large enough that the potential is small on the boundary in average. Nevertheless, the method has been generalized by Nonner and Gillespie [59] that makes it possible to directly impose arbitrary boundary conditions on the boundaries of the system. This generalization is not considered in this paper.

In the case of an inhomogeneous dielectric, we separate the total charge distribution into two parts. We treat only a fraction of the charges explicitly, this part of the charge density is called source (or external, or free) charge distribution, and it is denoted by $\rho(\mathbf{r})$. The rest of the degrees of freedom, corresponding usually to fast atomic and molecular motions, is replaced by their dielectric response using a constitutive relation. In the presence of an external electric field the dielectric matter containing the charges other than the source charges is polarized. If the dielectric response is linear, the polarization

$P(\mathbf{r})$ produced by an electric field $\mathbf{E}(\mathbf{r})$ can be given as,

$$\mathbf{P}(\mathbf{r}) = \varepsilon_0 \chi(\mathbf{r}) \mathbf{E}(\mathbf{r}) = -\varepsilon_0 \chi(\mathbf{r}) \nabla \psi(\mathbf{r}), \tag{1}$$

where $\chi(\mathbf{r})$ is the dielectric susceptibility. The dielectric susceptibility is space-dependent that characterizes an inhomogeneous dielectric in the domain $\mathcal{D}$. This corresponds to a local relative permittivity $\varepsilon(\mathbf{r}) = 1 + \chi(\mathbf{r})$. In general, this is a tensor, but in this work we restrict ourselves to a scalar relative permittivity ($\mathbf{P}$ and $\mathbf{E}$ have the same directions). For the case of a bulk system, this quantity is called dielectric constant; in this work we use the term dielectric coefficient to emphasize that it is not constant in space. The polarization charge density induced by the source charge distribution is associated with the potential through,

$$\rho_{\text{pol}}(\mathbf{r}) = -\nabla \cdot \mathbf{P}(\mathbf{r}) = \varepsilon_0 \nabla \cdot [\chi(\mathbf{r}) \nabla \psi(\mathbf{r})]. \tag{2}$$

Introducing the normalized versions of the source and the polarization charge densities,

$$g(\mathbf{r}) = \frac{\rho(\mathbf{r})}{\varepsilon_0}, \tag{3}$$

and

$$h(\mathbf{r}) = \frac{\rho_{\text{pol}}(\mathbf{r})}{\varepsilon_0}, \tag{4}$$

Poisson's equation can be given as,

$$\nabla^2 \psi(\mathbf{r}) = -[g(\mathbf{r}) + h(\mathbf{r})], \tag{5}$$

and the corresponding functional [47] is,

$$I[\psi] = \frac{1}{2} \int_{\mathcal{D}} \nabla \psi \cdot \nabla \psi d\mathbf{r} - \int_{\mathcal{D}} \psi \left[g + \frac{1}{2} \nabla \cdot (\chi \nabla \psi) \right] d\mathbf{r}. \tag{6}$$

In order to express $I[\psi]$ as a functional of the polarization charge density, the potential is also split into the "external" and the "induced" parts which are expressed in terms of $g(\mathbf{r})$ and $h(\mathbf{r})$ with the help of the Green's function as,

$$\begin{aligned} \psi(\mathbf{r}) &= \psi_e(\mathbf{r}) + \psi_i(\mathbf{r}) = \\ &= \int_{\mathcal{D}} G(\mathbf{r} - \mathbf{r}') g(\mathbf{r}') d\mathbf{r}' + \int_{\mathcal{D}} G(\mathbf{r} - \mathbf{r}') h(\mathbf{r}') d\mathbf{r}', \end{aligned} \tag{7}$$

where the Green's function satisfies,

$$\nabla^2 G(\mathbf{r} - \mathbf{r}') = -\delta(\mathbf{r} - \mathbf{r}'), \tag{8}$$

with $\delta(\mathbf{r} - \mathbf{r}')$ being the Dirac delta function. Substituting Eq. (7) into Eq. (6), the functional can be given as a function of $g(\mathbf{r})$, $h(\mathbf{r})$, and $\psi_e(\mathbf{r})$, e. g. $I = I[g, h, \psi_e]$. The task is to determine the polarization charge density $h(\mathbf{r})$ for a given external charge density $g(\mathbf{r})$ that satisfies Eq. (5), or, equivalently, minimizes $I[g, h, \psi_e]$. Determining $h(\mathbf{r})$ for a fixed $g(\mathbf{r})$ is equivalent to minimizing the h-dependent part of the functional $I[g, h, \psi_e]$, which is denoted by $I_2[h]$. Allen et al. [47] showed that the extremum condition,

$$\frac{\delta I_2[h]}{\delta h(\mathbf{r})} = 0, \tag{9}$$

leads back to the constitutive relation in Eq. (1), that the extremum is a minimum, and that the value of $I[h]$ at the minimum reduces to minus the electrostatic energy. Allen *et al.* [47] solved the variational problem (after discretization) with a steepest descent method. In our previous work [58], we have proposed a different route that results in the integral equation,

$$h(\mathbf{r})\varepsilon(\mathbf{r}) - \int_{\mathcal{D}} h(\mathbf{r}')\nabla_{\mathbf{r}}\varepsilon(\mathbf{r}) \cdot \nabla_{\mathbf{r}}G(\mathbf{r} - \mathbf{r}')d\mathbf{r}' =$$
$$= \nabla_{\mathbf{r}}\varepsilon(\mathbf{r}) \cdot \nabla_{\mathbf{r}}\psi_e(\mathbf{r}) - [\varepsilon(\mathbf{r}) - 1] g(\mathbf{r}). \tag{10}$$

Discretizing the central equation (Eq. (10)) of the ICC method leads to a matrix equation. In this work, we focus on the case of sharp dielectric boundaries, therefore, we will show the details of discretization for that case. To our knowledge, this equation for the general case $\varepsilon(\mathbf{r})$ has not been derived previously. Recently, Frediani et al. [11] has reported an integral equation for the case of a molecule solvated at a diffuse interface between two fluid phases (liquid/liquid or liquid/vapor). Their interface is described by a dielectric profile $\varepsilon(z)$ that is a continuous function of the z coordinate, while the charge distribution of the molecule is placed in a cavity formed in the diffuse interface. An integral equation has been developed by Frediani et al. by finding the appropriate Green's function through certain integral operators. The resulting equation is similar to Eq. (10), but is less general.

2.2   *Discrete, point source charges*

When the source charges are point charges in discrete locations, the source charge density is given by

$$g(\mathbf{r}) = \frac{e}{\varepsilon_0} \sum_k z_k \delta(\mathbf{r} - \mathbf{r}_k), \tag{11}$$

where source charge k with valence z_k is located at $\mathbf{r}_k$ and e is the elementary charge. Because these charges have no surface area, the induced charge around each point charge k is localized at its position $\mathbf{r}_k$. Assuming that the dielectric is locally uniform around the source charge $z_k e$ with dielectric coefficient $\varepsilon(\mathbf{r}_k)$, the magnitude of the induced charge is $-z_k e[\varepsilon(\mathbf{r}_k) - 1]/\varepsilon(\mathbf{r}_k)$ [60]. Therefore, the contribution to h from the induced charges around the source charges is,

$$h'(\mathbf{r}) = -\frac{e}{\varepsilon_0} \sum_k z_k \frac{\varepsilon(\mathbf{r}_k) - 1}{\varepsilon(\mathbf{r}_k)} \, \delta(\mathbf{r} - \mathbf{r}_k). \tag{12}$$

Let us consider the source point charges $g(\mathbf{r})$ and the induced charges $h'(\mathbf{r})$ localized on them together; and let us denote the sum of these two terms by $g(\mathbf{r})$ hereafter. In other words, we move $h'(\mathbf{r})$ from the group of induced charges to the group of source charges. Accordingly, the electric potential raised by $h'(\mathbf{r})$ also contributes to the potential of the source point charges; the sum of the two potentials is denoted by ψ_e henceforth. It can be shown that for the resulting potential,

$$\nabla^2 \psi_e(\mathbf{r}) = -\left[g(\mathbf{r}) + h'(\mathbf{r})\right] = -\frac{e}{\varepsilon_0} \sum_k \frac{z_k}{\varepsilon(\mathbf{r}_k)} \, \delta(\mathbf{r} - \mathbf{r}_k), \tag{13}$$

applies from which this potential is expressed as,

$$\psi_e(\mathbf{r}) = \frac{e}{4\pi\varepsilon_0} \sum_k \frac{z_k}{\varepsilon(\mathbf{r}_k)|\mathbf{r} - \mathbf{r}_k|}. \tag{14}$$

The dielectric coefficient $\varepsilon(\mathbf{r}_k)$ at the place of charge k appears in the denominator. This corresponds to a dielectric screening that is conventionally used in various descriptions of electrolytes where the solvent is interpreted as a dielectric continuum background (for instance, in the Debye-Hückel theory, in the Gouy-Chapman theory, or in the RPM of electrolyte solutions).

Substituting the redefined $g(\mathbf{r})$ and $\psi_e(\mathbf{r})$ into Eq. (10), we obtain,

$$h(\mathbf{r})\varepsilon(\mathbf{r}) - \int_D h(\mathbf{r}')\nabla_{\mathbf{r}}\varepsilon(\mathbf{r}) \cdot \nabla_{\mathbf{r}}G(\mathbf{r} - \mathbf{r}')\mathrm{d}\mathbf{r}' = \nabla_{\mathbf{r}}\varepsilon(\mathbf{r}) \cdot \nabla_{\mathbf{r}}\psi_e(\mathbf{r}), \tag{15}$$

where $h(\mathbf{r})$ refers solely to the induced charges other than $h'(\mathbf{r})$.

It is important to note that if the source charge were an ion represented, for example, as a point charge at the center of a hard dielectric sphere (with a dielectric coefficient different from that of the surrounding medium), then the induced charge on the ion surface must also be determined (as in the case of ASC methods); this would be another contribution to $h(\mathbf{r})$ on the LHS of Eq. (10). Since in a computer simulation the source charges are moving compared to the dielectric surfaces, the geometry of the dielectric pattern would constantly change, which, in turn, would make the computation time consuming. Therefore, in the simulations, we assume that the interior of the ion has the same dielectric coefficient as the surrounding medium. For the same reason, we assume that the ions move in regions of constant dielectric coefficient, namely, they do not overlap with dielectric boundaries and they are not displaced from one dielectric domain to another. In this work, we will show results for the cases where the interior of an ion has different dielectric coefficient than that of the surrounding medium.

2.3   Sharp dielectric boundaries

In the special case of sharp dielectric boundaries the dielectrics is separated into domains of uniform dielectric coefficients. The dielectric coefficient jumps from one value to another along a boundary. Let us denote the surface of the dielectric boundaries by $\mathcal{B}$. Then the induced charge is a surface charge on the dielectric interfaces (if the induced charges around the source charges are not considered), and the volume integral in Eq. (15) becomes a surface integral over the surface $\mathcal{B}$,

$$h(\mathbf{s})\varepsilon(\mathbf{s}) - \Delta\varepsilon(\mathbf{s}) \int_{\mathcal{B}} h(\mathbf{s}')\nabla_{\mathbf{s}}G(\mathbf{s} - \mathbf{s}') \cdot \mathbf{n}(\mathbf{s})\mathrm{d}\mathbf{s}' = \Delta\varepsilon(\mathbf{s})\nabla\psi_{\mathrm{e}}(\mathbf{s}) \cdot \mathbf{n}(\mathbf{s}) , \quad (16)$$

where the dielectric coefficient $\varepsilon(\mathbf{s})$ on the boundary is defined to be the arithmetic mean of the two dielectric coefficients on each side of the boundary. Furthermore, the dielectric jump $\Delta\varepsilon(\mathbf{s})$ is the difference of the two dielectric coefficients on each side of the boundary in the direction of the local unit normal of the surface $\mathbf{n}(\mathbf{s})$.

To solve Eq. (16) numerically, the surface $\mathcal{B}$ must be discretized; specifically, each discrete surface element $\mathcal{B}_\alpha$ of $\mathcal{B}$ is characterized by its center-of-mass $\mathbf{s}_\alpha$, area a_α, unit normal $\mathbf{n}_\alpha = \mathbf{n}(\mathbf{s}_\alpha)$, value of the mean dielectric coefficient $\varepsilon_\alpha = \varepsilon(\mathbf{s}_\alpha)$, and value of the dielectric jump $\Delta\varepsilon_\alpha = \Delta\varepsilon(\mathbf{s}_\alpha)$. Due to the assumption of the vanishingly small potential on $\mathcal{S}$, the Green's function simply is,

$$G(\mathbf{s} - \mathbf{s}') = \frac{1}{4\pi|\mathbf{s} - \mathbf{s}'|}. \quad (17)$$

Also, since the density of the discrete point charges is given, the resulting potential ψ_{e} is known from Eq. (14). Of course, $\nabla_{\mathbf{s}}G(\mathbf{s} - \mathbf{s}')$ and $\nabla\psi_{\mathrm{e}}(\mathbf{s})$ are also known. The otherwise continuous surface charge density $h(\mathbf{s})$ is then discretized into certain constant values $h_\alpha = h(\mathbf{s}_\alpha)$ taken on the surface element $\mathcal{B}_\alpha$.

Eq. (16) is valid for any vector $\mathbf{s}$. If we rewrite the equation for the discrete values of the centers of the surface elements $\mathbf{s}_\alpha$, the surface integral in Eq. (16) becomes a sum of surface integrals over the surface elements $\mathcal{B}_\beta$. Using the assumption that h_β is constant over $\mathcal{B}_\beta$, we obtain for a given α that,

$$\sum_\beta h_\beta \left[\varepsilon_\beta \delta_{\alpha\beta} - \Delta\varepsilon_\alpha \int_{\mathcal{B}_\beta} \nabla_{\mathbf{s}_\alpha} G(\mathbf{s}_\alpha - \mathbf{s}_\beta) \cdot \mathbf{n}_\alpha d\mathbf{s}_\beta \right] =$$

$$= \Delta\varepsilon_\alpha \nabla \psi_e(\mathbf{s}_\alpha) \cdot \mathbf{n}_\alpha, \qquad (18)$$

where $\delta_{\alpha\beta}$ is the Kronecker δ. This can be rewritten in a matrix form as,

$$A\mathbf{h} = \mathbf{c}, \qquad (19)$$

where each element of the matrix A is given by the expression in square brackets,

$$A_{\alpha\beta} = \varepsilon_\beta \delta_{\alpha\beta} - \Delta\varepsilon_\alpha I_{\alpha\beta}, \qquad (20)$$

where $I_{\alpha\beta}$ denotes the integral in Eq. (18). Each element of the column vector $\mathbf{h}$ is given by h_β and each element of the column vector $\mathbf{c}$ is given by the right hand side of Eq. (18),

$$c_\alpha = \Delta\varepsilon_\alpha \nabla \psi_e(\mathbf{s}_\alpha) \cdot \mathbf{n}_\alpha. \qquad (21)$$

For the calculation of the integral $I_{\alpha\beta}$, there are two levels of approximation to interpret the charge on a surface element. The first route is to consider the surface charge as a point charge of magnitude $h_\beta a_\beta$ localized at $\mathbf{s}_\beta$. We call this approach the *point charge* (PC) approximation. In this case the integral reduces to an interaction term between point charges,

$$I_{\alpha\beta} = \nabla_{\mathbf{s}_\alpha} G(\mathbf{s}_\alpha - \mathbf{s}_\beta) \cdot \mathbf{n}_\alpha a_\beta \qquad (22)$$

for $\beta \neq \alpha$ and 0 otherwise (an induced point charge does not polarize itself). This approach was used in our previous work [58], where we tested the method on planar dielectric interfaces.

On the higher level of approximation, the induced charge on the βth surface element is considered as a surface charge with the constant value h_β. This approach, which we call the *surface charge* (SC) approximation, was introduced in [59]. The main difference is geometrical: the values of the integrals in Eq. (18) do not depend on charges, they depend only on the geometry of the dielectric boundary surfaces and the way they are discretized. The integral represents the polarization of the induced charges on the surface element $\mathcal{B}_\beta$ by the induced charge at $\mathbf{s}_\alpha$. Practically, this approach means that we have to evaluate the surface integrals $I_{\alpha\beta}$. This is a numerical problem; a procedure to solve it was proposed in [59]. We parametrize the two-dimensional surface $\mathcal{B}_\beta$ by two variables u and v. There is a transformation that converts u and v into Cartesian coordinates: $x = X(u,v)$, $y = Y(u,v)$, and $z = Z(u,v)$.

Therefore, both $G(\mathbf{s}_\alpha - \mathbf{s}_\beta)$ and $\mathbf{n}(\mathbf{s}_\alpha)$ can be expressed in terms of the new parameters: $G(u_\alpha, v_\alpha, u_\beta, v_\beta)$ and $\mathbf{n}(u_\alpha, v_\alpha)$. Let us discretize the βth surface element into subelements by evenly dividing the variables u_β and v_β into subintervals of widths Δu_β and Δv_β. Then, the integral can be calculated as,

$$I_{\alpha\beta} = \sum_k \sum_l \nabla_\alpha G(u_\alpha, v_\alpha, u_{\beta,k}, v_{\beta,l}) \cdot \mathbf{n}(u_\alpha, v_\alpha) \, a(u_{\beta,k}, v_{\beta,l}) \, \Delta u_\beta \, \Delta v_\beta \,,$$

(23)

where $a(u_{\beta,k}, v_{\beta,l})$ denotes the area element and $(u_{\beta,k}, v_{\beta,l})$ gives the center of the k, lth subelement of the βth surface element. Also, care must be taken to ensure that $u_\alpha \neq u_{\beta,k}$ and $v_\alpha \neq v_{\beta,l}$.

It has been realized before that the convergence of the results with increasing the resolution of the grid is poor using the PC approximation. A curvature correction was used by many workers [1, 3, 7, 23, 24], where an empirical parameter was built into the diagonal elements of the matrix (for reviews, see [18, 61]). Instead of this uncertain parameter, it is more appropriate to evaluate the integrals numerically. In our SC approximation, the integrals are computed by assuming that the surface charge is constant within a surface element. A higher level approximation was used in several works [12–18] where the surface charge within a surface element is interpolated from the surface charges in the neighboring tiles. Recently, Chen and Chipman [18] have proposed a linear interpolation method using a triangulation of the surface. The sample points ($\mathbf{s}_\alpha$) are the corners of the triangles, and the integrals over the triangles are evaluated by interpolating the surface charge inside a triangle from those at the three corners. The development of a similar interpolation method for our parametrization procedure using the u, v parameters is under way.

The evaluation of the integrals in the SC approximation is quite time consuming. Nevertheless, the speed of filling and inverting the matrix is not an issue in computer simulations if the geometry of the dielectric boundaries does not change during the simulation. Thus, the inverse of A (or any factorization of A) need only be computed *once* for a given geometry and dielectric profile. The calculation of $I_{\alpha\beta}$ by subdiscretizing the surface elements does not increase the *size* of the matrix, it only influences the *fill time* of the matrix, which is also performed once at the beginning of a simulation.

Note that two points situated on a planar surface do not polarize each other. Therefore, it is expected that for planar interfaces the method of the calculation of the integral $I_{\alpha\beta}$ has smaller effect on the accuracy of the method. Indeed, in our previous work [58], we have obtained satisfactory results for one planar and two parallel planar interfaces using the PC approximation. In the case of curved surfaces, the accurate calculation of $I_{\alpha\beta}$ is very important as we will show for spherical and cylindrical geometries in the Results section.

2.4   *Calculation of the energy*

In an MC simulation the essential quantity is the change in the energy of the
system in an MC step. The MC step is normally a stochastic particle displacement, chosen from a uniform distribution, but biased moves are also possible
[62–65]. Some details of our MC simulations will be given in the Results
section, here we give the procedure with which the electrostatic energy is calculated in the framework the ICC method. We assume that the source charges
are point charges and that the dielectric boundaries are sharp as described in
the preceding two subsections.

The electrostatic energy can be split into two parts. The source charge –
source charge interaction energy is,

$$W_e = \frac{1}{2} \sum_j e z_j \psi_e(\mathbf{r}_j) \,, \qquad (24)$$

where the electrostatic potential of the source point charges is given by Eq. (14).
The source charge – induced charge interaction is given by,

$$W_i = \frac{e}{8\pi} \sum_j z_j \int_B \frac{h(\mathbf{s})}{|\mathbf{s} - \mathbf{r}_j|} d\mathbf{s} \,. \qquad (25)$$

After discretization, as in the case of filling the matrix, we have two choices
in the calculation of the above integral. In the PC approximation the integral
becomes a sum of point charge – point charge interactions,

$$W_i = \frac{e}{8\pi} \sum_j z_j \sum_\beta \frac{h_\beta a_\beta}{|\mathbf{s}_\beta - \mathbf{r}_j|} \,. \qquad (26)$$

In the above equations, the indices j and β range over the source point charges
and the surface elements, respectively. In the SC approximation the integral
expresses a point charge – surface charge interaction,

$$W_i = \frac{e}{8\pi} \sum_j z_j \sum_\beta h_\beta \int_{\mathcal{B}_\beta} \frac{d\mathbf{s}}{|\mathbf{s} - \mathbf{r}_j|} \,. \qquad (27)$$

The integral in Eq. (27) is a geometrical term similarly to $I_{\alpha\beta}$ in Eq. (18) and
it can be calculated similarly by the parametrization procedure outlined in the
previous subsection. The problem with the SC route is that the integral has
to be recalculated for the jth ion every time when it is displaced in the MC
step. This would slow the simulations down considerably. Fortunately, as we
will see from the results shown in the next section, this integral is important
only if the charge is very close to the dielectric boundary. Since we simulate

ions with a finite diameter and the ion centers (where the source charges are
located) cannot approach the surface closer than the half of the ionic diameter,
this problem does not arise in our simulations.

In a usual MC simulation we use single particle movements, namely, only
one particle is displaced in an MC step. This fact and the linearity of the matrix
equation $A\mathbf{h} = \mathbf{c}$ make it possible to decrease the computation time of the energy change ΔW. This is because the distances between particles that are not
moved do not change in the MC step and the distances between these particles
and the surface elements are also unchanged. Storing the intermolecular distances in an array, the computational burden can be decreased by saving time
consuming square roots. The details can be found in [58].

It is important to note that in our simulations we assume that the ions do
not cross the dielectric boundaries. If an ion were to cross a dielectric boundary in the MC step, the energy of interaction of the ion with the two different
dielectrics (that corresponds to the solvation energies) must be included. Calculating such energies is difficult. Instead, empirical parameters describing
this energy difference might be included in the calculation as in our previous
work [34] where we studied the selectivity of a calcium channel where the dielectric coefficient of the selectivity filter was different from that of the bulk
electrolyte. Nevertheless, estimating such empirical parameters is still problematic. Therefore, as a first approximation, we avoid this problem and assume
that the dielectric boundaries act as hard walls and the ions cannot leave their
host dielectrics.

3.   Results

3.1   *Planar geometry*

In the planar geometry we consider a dielectric slab shown in Fig. 1. Two
semi-infinite dielectrics of dielectric coefficients ε_1 and ε_3 are separated by a
dielectric slab of thickness D and with a dielectric coefficient ε_2. The boundaries of the slab are flat, sharp, and parallel. This can be regarded as a simple model of a membrane. This case has been studied in our previous paper
[58] where MC simulation results have been shown for the distribution of hard
sphere ions around a slab. Nevertheless, in our previous work, we did not use
the SC approximation. In the following, we will show that it is necessary only
if the width of the slab is small compared to the width of the surface elements.

When we solve the problem numerically, the number of surface elements,
and consequently, the size of the dielectric boundary surfaces must be finite.
This is in accordance with the practice in a simulation, where the simulation
cell is also finite. To approximate an infinite system in a simulation, periodic
boundary conditions are applied in the x and y directions. The closest image
convention is used not only for the ionic distances but any distances between

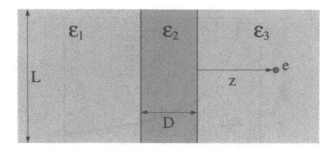

Figure 1. The slab geometry.

two physical points in the simulation cell including points on the boundary surfaces (s_β). Here we show results only for one source point charge. Nevertheless, to remain in connection with simulations, we use two square walls of dimensions $L \times L$, although using circular walls with the charge above the center is also possible. According to periodic boundary conditions, the grids on the surface of the two walls are constructed by evenly dividing them into square elements of width Δx (with $L/\Delta x$ being an integer). This means that the surface elements are parametrized by the x and y variables ($u = x$ and $v = y$). The value of Δx, namely the fineness of the grid, is a crucial point of the calculation (it was discussed in [58] in detail). Briefly, Δx must be small enough compared to the closest approach of a charge to the surface. Furthermore, the dimensions of the simulation cell have to be large enough, which is a usual criteria in computer simulations.

In this work, we concentrate on the issue of what happens if the width of the slab is small. In this case, the induced charges on the left wall are close to the induced charges on the right wall so the effect of their mutual polarization is large. Consequently, the accurate calculation of the $I_{\alpha\beta}$ integral, which represents this mutual polarization, is important. To investigate this effect, we calculate the polarization energy of a single point charge of magnitude e as a function of the distance of the charge from the slab. The polarization energy is the interaction energy between a charge and the polarization charges induced by this charge; this corresponds to W_i in Eq. (25) with $z_1 = 1$ ($j = 1$ because only one charge is present). This is a dominant term in the energy of a many particle system when an ion is close to the boundary, therefore, it is appropriate to test the accuracy of the method. Its precise value depends on the situation of the charge with respect to the surface element as we discussed in detail [58]. Here we calculate the polarization energy assuming that the charge is above the center of a square which corresponds to position 3 in Fig. 3 of [58]. The slab geometry has the advantage that analytical solutions are available in the

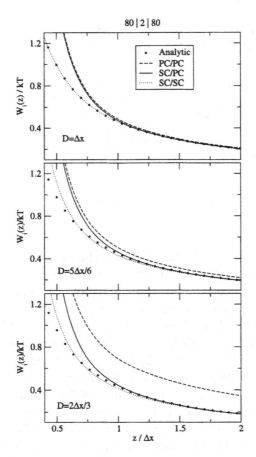

Figure 2.   The polarization energy W_i of a single charge of magnitude e as a function of the distance of the charge from the slab for different slab widths. The dielectric coefficients of the slab geometry are $\varepsilon_1 = 80; \varepsilon_2 = 2; \varepsilon_3 = 80$. The polarization energy is normalized by kT where $T = 300$ K. The ICC curves as obtained from different approaches (PC/PC, SC/PC, and SC/SC; the explanation of the abbreviations can be found in the main text) are compared to the analytical solution [66].

form of infinite series. Here, we use the formulas given by Allen and Hansen [66].

We consider three possibilities that differ whether we use the PC or the SC approximation in the calculation of $I_{\alpha\beta}$ and/or W_i :

(1) We can use Eq. (22) for the calculation of the matrix element and Eq. (26) for the calculation of W_i, namely, the PC approximation is used in both cases (PC/PC). This approach was used in [58].

(2) Using Eq. (23) for the calculation of the matrix element and Eq. (26) for the calculation of W_i means that the SC approximation is used only to fill the matrix (SC/PC).

(3) If we use Eq. (23) for the calculation of the matrix element and Eq. (27) for the calculation of W_i, the SC approximation is used in both cases (SC/SC). As mentioned before, using Eq. (27) significantly slows the simulation.

These notations will also be used in subsequent subsections.

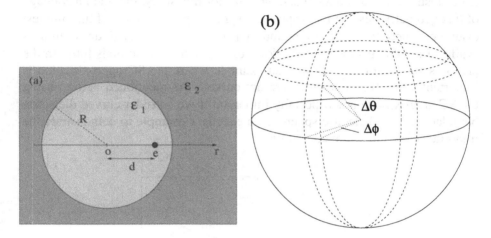

Figure 3. (a) The geometry of a dielectric sphere of dieletric coefficient ε_1 embedded in a mediun of dielectric coefficient ε_2 with a source point charge within the sphere. (b) The grid on the surface of a sphere is constructed by evenly dividing the spherical coordinates ϕ and θ into subintervals of widths $\Delta\phi$ and $\Delta\theta$.

Figure 2 shows the results for these cases for different values of D. For the width of walls the value $L = 20\Delta x$ is used. The temperature is $T = 300$ K in every calculation in this work. In the PC/PC case (dashed lines) the polarization energy is not reproduced even for large z when the width of the slab is small compared to Δx. In the case of $D = 2\Delta x/3$, the deviation is considerable. Using the SC approximation to fill the matrix (SC/PC), the magnitude of the energy for large z is recovered, but the behavior near the boundary is still bad (solid curves). Using the SC approximation to calculate the energy also (SC/SC), the behavior of the curves becomes much better even for small distances from the boundary. It is worth emphasizing that the SC/SC approach gives good results for $D = \Delta x$, which means that the errors due to different positions of the charge with respect to the grid [58] can be overcome (at least partly) by using the SC approximation to calculate W_i.

3.2   *One dielectric sphere*

A dielectric sphere of dielectric coefficient ε_1 embedded in an infinite dielectric of permittivity ε_2 is an important case from many points of view. The idea of a cavity formed in a dielectric is routinely used in the classical theories of the dielectric constant [67–69]. Such cavities are used in the studies of solvation of molecules in the framework of PCM [1–7] although the shape of the cavities mimic that of the molecule and are usually not spherical. Dielectric spheres are important in models of colloid particles, electrorheological fluids, and macromolecules just to mention a few. Of course, the ICC method is not restricted to a spherical sample, but, for this study, the main advantage of this geometry lies just in its spherical symmetry. This is one of the simplest examples where the dielectric boundary is curved; and an analytic solution is available for this geometry in the form of Legendre polynomials [60]. In the previous subsection, we showed an example where the SC approximation is important while the boundaries are not curved. As mentioned before, using the SC approximation is especially important if we consider curved dielectric boundaries. The dielectric sphere is an excellent example to demonstrate the importance of "curvature corrections".

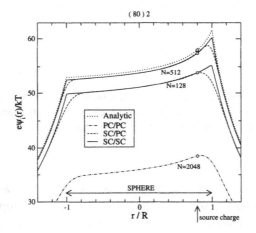

Figure 4.   The potential of the induced charges $e\psi_i(r)/kT$ as a function of r along the line crossing the center of the sphere and the source charge. The ICC curves as obtained from different approaches (PC/PC, SC/PC, and SC/SC) are compared to the analytical solution obtained from a series expansion of Legendre polynomials [60].

We will show results for the case where the dielectric coefficient is $\varepsilon_1 = 80$ inside, and $\varepsilon_2 = 2$ outside. For the reverse case, when the sourse charge is in the regime with the higher dielectric coefficient, we would obtain qualitatively similar, quantitatively even better converging results. Because of the spherical symmetry, the surface of the sphere is parametrized by the spherical coordinates ($u = \theta$ and $v = \phi$). The geometry is shown in Fig. 3a, while the corresponding grid is illustrated in Fig. 3b. A source charge is placed inside the sphere in a distance $d = 0.8R$ from the center of the sphere, where R is the radius of the sphere. Fig. 4 shows the potential produced by the polarization charges induced by the source charge on the surface of the sphere. The dimensionless potential $e\psi_i/kT$ is plotted as a function of r/R, where r is the distance from the center of the sphere along the line through the center and the source charge. The positive direction shows from the center to the source charge. In the case of the PC/PC approach there is a large difference between the analytic and the ICC solutions even for a very large number of surface elements $N = 2048$ (dot-dashed line). If we use the SC approximation to calculate the matrix elements (SC/PC), the agreement with the analytic solution becomes much better, but near the surface of the sphere the ICC curves (dashed lines) fail to reproduce the kinks in the analytic curve (dotted lines). Increasing the number of surface elements (from $N = 128$ to $N = 512$) the ICC curves approach the analytic curve. Using the SC/SC approach, the behavior of the ICC curves becomes satisfactory even in the vicinity of the boundary of the sphere (solid lines).

It is important to determine the centers of the surface elements correctly. If we parametrize the area element α by the variables u and v then the coordinates of the center are calculated as,

$$\bar{u}_\alpha = \frac{\iint_{B_\alpha} u\, a(u,v)\mathrm{d}u\mathrm{d}v}{\iint_{B_\alpha} a(u,v)\mathrm{d}u\mathrm{d}v}\,, \tag{28}$$

and similarly for $\bar{v}_\alpha$, where $\bar{s} = (\bar{u}_\alpha, \bar{v}_\alpha)$ is the center of the tile [59]. The centers of the subelements should be calculated similarly. If we determine the center by simply taking the centers of the Δu_α and Δv_α intervals instead of the above weighted average, we introduce a small error into our calculation. For the example of the one dielectric sphere, the polarization energy ($W_i/kT = e\psi_i(d)/kT$ denoted by open circles in Fig. 4) is 57.4 if we calculate $\bar{s}$ correctly from Eq. (28) (using 512 tiles), and 57.074 otherwise. The analytical value is 57.9616. Although it is small, this error might be important as in the case of the two spheres in the next subsection.

3.3 Two dielectric spheres

As mentioned before, the ions are modelled as point charges embedded in the center of a hard sphere where the sphere has the same dielectric coefficient as the surrounding medium. This approach replaces the polarization charges induced around the source charge from the surface of the sphere to the position of the source charge localized on it. The error made by this assumption was investigated by Allen and Hansen [70] who used their variational approach to calculate the effective interaction between charges within dielectric cavities. For a few special cases, they have developed solutions for the problems in the form of series expansions without using a grid. The general situation for spherical cavities can be seen in Fig. 5. Two dielectric spheres of radii R_1 and R_2 are immersed in a dielectric of dielectric coefficient ε_3. The dielectric coefficients in the spheres are ε_1 and ε_2. The distance of the centers of the spheres is r. Point charges of magnitudes q_1 and q_2 are placed at the centers of Sphere 1 and 2, respectively.

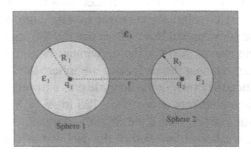

Figure 5. The geometry for two charges within dielectric spheres.

If the dielectric is homoegeneous ($\varepsilon_1 = \varepsilon_2 = \varepsilon_3$), the interaction potential between the charges is the Coulomb potential divided by ε_3. Introducing the effective dielectric coefficient $\varepsilon_{\text{eff}}(r)$ [70], the interaction potential can be written in the form of the Coulomb potential by

$$V_{\text{int}}(r) = \frac{q_1 q_2}{4\pi\varepsilon_0\, \varepsilon_{\text{eff}}(r)\, r} \tag{29}$$

for the case where dielectric boundaries are present. The interaction potential is defined by the difference,

$$V_{\text{int}}(r) = W(r) - W(\infty), \tag{30}$$

where $W(\infty)$ is the energy of the system when the spheres are infinitely far from each other. This term is the interaction energy between the ions and the polarization charges induced by the charges on their own spheres. This is an "intramolecular" term, while $V_{\text{int}}(r)$ is an "intermolecular" term which goes to zero increasing the distance of the spheres. For the former term an analytical expression exists,

$$W(\infty) = \frac{\varepsilon_1 - \varepsilon_3}{8\pi\varepsilon_0\varepsilon_1\varepsilon_3}\frac{q_1^2}{R_1} + \frac{\varepsilon_2 - \varepsilon_3}{8\pi\varepsilon_0\varepsilon_2\varepsilon_3}\frac{q_2^2}{R_2}. \tag{31}$$

Figure 6 shows results for the case $\varepsilon_1 = \varepsilon_2 = 1$, $\varepsilon_3 = 80$, $R = R_1 = R_2$ with charges of the opposite ($q_1 = -q_2 = e$, Fig. 6a) and the same ($q_1 = q_2 = e$, Fig. 6b) sign. This corresponds to two ions solvated in water, where the ions are modelled as vacuum spheres with point charges in the center. When the distance of the spheres is large, the solution of the problem converges to the case of point charges with the polarization charges localized on them (discussed in Sec. 2.2). In this limit, $\varepsilon_{\text{eff}}(r) \to \varepsilon_3$ for $r \to \infty$. We have calculated the energy of the two sphere systems for $r/R = 2.25, 2.5, 2.75$, and 3 using various numbers of surface elements ($N = 256, 484, 1024$, and 1600). The effective dielectric coefficient is plotted as a function of $1/N$ (open circles). In the limit of "infinitely fine" grid ($1/N \to 0$), the analytical results obtained from Figs. 5b and 5d of the paper of Allen and Hansen [70] are also shown (filled circles). Increasing the number of tiles, our results converge to the analytical data.

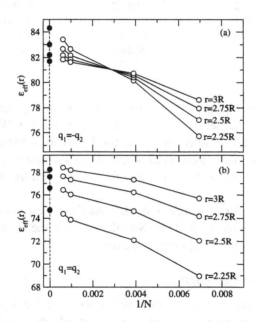

Figure 6. The effective dielectric coefficient as a function of $1/N$, where N is the number of tiles. Results are shown for charges of opposite (a) and equal (b) sign for different distances of the spheres. The filled circles are analytical results [70].

The convergence is quite slow, however. To explain this, the energies obtained for two specific cases with different fineness of grid are tabulated in Tab. 1. The small value of V_{int} is obtained from the sum of three large quantities: $V_{\text{int}} = W_e + W_i - W(\infty)$. For this reason, a small error in W_i or $W(\infty)$ causes a large error in V_{int}, and consequently, in ε_{eff}. The phenomenon of dielectric screening is well illustrated by these values. The large direct in-

Table 1. The various energies calculated for distance $r/R = 2.25$ for equal and opposite charges. The direct interaction between the charges is $W_e/kT = \pm 49.5140$. The analytical value for the energy of separated spheres is $W(\infty)/kT = -110.0139$ calculated from Eq. (31). The literature data for the effective dielectric coefficient is $\varepsilon_{\text{eff}} \sim 74.7$ and 84.3 for equal and opposite charges, respectively.

N	W_i/kT	$W(\infty)/kT$	V_{int}/kT	ε_{eff}
		$q_1 = q_2 = e$		
256	−158.8532	−110.0572	0.7181	68.95
484	−158.8739	−110.0466	0.6868	72.10
1024	−158.8799	−110.0362	0.6703	73.87
1600	−158.8801	−110.0318	0.6657	74.38
		$q_1 = -q_2 = e$		
256	−61.1972	−110.0572	−0.6540	75.71
484	−61.1504	−110.0466	−0.6176	80.15
1024	−61.1213	−110.0362	−0.5991	82.65
1600	−61.1114	−110.0318	−0.5937	83.4

teraction between the charges (W_e) is screened by a corresponding term raised by the polarization charges ($W_i - W(\infty)$). Note that for $W(\infty)$ the value obtained from our numerical method was used instead of the analytical value because the same numerical errors appear in both W_e and $W(\infty)$ which cancel each other. Using the analytical value for $W(\infty)$ results in a considerable overestimation of ε_{eff}.

In these two subsections, simple calculations for systems containing dielectric sphere(s) have been presented. Simulations for the distribution of ions around and within dielectric sphere(s) are under way.

3.4   *Ion channel geometry*

Ion channels are proteins with a narrow, highly selective pore in their center penetrating the cell membrane. They provide an effective and physiologically very important mechanism to control the pass of selected ions through the membrane. Calcium and sodium channels have crucial role in the function of nerve system, muscle contraction, and cell communication. The accurate 3D structure of these channels is unknown. However, the amino acid side chains that line the selectivity filter, which is a small and crowded region in the pore, are known. The selectivity of the channels is determined by these side chains which are 4 glutamate groups (EEEE locus) in the case of the calcium channel. Based on this minimal information a simple model for the calcium channel has been developed [37–43]. A small cylinder representing the filter is connected to baths that large enough for a bulk system to form in their middle. In this study, the filter is embedded in a membrane (Fig. 7a). The two baths are connected to the filter through two cone shaped vestibules at the entries of the filter (these vestibules were absent in our earlier studies). The system has a rotational symmetry: the surfaces in the simulation cell form a surface of revolution around the centerline of the pore.

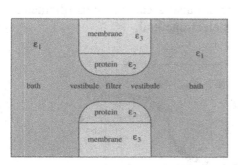 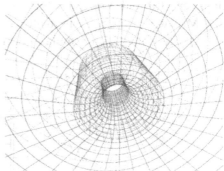

Figure 7. (a) The simulation cell for the ion channel geometry. (b) Illustration of the method to construct the grid on the dielectric boundary surfaces.

The dielectric coefficient was uniform ($\varepsilon = 80$) in our previous studies [39–42]. In this work, we present some MC results for the case where we allow various regions in the system to have different dielectric coefficient. Specifically, the membrane, the protein, and the electrolyte solution in which the ions move have dielectric coefficients $\varepsilon_3 = 2$, $\varepsilon_2 = 20$, and $\varepsilon_1 = 80$, respectively. The dielectric boundary surfaces that appear in the simulation cell have different geometries. Consequently, different u, v parameters are used for these surfaces. Due to the rotational symmetry, the variable ϕ is one of the parameters in all cases. The other parameter depends on the geometry of the various regions which are: (1) filter/protein and (2) protein/membrane boundaries – cylinders with z the second parameter, (3) vestibule/protein boundary – a cone shaped surface with a spherical curvature (an appropriate θ angle parametrize the curvature), and (4) bath/membrane boundary – planes perpendicular to the z-axis, the other parameter is r which is the distance from the rotational axis. The grid we constructed is illustrated in Fig. 7b; the grid is finest inside the filter (where important things happen) and it becomes coarser farther from the filter.

The whole simulation cell is confined by a large cylinder. The cell is finite, no periodic boundary conditions are applied. The radius of the cell is 60 Å, its length is 178.72 Å. These values are large enough so that the potential at the outer walls can be regarded as zero in average. The radius and length of the filter are 5 and 10 Å, respectively. The outer radius of the protein is 15 Å, its length (that equals the width of the membrane) is 30 Å. This means that the curvature radius of the vestibules is 10 Å.

There are 8 half charged oxygen ions (with diameters 2.8 Å) in the filter representing the 4 unprotonated structural groups of the EEEE locus of the calcium channel. These ions are assumed to be mobile inside the filter but they are restricted to the filter. There are 100 sodium and 100 chloride ions (diameters 1.9 and 3.62 Å) in the system in appropriate numbers to obtain a 0.1 M NaCl solution in the bath. There are also 2 calcium ions (diameter 1.98 Å). Thus, the whole simulation cell is electroneutral.

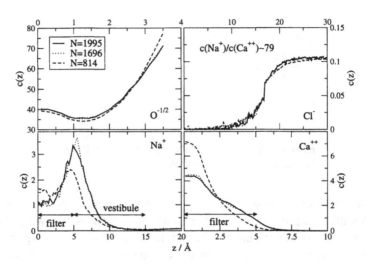

Figure 8. The concentration profiles for the various ionic species as obtained from MC simulations applying grids of different resolutions. The system is symmetric for the center plane of the membrane, so the results obtained for the left and right sides are averaged.

We present results of 3 simulations which differ from each other only in the resolution of the grid. The widths of the surface elements in the filter are 1.43, 1, and 0.91 Å, which correspond to total number of tiles $N = 814$, 1760, and 1995, respectively. The lengths of the simulations are 886 000, 339 000, and 308 000 MC cycles, respectively. Besides the usual particle displacements, biased particle exchanges between the channel and the baths were applied (for details, see [41, 42]).

The results for the density profiles (in mole/dm^3 unit) are shown in Fig. 8. Comparing the results to Fig. 7 of [41] (where a similar geometry was used but without dielectric boundaries), it can be seen that the presence of dielectric inhomogeneity has a large effect on the charge distribution in the channel. In our earlier study [41] both the Ca^{++} and Na$^+$ ions were concentrated in the center of the filter. On the contrary, our present results indicate that the Ca^{++}

ions tend to accumulate in the center of the filter, while the Na^+ ions are rather positioned at the entries of the filter and in the vestibules. This result clearly shows the importance of dielectric boundaries.

For this work, the important aspect of our simulation is the dependence of the results on the resolution of the grid. It is seen that our results converge as the number of surface elements is increased. For $N = 814$ the curves are quite different from those obtained for $N = 1696$ and 1995 although the qualitative behavior is the same. The curves obtained for the two better resolutions differ from each other only in minor details. This is in accordance with our earlier findings which showed that we can obtain accurate results from simulations if the dimensions of the surface elements are smaller than the closest approach of the ions to the surfaces. The SC/PC approximation was used in our simulations.

4.   Summary

We have presented the ICC method with developments in which the polarization charges are treated as surface charges with a constant value inside a boundary element (SC approximation). We have discussed the effect of using the SC approximation to calculate the matrix elements and polarization energy for various geometries. It was shown that this approach is important not only for curved dielectric boundaries but also for flat boundaries if the are close to each other. On the examples of dielectric spheres, it was shown that using the SC approximation (or "curvature corrections") is especially important for curved surfaces. If the geometry of dielectric interfaces does not change during a simulation, the method can efficiently been used in computer simulations, as our results for an ion channel geometry do show.

References

[1] Miertus, S., Scrocco, E., and Tomasi, J. *Chem. Phys.*, 1981, **55**, p. 117.

[2] Tomasi, J., and Persico, M. *Chem. Rev.*, 1994, **94**, p. 2027.

[3] Klamt, A., and Schüürmann, G. *J. Chem. Soc. Perkin. Trans.*, 1993, **2**, p. 799.

[4] Mennucci, B., Cancès, E., and Tomasi, J. *J. Phys. Chem. B*, 1997, **101**, p. 10506.

[5] Mennucci, B., and Cancès, E. *J. Math. Chem.*, 1998, **23**, p. 309.

[6] Mennucci, B., Cammi, R., and Tomasi, J. *J. Chem. Phys.*, 1998, **109**, p. 2798.

[7] Hoshi, H., Sakurai, M., Inoue, Y., and Chûjô, R. *J. Chem. Phys.*, 1987, **87**, p. 1107.

[8] Cammi, R., and Tomasi, J. *J. Chem. Phys.*, 1994, **100**, p. 7495.

[9] Cammi, R., and Tomasi, J. *J. Chem. Phys.*, 1994, **101**, p. 3888.

[10] Pomelli, C.S., Tomasi, J., and Barone, V. *Theor. Chem. Acc.*, 2001, **105**, p. 446, (and references therein).

[11] Frediani, L., Cammi, R., Corni, S., and Tomasi, J. *J. Chem. Phys.*, 2004, **120**, p. 3893, (and references therein).

[12] Zauhar, R.J., and Morgan, R.S. *J. Comput. Chem.*, 1988, **9**, p. 171.

[13] Yoon, B.J., and Lenhoff, A.M. *J. Comput. Chem.*, 1990, **11**, p. 1080.

[14] Juffer, A.H., Botta, E.F.F., van Keulen, B.A.M., van der Ploeg, A., and Berendsen, H.J.C. *J. Comp. Phys.*, 1991, **97**, p. 144.

[15] Fox, T., Rösch, N., and Zauhar, R.J. *J. Comput. Chem.*, 1993, **14**, p. 253.

[16] Liang, J., and Subramaniam, S. *Biophys. J.*, 1997, **73**, p. 1830.

[17] Bordner, A.J., and Huber, G.A. *J. Comput. Chem.* **24**, 353 (2003).

[18] Chen, F., and Chipman, D.M. *J. Chem. Phys.*, 2003, **119**, p. 10289.

[19] Torrie, G.M., Valleau, J.P., and Patey, G.N. *J. Chem. Phys.*, 1982, **76**, p. 4615.

[20] Jacoboni, C., and Luigi, P. (1989). *The Monte Carlo Method for Semiconductor Device Simulation*. New York: Springer Verlag.

[21] Parsegian, V.A. *Nature (Lond.)*, 1969, **221**, p. 844.

[22] Neumcke, B., and Läuger, P. *Biophys. J.*, 1969, **9**, p. 1160.

[23] Hoyles, M., Kuyucak, S., and Chung, S.-H. *Phys. Rev. E*, 1998, **58**, p. 3654.

[24] Hoyles, M., Kuyucak, S., and Chung, S.-H. *Comput. Phys. Commun.*, 1998, **115**, p. 45.

[25] Chung, S.-H., Hoyles, M., Allen, T., and Kuyucak, S. *Biophys. J.*, 1998, **75**, p. 793.

[26] Chung, S.-H., Allen, T., Hoyles, M., and Kuyucak, S. *Biophys. J.*, 1999, **77**, p. 2517.

[27] Graf, P., Nitzan, A., Kurnikova, M., and Coalson, R. *J. Phys. Chem. B*, 1997, **101**, p. 5239.

[28] Levitt, D.G. *Biophys. J.*, 1978, **22**, p. 209; *ibid*, 1978, **22**, p. 221.

[29] Corry, B., Allen, T., Kuyucak, S., and Chung, S.-H. *Biophys. J.*, 2001, **80**, p. 195.

[30] Baştuğ, T., and Kuyucak, S. *Biophys. J.*, 2003, **84**, p. 2871.

[31] Im, W., and Roux, B. *J. Chem. Phys.*, 2001, **115**, p. 4850.

[32] Schuss, Z., Nadler, B., and Eisenberg, R.S. *Phys. Rev. E*, 2001, **64**, p. 036116.

[33] Nadler, B., Hollerbach, U., and Eisenberg, R.S. *Phys. Rev. E*, 2003, **68**, p. 021905.

[34] Boda, D., Varga, T., Henderson, D., Busath, D.D., Nonner, W., Gillespie, D., and Eisenberg, B. *Mol. Sim.*, 2004, **30**, p. 89.

[35] Moy, G., Corry, B., Kuyucak, S., and Chung, S.-H. *Biophys. J.*, 2000, **78**, p. 2349.

[36] Corry, B., Kuyucak, S., and Chung, S.-H. *Biophys. J.*, 2000, **78**, p. 2364.

[37] Nonner, W., Catacuzzeno, L., and Eisenberg, B. *Biophys. J.*, 2000, **79**, p. 1976.

[38] Nonner, W., Gillespie, D., Henderson, D., and Eisenberg, B. *J. Phys. Chem. B*, 2001, **105**, p. 6427.

[39] Boda, D., Busath, D.D., Henderson, D., and Sokołowski, S. *J. Phys. Chem. B*, 2000, **104**, p. 8903.

[40] Boda, D., Henderson, D., and Busath, D.D. *J. Phys. Chem. B*, 2001, **105**, p. 11574.

[41] Boda, D., Henderson, D., and Busath, D.D. *Mol. Phys.*, 2002, **100**, p. 2361.

[42] Boda, D., Busath, D.D., Eisenberg, B., Henderson, D., and Nonner, W. *Phys. Chem. Chem. Phys.*, 2002, **4**, p. 5154.

[43] Gillespie, D., Nonner, W., and Eisenberg, R.S. *J. Phys.: Condens. Matt.*, 2002, **14**, p. 12129.

[44] Lu, J., and Green, M.E. *Progr. Colloid Polim. Sci.*, 1997, **103**, p. 121.

[45] Green, M.E., and Lu, J. *J. Phys. Chem. B*, 1997, **101**, p. 6512.

[46] Lu, J., and Green, M.E. *J. Phys. Chem. B*, 1999, **103**, p. 2776.

[47] Allen, R., Hansen, J.-P., and Melchionna, S. *Phys. Chem. Chem. Phys.*, 2001, **3**, p. 4177.

[48] Allen, R., Melchionna, S., and Hansen, J.-P. *Phys. Rev. Letters*, 2002, **89**, p. 175502.

[49] Allen, R., Melchionna, S., and Hansen, J.-P. *J. Phys.: Condens. Matter*, 2003, **15**, p. S297.

[50] Allen, R., Hansen, J.-P., and Melchionna, S. *J. Chem. Phys.*, 2003, **119**, p. 3905.

[51] Marcus, R.A. *J. Chem. Phys.*, 2956, **24**, p. 966; *ibid*, 1956, **24**, p. 979.

[52] Felderhof, B.U. *J. Chem. Phys.*, 1977, **67**, p. 493.

[53] Attard, P. *J. Chem. Phys.*, 2003, **119**, p. 1365.

[54] Löwen, H., Hansen, J.-P., and Madden, P.A. *J. Chem. Phys.*, 1993, **98**, p. 3275.

[55] York, D.M., and Karplus, M. *J. Phys. Chem. A*, 1999, **103**, p. 11060.

[56] Marchi, M., Borgis, D., Levy, N., and Ballone, P. *J. Chem. Phys.*, 2001, **114**, p. 437.

[57] HaDuong, T., Phan, S., Marchi, M., and Borgis, D. *J. Chem. Phys.*, 2002, **117**, p. 541.

[58] Boda, D., Gillespie, D., Nonner, W., Henderson, D., and Eisenberg, B. *Phys. Rev. E*, 2004, **69**, p. 046702.

[59] Nonner, W., and Gillespie, D. *Biophys. J.*, 2004, (in preparation).

[60] Jackson, J.D. (1999). *Classical Electrodynamics*, 3rd ed. New York: Wiley.

[61] Chipman, D.M., and Dupuis, M. *Theor. Chem. Acc.*, 2002, **107**, p. 90.

[62] Allen, M.P., and Tildesley, D.J. (1987). *Computer Simulation of Liquids*. New York: Oxford.

[63] Frenkel, D., and Smit, B. (1996). *Understanding Molecular Simulations*. San Diego: Academic Press.

[64] Sadus, R.J. (1999). *Molecular Simulation of Fluids; Theory, Algorithms, and Object-Orientation*. Amsterdam: Elsevier.

[65] Schlick, T. (2002). *Molecular Modeling and Simulation*. New York: Springer Verlag.

[66] Allen, R., and Hansen, J.-P. *Mol. Phys.*, 2003, **101**, p. 1575.

[67] Born, M. *Z. Phys.*, 1920, **1**, p. 45.

[68] Kirkwood, J.G. *J. Chem. Phys.*, 1934, **2**, p. 351.

[69] Onsager, L. *J. Am. Chem. Soc.*, 1936, **58**, p. 1586.

[70] Allen, R., and Hansen, J.-P. *J. Phys.: Condens. Matter*, 2002, **14**, p. 11981.

CONCEPT OF ION ASSOCIATION IN THE THEORY OF ELECTROLYTE SOLUTIONS

Application of the multidensity integral equation theory of associating fluids

M. Holovko

Institute for Condensed Matter Physics
of the National Academy of Sciences of Ukraine,
1 Svientsitskii Str., 79011 Lviv, Ukraine

Abstract   Analytical solution of the associative mean spherical approximation (AMSA) and the modified version of the mean spherical approximation – the mass action law (MSA-MAL) approach for ion and ion-dipole models are used to revise the concept of ion association in the theory of electrolyte solutions. In the considered approach in contrast to the traditional one both free and associated ion electrostatic contributions are taken into account and therefore the revised version of ion association concept is correct for weak and strong regimes of ion association. It is shown that AMSA theory is more preferable for the description of thermodynamic properties while the modified version of the MSA-MAL theory is more useful for the description of electrical properties. The capabilities of the developed approaches are illustrated by the description of thermodynamic and transport properties of electrolyte solutions in weakly polar solvents. The proposed theory is applied to explain the anomalous properties of electrical double layer in a low temperature region and for the treatment of the effect of electrolyte on the rate of intramolecular electron transfer. The revised concept of ion association is also used to describe the concentration dependence of dielectric constant in electrolyte solutions.

Keywords: Electrolyte solutions, ion association, associative mean spherical approximation, electron transfer, electrical double layer, ion-dipole model, dielectric constant

1.   Introduction

A concept of ion association in electrolyte solutions was introduced about eighty years ago by Bjerrum [1] in order to improve the Debye-Hückel (DH) theory [2]. In accordance with this concept an electrolyte solution is considered to be a mixture of free ions and ion clusters (usually ion pairs and some-

D. Henderson et al. (eds.), Ionic Soft Matter: Modern Trends in Theory and Applications, 45–81.
© 2005 *Springer. Printed in the Netherlands.*

times also trimers, tetramers or more complex clusters), which are assumed to take part in chemical equilibrium according to the corresponding mass action law(MAL). The phenomenon of ion association has an important effect on thermodynamic, transport, dielectric and other properties of electrolyte solutions [3]. For example, the formation of ion pairs in a symmetrical electrolyte cancels the charges of the ions by forming neutral polar dumbbells. These dumbbells do not contribute to the electric conductivity but its contribution can be important for the description of thermodynamic properties. In the DH theory ions are treated as point particles. Since the point ion pairs do not give any electrostatic contribution to thermodynamic properties in the original Bjerrum approach, the DH theory was modified by correcting the ion concentration in electrostatic contribution using only the concentrations of free ions obtained from the mass action law (MAL). The second modification concerns the ideal terms of thermodynamic properties in which a certain number of ions has been substituted by the number of ion pairs.

A natural extension of the DH theory to the non-point ions was given in the framework of the mean spherical approximation (MSA) [4, 5]. However the direct combination of the Bjerrum approach and MSA theory (MSA-MAL approximation [3, 6]) is correct only in the regime of weak association where the contributions of electrostatic interaction from ion clusters can be neglected. At the increasing association, the electrostatic contributions of the ion clusters increase and are not negligible for non-point ion clusters. More correct extension of MSA theory to the ion association is due to the associative mean spherical approximation (AMSA) [7, 8], which is based on the modern theory of associating fluids [9–12]. This approach coincides with the traditional MSA-MAL approach in the case of weak association. The AMSA is also correct in the regime of strong association due to the electrostatic contributions from ion pairs being taken into account. Moreover, the AMSA theory was modified in order to take into account the formation of ion trimers and tetramers [13, 14]. The AMSA theory was also generalized for the ion-dipole fluids [15, 16] that makes it possible to describe solvation effects together with association phenomena.

The DH and MSA theory, that are linear in charge can be considered in the framework of linearized Poisson-Boltzmann (PB) equation. The concept of ion association entails nonlinearity in the treatment of electrostatic interactions by the formulation of appropriate thermodynamic equilibrium constants between free ions and ion clusters [14]. In general, this formulation can be considered as the division of ion-ion interaction potentials into an associative part responsible for the ion association, and nonassociative part which is more or less arbitrary. In order to optimize this division in the framework of associative hypernetted chain approximation (AHNC), the division of energy and distance were considered [17] with the parameters calculated from the condition of stationarity of the system free energy. In a similar way the ion association was also treated using the AMSA [18]. In AMSA description of ion association, the association constant can be also given in the Ebeling form [19, 20] which fixes the value of the electrostatic part of the second virial coefficient and in general can be considered as the way of summation of optimized cluster expansions for the fluids with a strongly attractive interaction.

In this chapter some aspects of the present state of the concept of ion association in the theory of electrolyte solutions will be reviewed. For simplification our consideration will be restricted to a symmetrical electrolyte. It will be demonstrated that the concept of ion association is useful not only to describe such properties as osmotic and activity coefficients, electroconductivity and dielectric constant of nonaqueous electrolyte solutions, which traditionally are explained using the ion association ideas, but also for the treatment of electrolyte contributions to the intramolecular electron transfer in weakly polar solvents [21, 22] and for the interpretation of specific anomalous properties of electrical double layer in low temperature region [23, 24]. The majority of these properties can be described within the McMillan-Mayer or ion approach when the solvent is considered as a dielectric continuum and only ions are treated explicitly. However, the description of dielectric properties also requires the solvent molecules being explicitly taken into account which can be done at the Born-Oppenheimer or ion-molecular approach. This approach also leads to the correct description of different solvation effects. We should also note that effects of ion association require a different treatment of the thermodynamic and electrical properties. For the thermodynamic properties such as the osmotic and activity coefficients or the adsorption coefficient of electrical double layer, the ion pairs give a direct contribution and these properties are described correctly in the framework of AMSA theory. Since the ion pairs have no free electric charges, they give polarization effects only for such electrical properties as electroconductivity, dielectric constant or capacitance of electrical double layer. Hence, to describe the electrical properties, it is more convenient to modify MSA-MAL approach by including the ion pairs as new polar entities.

2.   Ion approach

In this section we consider the application of the concept of ion association to describe the properties of electrolyte solutions within the ion or McMillan-Mayer level approach. In this approach the effects of solvent molecules are taken into account by introducing the dielectric constant into Coulomb interaction law and by appropriately choosing the short-range part of ion-ion interactions. To simplify, we consider here the restrictive primitive model (RPM) given by the following interactions,

$$
U_{ij}(r) = \begin{cases} \infty, & r < R_i, \\ \\ \frac{Z_i Z_j e^2}{\epsilon r}, & r > R_i, \end{cases} \tag{1}
$$

where $Z_+ = -Z_- = 1$, R_i is the diameter of ions, e is the elementary charge, ϵ is the solvent dielectric constant.

In the beginning of this section the AMSA approach will be applied to the description of this model of electrolyte solution. The obtained results will be applied to describe the thermodynamic properties of electrolyte solution and to study the effect of electrolyte solution on intramolecular transfer reactions. Finally, the specific features of the effect of ion association on the properties of electrical double layer will be discussed.

2.1 Associative MSA theory

Modern theory of associative fluids is based on the combination of the activity and density expansions for the description of the equilibrium properties. The activity expansions are used to describe the clusterization effects caused by the strongly attractive part of the interparticle interactions. The density expansions are used to treat the contributions of the conventional nonassociative part of interactions. The diagram analysis of these expansions for pair distribution functions leads to the so-called multidensity integral equation approach in the theory of associative fluids. The AMSA theory represents the two-density version of the traditional MSA theory [4, 5] and will be used here for the treatment of ion association in the ionic fluids.

According to the MAL, the formation of ion pairs determines the concentration of free ions $c_+ = c_- = \alpha c$,

$$
\frac{1-\alpha}{\alpha^2} = cK^{\text{as}} \frac{(y'_{\pm})^2}{y'_0}, \tag{2}
$$

where α is the degree of dissociation, $c = \rho_i/2N_A$ is the electrolyte concentration evaluated within the theory, $\rho_i = \rho_+ + \rho_-$ is the total number density, N_A is the Avogadro number, K^{as} is the equilibrium constant of the formation of ion pairs, $y'_{\pm}$ is the mean activity coefficient of the free ions in solution, and y'_0 is the activity coefficient of the ion pairs.

Within the multidensity approach the diagrams that appear in the activity expansions for the one-point density are classified with respect to the number of associative bonds incident with the labelled white circle. Thus, the MAL is written in the form [9–12],

$$\frac{1-\alpha}{\alpha^2} = 4\pi c N_A \int_0^\infty f^{as}(r) g_{+-}^{00}(r) r^2 dr, \tag{3}$$

where $f^{as}(r) = \exp(-\beta U_{+-}^{as}(r)) - 1$ is the Mayer function for the associative interaction, $\beta = 1/kT$, T is the temperature, k is the Boltzmann constant.

In the sticky limit it follows,

$$f^{as}(r) = B\delta(r - R_i), \tag{4}$$

and Eq. (3) can be rewritten in the form,

$$\frac{1-\alpha}{\alpha^2} = 4\pi c N_A B R_i^2 g_{+-}^{00}(R_i), \tag{5}$$

where $g_{+-}^{00}(R_i)$ is the contact value, i.e. the value at $r = R_i$ of the pair distribution function of the nonbonded oppositely charged ions.

From the comparison of Eqs. (5) and (2) for charged hard spheres it follows,

$$K^{as} = 4\pi N_A B R_i^2 e^{b_i}, \qquad \frac{(y_\pm')^2}{y_0'} = e^{-b_i} g_{+-}^{00}(R_i), \tag{6}$$

where $b_i = e^2/(\epsilon k T R_i)$ is the Bjerrum parameter characterizing the Coulomb interactions between ions at contact distance.

A formal short-range cation-anion interaction, $U_{+-}^{as}(r)$, which is responsible for the formation of ion pairs, is fixed by the association constant in Ebeling form [19, 20],

$$\begin{aligned}
K^{as} &= 8\pi N_A R_i^3 \int_1^\infty r^2 dr \left[\mathrm{ch}\left(\frac{b_i}{r}\right) - 1 - \frac{1}{2}\frac{b_i^2}{r^2} \right] \\
&= 8\pi N_A R_i^3 \sum_{m \geqslant 2} \frac{b_i^{2m}}{(2m)!(2m-3)}.
\end{aligned} \tag{7}$$

Similarly to the one-point density in the multidensity approach, the total pair correlation function $h_{ij}(r)$ between two ions i and j is represented as a sum of four terms [9–13],

$$h_{ij}(r) = h_{ij}^{00}(r) + h_{ij}^{01}(r) + h_{ij}^{10}(r) + h_{ij}^{11}(r), \tag{8}$$

where the superscript 0 announces that the corresponding ion is free, and the superscript 1 indicates, when it is bonded into an ion pair, and the function $g_{ij}^{00}(r) = h_{ij}^{00}(r) + 1$.

The classification and topological reduction of the diagrams for pair correlation functions leads to the Wertheim modification of the Ornstein-Zernike (WOZ) equations [9],

$$\mathbf{h}_{ij}(r_{12}) = \mathbf{C}_{ij}(r_{12}) + \sum_{l} \rho_l \int \mathbf{C}_{il}(r_{13}) \mathbf{x} \mathbf{h}_{lj}(r_{32}) \, d\bar{r}_3, \qquad (9)$$

where the matrices $\mathbf{h}_{ij}(r)$ and $\mathbf{C}_{ij}(r)$ are composed of the elements $h_{ij}^{\alpha\beta}(r)$ and $C_{ij}^{\alpha\beta}(r)$, and

$$\mathbf{x} = \begin{pmatrix} 1 & 1 \\ 1 & 0 \end{pmatrix}. \qquad (10)$$

As usual, due to the symmetry of the RPM it is possible to define the sum (s) and the difference (d) combinations for h functions,

$$\begin{aligned}
h_{\mathrm{s}}^{\alpha\beta}(r) &= \frac{1}{2} \left(h_{++}^{\alpha\beta}(r) + h_{+-}^{\alpha\beta}(r) \right), \\
h_{\mathrm{d}}^{\alpha\beta}(r) &= \frac{1}{2} \left(h_{++}^{\alpha\beta}(r) - h_{+-}^{\alpha\beta}(r) \right)
\end{aligned} \qquad (11)$$

and corresponding similar expressions for C functions. Then, the WOZ Eq. (9) decouples into a set of two matrix equations,

$$\mathbf{h}_{\mathrm{s}}(r_{12}) = \mathbf{C}_{\mathrm{s}}(r_{12}) + \rho_i \int \mathbf{C}_{\mathrm{s}}(r_{13}) \mathbf{x} \mathbf{h}_{\mathrm{s}}(r_{32}) \, d\bar{r}_3, \qquad (12)$$

$$\mathbf{h}_{\mathrm{d}}(r_{12}) = \mathbf{C}_{\mathrm{d}}(r_{12}) + \rho_i \int \mathbf{C}_{\mathrm{d}}(r_{13}) \mathbf{x} \mathbf{h}_{\mathrm{d}}(r_{32}) \, d\bar{r}_3. \qquad (13)$$

The AMSA closure for the electroneutrality sum problem (subscript s) is the same as for the associative Percus-Yevick (APY) approximation, [25]

$$\begin{aligned}
h_{\mathrm{s}}^{00}(r) &= -1, h_{\mathrm{s}}^{10}(r) = h_{\mathrm{s}}^{11}(r) = 0, \quad r < R_i, \\
C_{\mathrm{s}}^{00}(r) &= C_{\mathrm{s}}^{01}(r) = C_{\mathrm{s}}^{10}(r) = 0, \\
C_{\mathrm{s}}^{11}(r) &= \frac{(1-\alpha)}{4\pi\rho_i} \delta(r - R_i), \qquad r \geqslant R_i.
\end{aligned} \qquad (14)$$

Similarly, for the difference combination (subscript d),

$$\begin{aligned}
h_{\mathrm{d}}^{00}(r) &= h_{\mathrm{d}}^{01}(r) = h_{\mathrm{d}}^{10}(r) = h_{\mathrm{d}}^{11}(r) = 0, \quad r < R_i, \\
C_{\mathrm{d}}^{00}(r) &= -\frac{b_i R_i}{r}, \; C_{\mathrm{d}}^{01}(r) = C_{\mathrm{d}}^{10}(r) = 0, \\
C_{\mathrm{d}}^{11}(r) &= -\frac{(1-\alpha)}{4\pi\rho_i} \delta(r - R_i), \qquad r \geqslant R_i.
\end{aligned} \qquad (15)$$

The function $g_{+-}^{00}(R_i)$, introduced into the MAL, Eq. (5), is defined by exponential approximation, [26]

$$g_{+-}^{00}(R_i) = \left[1 + h_s^{00}(R_i)\right] \exp\left[-h_d^{00}(R_i)\right].$$ (16)

The solution for $h_s(r)$ formally coincides with the Wertheim solution [25] for dimerizing hard spheres with α calculated from Eq. (5). The solution for $h_d(r)$ was also obtained [6, 7] for the RPM and for the asymmetrical ionic fluids [8] using the Wertheim-Baxter factorization technique.

The solution of AMSA for Eq. (13) includes the single parameter,

$$J_0 = 2\pi\rho_i \int_0^\infty r \, dr \sum_\beta h_d^{0\beta}(r)$$ (17)

connected with the fraction of free ions α and the Debye screening parameter,

$$\kappa_i = \left[4\pi\beta \, e^2 \rho_i/\epsilon\right]^{1/2}$$

by the nonlinear equation,

$$(\kappa_i R_i)^2 (1 + J_0 R_i)^4 \left[\alpha - J_0 R_i(1-\alpha)\right] = 4 (J_0 R_i)^2.$$ (18)

This equation can be written in another form, if the Blum's screening parameter Γ is introduced instead of J_0,

$$J_0 R_i = -\frac{\Gamma R_i}{1 + \Gamma R_i}.$$ (19)

Then, a simple nonlinear equation is derived for the scaling parameter Γ first given by Bernard and Blum [26],

$$4 (\Gamma R_i)^2 (1 + \Gamma R_i)^2 = (\kappa_i R_i)^2 \frac{\alpha + \Gamma R_i}{1 + \Gamma R_i}.$$ (20)

Equations (18) and (20) containing the degree of dissociation α are considered together with the MAL equation (5). The contact value, $h_d(R_i)$ is needed to calculate α and is equal,

$$h_d^{00}(R_i) = -b_i(1 + J_0 R_i)^2 = -\frac{b_i}{(1 + \Gamma R_i)^2}$$ (21)

and using the exponential form (16) for $g_{+-}(R_i)$, the Eq. (5) can be written as,

$$\frac{1-\alpha}{\alpha^2} = c 4\pi N_A B R_i^2 \left[1 + h_s^{00}(R_i)\right] \exp\left[\frac{b_i}{(1 + \Gamma R_i)^2}\right].$$ (22)

In the AMSA, the contact value of $1+h_s^{00}(R_i)$ is equal to the Percus-Yevick result for the hard-sphere fluid [25],

$$1 + h_s^{00}(R_i) = \frac{1 + \frac{1}{2}\eta_i}{(1 - \eta_i)^2},$$
(23)

where $\eta_i = (1/6)\pi\rho_i R_i^3$.

The analysis of Eq. (18) for J_0 or Eq. (20) for Γ, suggests the existence of two different regimes, namely, of the weak association $(\alpha \to 1)$ and the strong association $(\alpha \to 0)$. In the regime of weak association Eq. (20) for Γ reduces to Γ_α,

$$4(\Gamma_\alpha R_i)^2(1 + \Gamma_\alpha R_i)^2 = (\kappa_i R_i)^2\alpha$$
(24)

and in this regime only the electrostatic contribution due to free ions is important; contributions due to ion pairs can be neglected. The regime of weak association corresponds to the traditional MSA-MAL description of the ion association [3] and is realized for,

$$1 \geqslant \alpha \gg \frac{\Gamma R_i}{(1 + 2\Gamma R_i)}.$$
(25)

In the regime of strong association Eq. (20) for Γ reduces to

$$4\Gamma R_i (1 + \Gamma R_i)^3 = (\kappa_i R_i)^2 (1 - \alpha)$$
(26)

and only the electrostatic contribution due to the ion pairs is important; the contribution due to the free ions can be neglected. This regime is realized for,

$$0 \leqslant \alpha \ll \frac{\Gamma R_i}{1 + 2\Gamma R_i}.$$
(27)

In the AMSA the thermodynamic properties contain three different contributions: the hard-sphere (HS) contribution, the contribution from the mass action law (MAL) and the contribution from electrostatic ion-ion interaction (EL) [6, 13, 14]. For example, the osmotic coefficient, $\Phi = p/(\rho_i kT)$, can be represented in the form,

$$\Phi = \frac{p^{HS}}{\rho_i kT} + \frac{p^{MAL}}{\rho_i kT} + \frac{p^{EL}}{\rho_i kT},$$
(28)

where the hard-sphere contribution may be calculated using the Carnahan-Starling expression for hard spheres [27],

$$\frac{p^{HS}}{\rho_i kT} = \frac{1 + \eta_i + \eta_i^2 - \eta_i^3}{(1 - \eta_i)^3}.$$
(29)

The effect of ion pair formation to the hard-sphere contributions can be evaluated with the help of thermodynamic perturbation theory [25] and is included into the MAL terms,

$$
\begin{aligned}
\frac{p^{\mathrm{MAL}}}{\rho_i kT} &= -\frac{1}{2}(1-\alpha)\left[1+\rho_i\frac{\partial \ln h_{\mathrm{s}}(R_i)}{\partial}\rho_i\right] \\
&= -\frac{1}{2}(1-\alpha)\frac{1+\eta_i-\frac{1}{2}\eta_i^2}{(1-\eta_i)\left(1-\frac{1}{2}\eta_i\right)} \, .
\end{aligned}
\tag{30}
$$

The electrostatic contribution was obtained by Bernard and Blum [26] by the direct integration of the internal energy and is of the same form as in the ordinary MSA [28],

$$
\frac{p^{\mathrm{EL}}}{\rho_i kT} = -\frac{(\Gamma R_i)^3}{3\pi\rho_i R_i^3} \, .
\tag{31}
$$

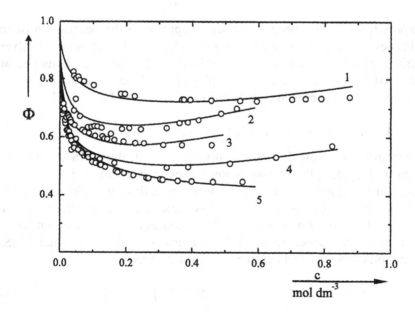

Figure 1. Comparison of osmotic coefficients from the AMSA calculations [6, 14] with experimental data [29,30]: 1. LiClO$_4$ in acetonitrile ($\epsilon = 35.95$); 2. Bu$_4$NBr in ethanol ($\epsilon = 24.36$); 3. LiClO$_4$ in acetone ($\epsilon = 20.70$); 4. LiClO$_4$ in 2-propanol ($\epsilon = 19.39$); 5. Bu$_4$NBr in acetone ($\epsilon = 20.70$).

For the activity coefficient of the free ions one has,

$$
\ln y_\pm = \ln y_\pm^{\mathrm{HS}} + \ln y_\pm^{\mathrm{MAL}} + \ln y_\pm^{\mathrm{EL}} \, ,
\tag{32}
$$

where in the same approximation as for osmotic coefficient,

$$\ln y_{\pm}^{\mathrm{HS}} = \frac{(1 + 2\eta_i)^2}{(1 - \eta_i)^4}, \tag{33}$$

$$\ln y_{\pm}^{\mathrm{MAL}} = \ln \alpha - \frac{1}{4}(1 - \alpha)\rho_i \frac{\partial \ln h_{\mathrm{s}}(R_i)}{\partial \rho_i}$$

$$= \ln \alpha - \frac{1}{4}(1 - \alpha)\frac{5\eta_i - 2\eta_i^2}{(1 - \eta_i)(1 - \frac{1}{2}\eta_i)}, \tag{34}$$

$$\ln y_{\pm}^{\mathrm{EL}} = -b_i \frac{\Gamma R_i}{1 + \Gamma R_i}. \tag{35}$$

From the comparison of Eqs. (5), (6), (22) and (35), the electrostatic part of the activity coefficient of ion pairs is given by equation,

$$\ln y_0^{\mathrm{EL}} = -b_i \frac{(\Gamma R_i)^2}{(1 + \Gamma R_i)^2}. \tag{36}$$

The expression (28) has been successfully applied to the description of the osmotic coefficients for different nonaqueous electrolyte solutions with solvent dielectric constant in the range $20 < \epsilon < 36$ [6, 14], when ion diameters are treated as adjustable parameters. Examples of the AMSA calculations of the osmotic coefficients for electrolyte solutions are given in Fig. 1 [14].

2.2   *Modified associative MSA theory*

For electrolytes of low dielectric constant, $\epsilon < 20$, the AMSA theory was modified by including the effects of ion trimers and tetramers [13, 14]. The formation of ion clusters which are more complex than ion pairs was first introduced to explain the conductivity of low dielectric constant electrolyte solutions [31]. Later on it was shown that this concept is also important for the explanation of other phenomena [3]. We suppose that in the modified AMSA the formation of trimers and tetramers follows the reaction,

$$(C^+ A^-)^0 + C^+ (A^-) \leftrightarrow \left[(C^+ A^-)^0 C^+ (A^-) \right], \tag{37}$$

where the notation $C^+(A^-)$ means that the cation C^+ can either be a free ion or is bonded with another ion pair. In the first case we have the formation of ion trimer while in the second case we have the formation of a tetramer. The relation corresponding to (37) holds for the anion A^-. As a simplification we consider bilateral triplet formation [3, 31], assuming that the formation of a trimer is characterized by only one association constant K_2. This means that the concentration of negative trimers $(ACA)^-$ equals that of positive trimers $(CAC)^+$.

We introduce γc as the concentration of ions of a species which are not bonded neither in a trimer or in a tetramer. Correspondingly $(1 - \gamma)c$ will be the concentration of ions of each species bonded in trimers and tetramers. Then the ion solute is assumed to exist in the form of single ions of concentration $\alpha\gamma c$, pairs, $(1 - \alpha)\gamma c$, trimers, $\alpha(1 - \gamma)c$, and tetramers, $(1 - \alpha)(1 - \gamma)c$. With this definition, the MAL of the formation of ion pairs that is given by Eq. (22), can be rewritten in the form,

$$\frac{1 - \alpha}{\alpha^2} = c\gamma K^{\text{as}} \left(1 + h_{\text{s}}^{00}(R_i)\right) \frac{\left(y_{\pm}^{\text{EL}}\right)^2}{y_0^{\text{EL}}}. \tag{38}$$

Similar as before, K^{as} is the association constant of ion pairs, for which we can use Eq. (7). The hard-sphere contribution $1 + h_{\text{s}}^{00}(R_i)$ is given by Eq. (23). For the electrostatic contributions to the activity coefficients of free ions, $y_{\pm}$, and of the ion pairs, y_0', we can use the expressions (35) and (36) with correction of Eq. (20) for Γ. For this purpose, we can neglect the electrostatic contribution from ion trimers and tetramers in $y_{\pm}^{\text{EL}}$ and y_0^{EL} as for Eq. (24) in the regime of a weak association, $\alpha \to 1$. For $\gamma \to 1$, we generalize the equation for Γ, in the following way,

$$4 \left(\Gamma R_i\right)^2 \left(1 + \Gamma R_i\right)^2 = \gamma \left(\kappa_i R_i\right)^2 \frac{\alpha + \Gamma R_i}{1 + \Gamma R_i}. \tag{39}$$

The second MAL relation for the ion trimer and tetramer formations can be written in two different forms. For the regime of a weak ion pairing ($\alpha \to 1$) the trimer formation process is the dominating process and,

$$\frac{1 - \gamma}{\gamma^2} = (1 - \alpha)y_3^{\text{HS}}(R_i)cK_2^{\text{as}} \frac{y_{\pm}^{\text{EL}} y_0^{\text{EL}}}{y_3^{\text{EL}}}. \tag{40}$$

For the regime of a strong ion pairing ($\alpha \to 0$) the tetramer formation process is the dominating process and,

$$\frac{1 - \gamma}{\gamma^2} = (1 - \alpha)cy_4^{\text{HS}}(R_i)K_2^{\text{as}} \frac{(y_0^{\text{EL}})^2}{y_4^{\text{EL}}}. \tag{41}$$

The association constant K_2^{as} is related to the formation of ion trimers and tetramers and by analogy with the Ebeling association constant can be expressed as a part of the third ion virial coefficient in the case of Eq. (40), or as a part of the fourth ion virial coefficient in the case of Eq. (41). Third and fourth ion virial coefficients can be approximated as the electrostatic parts of the second virial coefficient for ion-dipole and dipole-dipole interactions, respectively. For the sake of simplicity, we consider K_2 as the adjustable parameter and neglect the hard-sphere contribution of Eqs. (40) and (41), i.e. $y_3^{HS}(R_i)$ and $y_4^{HS}(R_i)$. For the electrostatic part of the activity coefficient of the ion trimers, y_3^{EL}, and ion tetramers, y_4^{EL}, we assume that by analogy with the relations (34) and (35) we can write,

$$\ln y_3^{EL} = -b_i \frac{(\Gamma R_i)^3}{(1+\Gamma R_i)^3}, \qquad \ln y_4^{EL} = -b_i \frac{(\Gamma R_i)^4}{(1+\Gamma R_i)^4}. \qquad (42)$$

The formation of trimers and tetramers is the process that follows the formation of ion pairs. Usually $K^{as} \gg K_2^{as}$ [3]. This leads to the following order of the quantities α and γ,

$$0 \leqslant \alpha \leqslant \gamma \leqslant 1. \qquad (43)$$

Due to this fact, Eq. (41) is preferable.

After including trimers and tetramers, the osmotic coefficient, Φ, has the form similar to Eq. (28), where only the MAL term is modified from Eq. (30) to,

$$\begin{aligned}
\frac{p^{MAL}}{\rho_i kT} &= -\gamma \frac{(1-\alpha)}{2}\left[1 + \rho_i \frac{\partial \ln y_2^{HS}(R_i)}{\partial \rho_i}\right] \\
&\quad - \frac{2}{3}\alpha(1-\gamma)\left[1 + \rho_i \frac{\partial \ln y_3^{HS}(R_i)}{\partial \rho_i}\right] \\
&\quad - \frac{3}{4}(1-\alpha)(1-\gamma)\left[1 + \rho_i \frac{\partial \ln y_4^{HS}(R_i)}{\partial \rho_i}\right]. \qquad (44)
\end{aligned}$$

The results obtained also are useful for the calculation of the ionic conductivity of nonaqueous electrolyte solutions. Several attempts exist for the calculation of the molar conductivity of associating electrolytes beyond the limiting law at the level of the MSA [3, 32, 33], where, however, only ion pairs were taken into account. Ion pairs and tetramers are electrically neutral, nonconducting species in the solution, by contrast to the ion trimers. The total concentration of charged particles is given by,

$$\alpha\gamma c + \frac{1}{3}\alpha(1-\gamma)c = \frac{2\gamma+1}{3}\alpha c. \qquad (45)$$

Following the concept underlying the MSA-MAL conductivity equation [3, 32, 33] and by taking into account that the total concentration of electrically conducting particles is $\alpha(2\gamma+1)/3$ instead of αc, yields an appropriate equation to express the molar conductivity in the AMSA for symmetrical electrolytes [13]. The possibility of such modification of the AMSA theory is quite promising for the description and interpretation of thermodynamic and transport properties of electrolyte solutions in a weakly polar solvent.

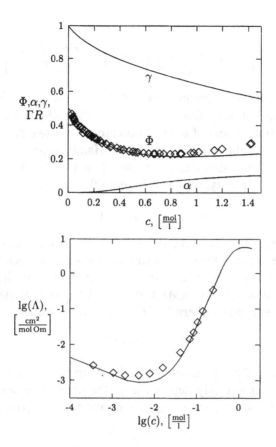

Figure 2. Osmotic coefficient Φ and equivalent conductivity Λ of LiClO$_4$ solution in DMC as the functions of electrolyte concentration c. $\square$ – experimental values [29, 33], solid line – theoretical prediction from the AMSA theory [13, 14].

As an example, in Fig. 2 the comparison of the experimental [30, 34], and calculated data [13, 14], for the solution of LiClO$_4$ in dymethyl carbonate (DMC) ($\epsilon = 3.1$) at 25°C is shown.

2.3 *Electrolyte effects in the intramolecular electron transfer*

The nontraditional example of applying the AMSA theory is connected with the treatment of electrolyte effects in intramolecular electron transfer (ET) reactions [21, 22]. Usually the process of the transfer of the electron from donor (D) to acceptor (A) in solutions is strongly nonadiabatic. The standard description of this process in connected with semiclassical Marcus theory [35], which reduces a complex dynamical problem of ET to a simple expression of electron transfer rate,

$$k_{ET} = \frac{2\pi}{\hbar} |V_{AD}|^2 \frac{1}{\sqrt{4\pi\lambda kT}} \exp\left\{-\frac{(\lambda + \Delta G^0)^2}{4kT\lambda}\right\}, \qquad (46)$$

which comprises just three parameters ΔG^0, λ, and V_{AD}. The parameter ΔG^0 is the reaction driving force, λ is the reorganization energy measured by the energy rearrangement of the solution around the two reactants, V_{AD} is the matrix element of the effective electronic Hamiltonian which couples the electronic states between donor and acceptor, $\hbar$ is the Planck's constant.

For the sake of simplicity, we focus here on the univalent charge shift type reactions,

$$D^- + A \rightarrow D + A^-, \qquad (47)$$

and we assume that the donor and acceptor are of the same size, $R_D = R_A$, and following the reaction these sizes and the distance between the donor and the acceptor do not change. The electrolyte is modelled by the RPM model which is described within the AMSA theory. We are restricted here to the consideration of only the exothermic ET reactions, for which

$$\lambda \gg -\Delta G^0. \qquad (48)$$

As a result, from all three parameters ΔG^0, λ and V_{AD} in the Marcus formula (46), only λ is strongly dependent on the electrolyte solution. Within this dependence the solvent and ion contributions can be separated,

$$\lambda = \lambda^{sol} + \lambda^{ion}. \qquad (49)$$

The molecular solvent contribution, λ^{sol} is often described by modelling the solvent as a dielectric continuum which leads to the standard Marcus expression [36] or by a more up-to-date liquid state theory such as integral equation technique [37] or computer simulations [38].

In the limit of slow ET, the ion contribution, λ^{ion}, can be presented as the sum of the reorganization energy of the donor, λ_D^{ion}, of the acceptor, λ_A^{ion}, and of the changes in the ion-ion potential between the donor and acceptor due to the presence of the electrolyte,

$$\lambda^{ion} = \lambda_D^{ion} + \lambda_A^{ion} + kT h_{DA}^d(L) + \frac{e^2 \Delta Z_D \Delta Z_A}{\epsilon L}, \qquad (50)$$

where L is the center-to-center distance between the donor and acceptor which in the course of the ET reaction change their charges by ΔZ_D and ΔZ_A, respectively; since the total charge of the system does not change, $\Delta Z_D = -\Delta Z_A$. The two last terms are the corrections for the finite donor-acceptor separation. $h_{\mathrm{DA}}^d(L)$ is the electrostatic part of the donor-acceptor pair correlation function.

The donor and acceptor contributions to λ_p^{ion}, where $p = A$ or D, are defined as $1/2$ of the corresponding Stokes shift,

$$\lambda_p^{\mathrm{ion}} = \frac{1}{2} \sum_{i=+,-} \rho_i \int \Delta U_{ip}(r) \Delta g_{ip}(r) \mathrm{d}\vec{r}, \tag{51}$$

with the summation performed over all anions and cations of the electrolyte, and where

$$\Delta U_{ip}(r) = U_{ip}^{(\mathrm{initial})}(r) - U_{ip}^{(\mathrm{final})}(r) = \frac{e^2 Z_i \Delta Z_p}{\epsilon r}, \tag{52}$$

describes the changes of the donor-ion and acceptor-ion pair potentials in the course of the ET event. The corresponding change of the donor-ion or acceptor-ion pair correlation functions (i.e. the so-called screening potentials) is given by

$$\Delta g_{ip}(r) = g_{ip}^{\mathrm{final}}(r) - g_{ip}^{\mathrm{initial}}(r), \tag{53}$$

where the sign conventionally adopted in Eqs. (52) and (53) ensures that the value of λ^{ion} defined by Eq. (51) is always positive regardless of the sign of ΔZ_p.

In the case that we have considered, two types of ion association will take place under most conditions: the pairing between ions in the electrolyte described by Eq. (2) and the pairing between donor and counterions and acceptor and counterions, the so-called specific ion pairs. The latter clusterization is also described by the MAL with donor and acceptor being treated separately,

$$\frac{1 - \alpha_p}{\alpha_p} = \alpha c K_p^{\mathrm{as}} \frac{y_p y_i}{y_{pi}}, \tag{54}$$

where $1 - \alpha_p$ is the fraction of the donor-ion or acceptor-ion pairs, $(p = A$ or $D)$, K_p^{as} is equilibrium constant for the formation of corresponding ion pairs, y_p is the activity coefficient for the donor or the acceptor, y_i is the activity coefficient for the respective counterions, y_{pi} is the activity coefficient of the donor-ion or acceptor-ion pairs.

For the symmetric electrolyte, $y_i = y_\pm$, which is given by Eq. (35). For the equilibrium constant, K_p^{as}, in analogy to the electrolyte we also have chosen the Ebeling expression,

$$K_p^{\mathrm{as}} = 8\pi N_A R_{ip}^3 \sum_{m \geqslant 2} \frac{b_{ip}^{2m}}{(2m)!(2m-3)}, \tag{55}$$

which sets the value of the second virial coefficient for the DA-electrolyte interaction and where $b_{ip} = (|Z_i \Delta Z_p| e^2)/(\epsilon k T R_{ip})$ is the Bjerrum parameter describing the Coulomb interaction between either donor or acceptor and counterions at the contact distance $R_{ip} = (R_i + R_p)/2$, R_p is the diameter of a donor or acceptor. In the AMSA treatment, the MAL-relations (54) can be written in the form similar to Eq. (5),

$$\frac{1 - \alpha_p}{\alpha_p} = 4\pi c \alpha N_A B_{ip} R_{ip}^2 g_{ip}^{oo}(R_{ip}), \qquad (56)$$

where $g_{ip}^{00}(R_{ip})$ is the contact value at $r = R_{ip}$ of the pair distribution function of the nonassociated donor or acceptor with the corresponding counterions, B_{ip} is the strength of the corresponding pairing. Again, from comparing Eqs. (54) and (56),

$$K_p^{as} = 4\pi N_A B_{ip} R_{ip}^2 \exp(b_{ip}), \qquad \frac{y_p y_i}{y_{ip}} = \exp(-b_{ip}) g_{ip}^{00}(R_{ip}). \qquad (57)$$

As a result of specific pairing, $\Delta g_{ip}(r)$ can be divided into two terms corresponding to the direct ion pairing and to the continuum-like ion atmosphere [12],

$$\Delta g_{ip}(r) = h_{ip}^{\text{ion pair}}(r) + h_{ip}^{\text{ion atm}}(r), \qquad (58)$$

where

$$h_{ip}^{\text{ion pair}}(r) = \frac{1 - \alpha_p}{4\pi c R_{ip}^3} \delta(r - R_{ip}), \qquad (59)$$

respectively. As a result, λ_p^{ion} also consists of two terms,

$$\lambda_p^{\text{ion}} = \frac{e^2 Z |\Delta Z_p|}{2\epsilon R_{ip}}(1 - \alpha_p) + 2\pi c e^2 \Delta Z_p \frac{1}{\epsilon} \sum_{i=+,-} Z_i \int h_{ip}^{\text{ion atm}}(r) r \mathrm{d}r, \qquad (60)$$

the first of which describes the direct interaction between the donor (or acceptor) and the counterion while the second term contains the ion atmosphere contribution to λ. Both terms are positive regardless of the sign of ΔZ_p. The specific form of $h_{ip}^{\text{ion atm}}(r)$ depends on the selected model of electrolyte which was obtained in the framework of the AMSA approach. For this purpose the AMSA Eqs. (13) and (15) for electrolyte were expanded by the inclusion of a corresponding equation for the correlation functions of donor-electrolyte, acceptor-electrolyte and donor-acceptor. We omit the details of these calculations and reproduce here only the final results needed to evaluate λ^{ion} [22],

$$y_p = \exp\left(-\frac{e^2(\Delta Z_p)^2}{\epsilon kT}\frac{\Gamma}{1+\Gamma R_p}\right),$$

$$h_{ip}^{00}(R_{ip}) = \frac{b_{ip}}{(1+\Gamma R_i)(1+\Gamma R_p)}. \tag{61}$$

Here, according to Eq. (57),

$$y_{ip} = y_p y_i \exp\left[b_{ip} - h_{ip}^{00}(R_{ip})\right]. \tag{62}$$

Finally, the ion contribution to the reorganization energy of the donor and acceptor can be written in the form,

$$\lambda_p^{\text{ion}} = \frac{1}{2}\frac{e^2 Z\Delta Z_p}{\epsilon R_{ip}}\left[(1-\alpha_p) + \frac{2\Gamma R_{ip}}{1+\Gamma R_{ip}}\right]. \tag{63}$$

The calculation of $G_{\text{AD}}(L)$ is more complicated and, in general, $G_{\text{AD}}(L)$ cannot be presented in the explicit form. However, in a wide range of electrolyte concentration $G_{\text{AD}}(L)$ can be presented with a considerable accuracy in the Debye-Hückel-like form [22],

$$G_{\text{AD}}(L) = \frac{\Delta Z_A \Delta Z_D e^2 \exp(-2\Gamma(L - R_{\text{AD}}))}{\epsilon kTL(1+\Gamma R_A)(1+\Gamma R_D)}. \tag{64}$$

Finally, for the symmetrical case, $R_A = R_D, \alpha_A = \alpha_D, |\Delta Z_p| = 1$, the ion contribution to the reorganization energy can be presented in the form,

$$\lambda^{\text{ion}} = \frac{e^2}{\epsilon R_{iD}}\left[(1-\alpha_D) + \frac{2\Gamma R_{iD}}{1+\Gamma R_{iD}}\right] +$$
$$\frac{e^2}{\epsilon L}\left[\frac{\exp(-2\Gamma(L-R_{\text{AD}}))}{(1+\Gamma R_A)^2} - 1\right]. \tag{65}$$

The capability of the developed theoretical model is illustrated in Fig. 3, where the comparison of experimental data [39] with theoretical prediction for ET rates in trans-1-(4-biphenylyl)-4-(2-naphthyl)cyclohexane (C-1,4) in tetrahydrofuran (THF) ($\epsilon = 7.58$) as function of sodium tetraphenylboron (TPB) concentration is presented. The calculations are performed with ion size $R_i = 7.6$ Å which, according to the Ebeling expression for association constant, reproduced the experimental value of the dissociation constant of the considered electrolyte, $L = 11.8$ Å. The sizes of donor and acceptor, according to [40] are: $R_D = R_A = 5.9$ Å. As we can see from Fig. 3, the theory quite accurately reproduces the dramatic reduction of ET rates in the

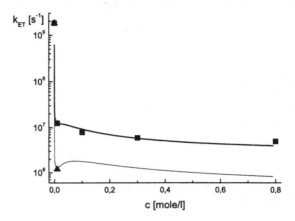

Figure 3.   Electron transfer rate in C-1,4 in THF as the function of concentration of tetraphenylboron salts (■ and ▲) – experimental values for TPB-Na and TPB-TBA, respectively [39] (solid line) – theoretical values from the AMSA theory [22].

presence of 10 mM of electrolyte. Such a behavior is connected with the similar concentration dependence of the dissociation degree of specific ion pairs α_D that leads to the strong increase of reorganization energy in this concentration region due to specific ion pair contribution. For higher concentrations, the theory also reproduces the concentration dependence of ET rate that is rather slow due to saturation of specific ion pair contribution. Figure 3 also shows the concentration dependence of ET rate for the same compound in the presence of TPB tetrabutyl ammonium (TBA^+) solution in THF. The decrease of the ion size by 0.6 Å that reproduces the decrease of the dissociation constant of this electrolyte by two times is sufficient for being reproduced in order to produce a larger decrease of ET rate compared with the previous electrolyte.

The specific ion pairing contribution is also important for the solvent with ϵ significantly larger than for THF [41]. Such a situation is illustrated in Fig. 4 where the electrolyte effect on the activation energy of electron exchange between ferrocene and ferrocenium ion $(C_{p2}Fe^{0/+})$ in acetone $(\epsilon = 20.70)$ versus ion concentration is presented.

The electrolyte contribution to the activation energy in the framework of the Marcus theory is given by equation,

$$\Delta G^\star = \frac{(\lambda + \Delta G^0)^2}{4\lambda}.$$ (66)

In the considered case of a weakly exothermic intramolecular ET shift reactions $\Delta G^\star$ are connected with ion contribution to the reorganization energy

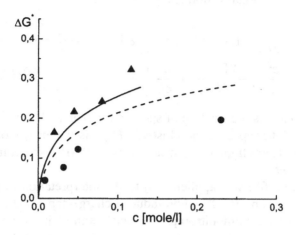

Figure 4. The activation energy of ferrocenium-ferrocene ($C_{p2}Fe^{+/0}$) self-exchange in acetone ($\epsilon = 20.70$) as the function of electrolyte concentration. ($\blacktriangle$ *and* $\bullet$) – experimental values for tetraethylammoniym perchlorate and tetraethylammoniym hexafluorophosphate correspondingly [42], (solid and dashed line)-theoretical prediction from the AMSA theory with parameters $R_d = R_a = L = 8$ Å, $R_i = 6$ Å, and $R_i = 8$ Å, respectively [41].

and in the simplest case, $R_{iD} = R_i = L$, is reduced to the expression,

$$\Delta G^{\star}_{\text{ion}} = \frac{1}{4}\lambda^{\text{ion}} = \frac{e^2}{4\epsilon R_i}\left[(1 - \alpha_D) + \frac{\Gamma^2 R_i^2}{(1 + \Gamma R_i)^2}\right], \qquad (67)$$

where the first term, $(e^2/\epsilon R_i)(1 - \alpha_D)$, describes the specific ion pair contribution while the second one, $\ln y_0 = (e^2/\epsilon R_i)\Gamma^2 R_i^2(1 + \Gamma R_i)^{-2}$, is connected with the activity coefficient y_0 of the ion pair (36). We can see that ion atmosphere effect in the considered case is not important and, in the framework of the Marcus theory, the contribution of ion pairing provides for the correct description of the experimental data [42] of electrolyte effects on the kinetics of ferrocenium-ferocene self-exchange and any speculations about the deviation from Marcus theory [43] are irrelevant.

In order to treat the electrolyte effects in ET reactions in very weakly polar solvents with $\epsilon < 5$, the possibility of forming the clusters of the order higher than ion pairs should be included into the theoretical model [44]. The introduction of such ion clusters is founded on the two additional MALs, one of which describes the clusterization in electrolyte while the second - the clusterization between donor or acceptor of a probe molecule and electrolyte. The modification of the AMSA treatment leads to the following expression for ion contribution to the reorganization energy,

$$\lambda^{\text{ion}} = \frac{e^2}{\epsilon R_{i\text{D}}} \gamma_{\text{D}} (1 - \alpha_{\text{D}}) + E_{\text{h.c.}} (1 - \gamma_{\text{D}}) (1 - \alpha_{\text{D}}) +$$
$$\frac{e^2}{\epsilon} \frac{2\Gamma}{(1 + \Gamma R_{i\text{D}})} + \frac{e^2}{\epsilon L} \left[\frac{\exp(-2\Gamma(L - R_{\text{AD}}))}{(1 + \Gamma R_{\text{A}})^2} - 1 \right], \quad (68)$$

where $\gamma_{\text{D}}(1 - \alpha_{\text{D}})$ is the fraction of specific ion pairs, $(1 - \gamma_{\text{D}})(1 - \alpha_{\text{D}})$ is the fraction of higher specific ion clusters, $E_{\text{h.c.}}$ is the energy of bonding of donor (or acceptor) in a higher ion cluster, Γ is the AMSA screening parameter defined by Eq. (39).

The expression (68) was applied [44] to the interpretation of the effect of electrolyte on the stability of a photoinduced charge-separated states of such probe molecules as p-aminonitrobiphenyl and p-aminonitroterphenyl in solution of different tetrabutulammonium salts in nonpolar solvents such as benzene ($\epsilon = 2.28$) and toluene ($\epsilon = 2.38$). According to the experimental data [45, 46], the observed spectral shifts in toluene are approximately $\lambda^{\text{ion}}_{\text{toluene}} / \lambda^{\text{ion}}_{\text{THF}} \approx \gamma_{\text{D}} \epsilon_{THF} / \epsilon_{toluene} \approx 3.2\gamma_{\text{D}}$ times larger than in THF.

2.4   *Concept of ion association in the theory of electrical double layer*

Concept of ion association is also important in explaining the specific properties of the electrical double layer, especially in the regime of strong ion-ion interaction when ion clusterization occurs. For the sake of simplicity, we consider here the RPM model near a charged hard wall. The interaction between ions and surface is given by the potential,

$$U_i(z) = \begin{cases} \infty, & z < R_i/2, \\ \frac{Z_i e E z}{\epsilon}, & z > R_i/2, \end{cases} \quad (69)$$

where $E/4\pi = q_s$ is the surface charge per unit area, z is the normal distance of an ion from the wall.

The equilibrium properties of the double layer can be described by the ion density profiles, $\rho_+(z)$ and $\rho_-(z)$, from which the charge and density profiles can be organized,

$$q(z) = e(\rho_+(z) - \rho_-(z)), \quad (70)$$

$$\rho_i(z) = \rho_+(z) + \rho_-(z), \quad (71)$$

respectively. The charge density profile satisfies the condition of local electroneutrality,

$$\int_0^\infty dz q(z) = -q_{\rm s}. \qquad (72)$$

The density profile satisfies the so-called contact theorem [47], according to which,

$$kT\rho_i\left(\frac{R_i}{2}\right) = p + \frac{E^2}{8\pi\epsilon}, \qquad (73)$$

where p is the pressure of the bulk electrolyte given by expression (28).

The electrical properties of the double layer are described by the charge profile. In particular, the double layer potential,

$$\Psi = -\frac{4\pi}{\epsilon}\int_0^\infty z dz q(z), \qquad (74)$$

and the capacitance of the double layer,

$$C = \frac{q_{\rm s}}{\Psi}. \qquad (75)$$

According to [23], the formation of ion pairs leads to the division of the charge profile into two parts,

$$q(z) = q_0(z) + q_{\rm pairs}(z), \qquad (76)$$

where $q_0(z)$ is the charge profile of the free ions with density $\alpha\rho_i$ while $q_{\rm pairs}(z)$ is the charge profile of the ions bonded in ion pairs with the density $1/2 \cdot (1 - \alpha)\rho_i$; $q_{\rm pairs}(z)$ describes the bonded ions and is related to the polarization profile [48],

$$q_{\rm pairs}(z) = -{\rm div} P(z). \qquad (77)$$

Due to this, the contribution of $q_{\rm pairs}(z)$ into electrical properties is totally different in comparison with $q_0(z)$. Usually the polarization effects only modify the dielectric properties of double layer and hence its contribution for a low ion concentration can be neglected after all,

$$q(z) \approx q_0(z). \qquad (78)$$

For a weakly charged surface such consideration of q(z) reduces to the MSA-MAL description in the framework of which the MSA theory for double layer [49] should be modified by correcting the ion density using the density of free ions, $\alpha\rho_i$, obtained from Eq. (2). However, for the activity coefficients in the form (35) and (36), we should change Γ to Γ_α defined by Eq. (24),

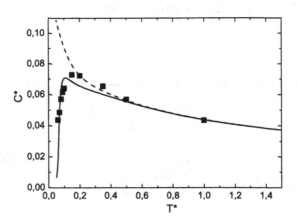

Figure 5. The reduced capacitance, C^*, as the function of reduced temperature, T^*, and obtained from Monte Carlo simulation [50, 51] (points) and theory (curves) [23], $\eta_i = 0.021$. The solid and dashed lines correspond to the MSA and MSA-MAL theory, respectively.

since, we neglect the polarization effects for electrical properties. The Ebeling expression for the association constant remains in the form given by Eq. (7).

Hence, the MSA-MAL approach leads to the following expression for the capacitance of the double layer,

$$C = \frac{\epsilon}{2\pi} \Gamma_\alpha , \tag{79}$$

that is different from the MSA result [49] by the switch from Γ to Γ_α. This change is crucial for the temperature dependence of C. In the limit of low electrolyte concentration $2\Gamma_\alpha \to \kappa_\alpha = \kappa_i \sqrt{\alpha}$. Since $\kappa_\alpha \sim \sqrt{\alpha/T}$, the MSA-MAL theory gives a correct explanation of the temperature dependence of the capacitance at high and at low temperatures. At high temperature, in accordance with (2) $\alpha \to 1$, and we have the classical $1/\sqrt{T}$ behavior of the capacitance. At low temperature, $K^{as} \to \infty$, $\alpha \to 0$ and the capacitance $C \to 0$; this implies a change in the slope of the temperature dependence, in agreement with computer simulations [50–52]. Figure 5 illustrates the reduced values of the capacitance $C^* = CR_i/\epsilon$ calculated in the MSA and MSA-MAL approaches as a function of the reduced temperature $T^* = 1/b_i$ at a volume fraction, $\eta_i = 0.021$. For comparison, the data obtained from computer simulations [50, 51] are also presented.

The thermodynamic properties of double layer are described by the density profile $\rho_i(z)$. In particular, the adsorption coefficient is equal to

$$\Gamma_a = \int_{R/2}^{\infty} (\rho_i(z) - \rho_i)\mathrm{d}z . \tag{80}$$

The density profile $\rho_i(z)$ for the uncharged surface has three different terms according to the contact theorem (73) and according to the expression (28) for the pressure. The first two of them describe the hard-sphere and ion-pairing contributions, respectively. They can be obtained in the framework of the associative version of the Henderson-Abraham-Barker (HAB) theory [53, 54]. According to the obtained results [54],

$$\frac{1}{\rho_i}\rho_i^{\mathrm{NONEL}}(R_i/2) = \frac{1+2\eta_i}{(1-\eta_i)^2} - \frac{1-\alpha}{2(1-\eta_i)}, \tag{81}$$

$$\pi R_i^2 \Gamma_a^{\mathrm{NONEL}} = \frac{9\eta_i^2 + \frac{3}{2}\eta_i[4\eta_i - 1](1-\alpha)}{1+2\eta_i+\frac{3}{2}\eta_i(1-\alpha)}. \tag{82}$$

The two terms for the non-electrostatic contribution for the contact value $\rho_i(R/2)$ are slightly different from the corresponding expressions (29) and (30). However, for the low ion concentrations this difference is not so important. The principal defect of the HAB approach is connected with the difficulty to reproduce the electrostatic contribution which for the contact value should give the expression corresponding to (31).

The solution of this problem can be obtained out in the framework of inhomogeneous integral equation approach [55]. Some other ways exist as well [56–58]. One of them is related with the application of the field theoretical approach to the solution of this problem and is discussed by Badiali et al. in the first chapter of this volume. However, the expression for $\rho^{\mathrm{EL}}(z)$ in an explicit form was obtained only for the point ions,

$$\rho_i^{\mathrm{EL}}(z) = -\frac{e^2\rho_i}{2\epsilon kT}\int_0^{\infty} \frac{pdp}{\sqrt{p^2+\kappa_i^2}} \cdot \frac{\sqrt{p^2+\kappa_i^2}-p}{\sqrt{p^2+\kappa_i^2}+p} \exp\left[-2z\sqrt{p^2+\kappa_i^2}\right],$$
$$z > 0, \tag{83}$$

where p is a plane wave number.

The important feature of this expression is the reproducibility of the electrostatic term in the Debye-Hückel form by the contact value, $\rho_i^{\mathrm{el}}(0)$ [58]. Since the bulk correlation functions within the AMSA theory at a low ion concentration can be presented in the Debye-Hückel-like form (64), the modified version of the expression (83) can be proposed [24],

$$
\rho_i^{\mathrm{EL}}(z) = -\frac{e^2 \rho_i}{2\epsilon kT} \left(\frac{2\Gamma}{\kappa_i}\right)^2 \left[\alpha \int_0^\infty \frac{p\,dp}{\sqrt{p^2+4\Gamma^2}} \cdot \frac{\sqrt{p^2+4\Gamma^2}-p}{\sqrt{p^2+4\Gamma^2}+p} \right.
$$

$$
\times \exp\left[-2\left(z-\frac{1}{2}R_i\right)\sqrt{4\Gamma^2+p^2} \right]
$$

$$
\left. + (1-\alpha)\frac{2}{3}\Gamma \exp\left[-2\Gamma_\alpha\left(z-\frac{1}{2}R_i\right) \right] \right],
$$

$$
z > \frac{1}{2}R_i,
\tag{84}
$$

where the last term corrects z-dependence of $\rho_i^{\mathrm{EL}}(z)$, while Γ and Γ_α are given by Eqs. (20) and (24), respectively.

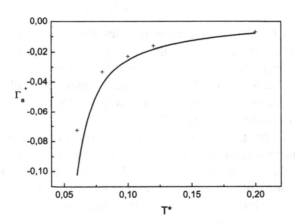

Figure 6.   The reduced value of the adsorption coefficient, $\Gamma_a^\star$, as the function of reduced temperature, $T^\star$ for the RPM model with a bulk volume fraction, $\eta_i = 0.00785$ near to uncharged hard wall. The solid curve gives theoretical prediction [24] while the symbols correspond to Monte Carlo data [51, 52].

According to the contact theorem, the expression (84) reproduces exactly the AMSA result (31) for the electrostatic part of the pressure,

$$
kT\rho_i^{\mathrm{EL}}\left(\frac{1}{2}R_i\right) = -\frac{\Gamma^3}{3\pi},
\tag{85}
$$

and Eq. (84) gives the following contribution to the adsorption coefficient,

$$\Gamma_a^{EL} = -\frac{\Gamma^2}{2\pi}\left[\frac{1}{8}\alpha(2\ln 2 - 1) + \frac{1}{3}(1-\alpha)\frac{\Gamma}{\Gamma_\alpha}\right]. \tag{86}$$

The total adsorption coefficient is equal to

$$\Gamma_a = \Gamma_a^{NONEL} + \Gamma_a^{EL}, \tag{87}$$

where Γ_a^{NONEL} is given by the expression (82). The dimensionless adsorption coefficient for the uncharged surface, $\Gamma_a^\star = \Gamma_a R^2$, calculated according to Eqs. (82), (86) and (87) as the function of temperature $T^\star = 1/b_i$, is presented in Fig. 6. As we can see, the theory quite accurately reproduces the decrease of Γ_a with decrease of the temperature.

Finally, the expression (78) for $q(z)$ can be used to modify the nonlinear Poisson-Boltzmann theory in order to consider a highly charged surface [59, 60]. In this case, for the profile $\rho_i(z)$ the new term appears which exactly reproduces the last electrostatic term in the contact theorem (73).

3. Ion-molecular approach

In this section we consider the possibility of applying the ion association concept to the description of the properties of electrolyte solutions in the ion-molecular or Born-Oppenheimer level approach. The simplest ion-molecular model for electrolyte solution can be represented by the mixture of charged hard spheres and hard spheres with embedded dipoles, the so-called ion-dipolar model. For simplification we consider that ions and solvent molecules are characterized by diameters R_i and R_s, correspondingly. The model is given by the pair potentials,

$$
\begin{aligned}
U_{ij}(r) &= \begin{cases} \infty, & r < R_i, \\ \frac{Z_i Z_j e^2}{r}, & r > R_i, \end{cases} \\
U_{is}(1,2) &= \begin{cases} \infty, & r < R_{is}, \\ \frac{Z_i e p_s}{r^2}(\hat{r}\hat{p}), & r > R_{is}, \end{cases} \\
U_{ss}(1,2) &= \begin{cases} \infty, & r < R_s, \\ (p_s^2/r^3)\cdot[3(\hat{r}\hat{p}_1)(\hat{r}\hat{p}_2) - (\hat{p}_1\hat{p}_2)], & r > R_s, \end{cases}
\end{aligned} \tag{88}
$$

where p_s is the dipole moment of the solvent molecule, $R_{is} = 1/2\cdot(R_i + R_s)$, $\hat{r}, \hat{p}$ are the unit vectors of $\vec{r}_{12}$ and the dipole moment, respectively. The ion-dipole model has been solved in the MSA in the simplest case when all species have the same diameters [61–64]. This result was also generalized for the ion-dipole model with the arbitrary particle sizes [65, 66]. Recently, the AMSA solution for ion-dipole model was reported [15, 16]

In this section the results of the associative MSA theory for the ion-dipole model will be reviewed. Moreover, the modified MSA-MAL version for the ion-dipole system will be applied to describe the dielectric constant of electrolyte solutions.

3.1   *Associative MSA theory*

The effects of association in the ion-dipole model are more rich than in the RPM. In this model in addition to the association between ions it is also possible to consider the clusterization between solvent molecules as well as between ions and solvent molecules; the later can describe the formation of chains or network due to the bonds between solvent molecules, and the formation of specific solvation clusters around the ions. For the sake of simplicity, we focus here only on the case of ion pairs that is characterized by the MAL (2) or in the form of Eq. (5) for the degree of dissociation, α.

In the multidensity approach the pair correlation functions $h_{ij}(r)$, have the form (8),

$$h_{is}(12) = h_{is}^0(12) + h_{is}^1(12) . \tag{89}$$

Similarly to the MSA [61–66], the pair correlation functions, $h_{is}(12)$ and $h_{ss}(12)$, can be presented in the form,

$$
\begin{aligned}
h_{is}(12) &= h_{is}^s(r) + h_{is}^{011}(r)(\hat{r}\hat{p}), \\
h_{ss}(12) &= h_{ss}^s(r) + h_{ss}^{110}(r)(\hat{p}_1\hat{p}_2) + \\
&\quad h_{ss}^{112}(r)\left(3(\hat{r}\hat{p}_1)(\hat{r}\hat{p}_2) - (\hat{p}_1\hat{p}_2)\right) .
\end{aligned}
\tag{90}
$$

Due to the symmetry of the present ion-dipole model, the WOZ equation decouples into electrostatic and nonelectrostatic parts. The nonelectrostatic part has the form of WOZ equation for the two-component dimerizing-nondimerizing mixture [67]. Using the Wertheim-Baxter factorization technique, the solution of WOZ equation for the electrostatic part was performed in terms of energy parameters [68, 69],

$$J_\alpha = 2\pi\rho_i R_i \int_0^\infty r\,dr \sum_\beta h_{ii}^{(d)\alpha\beta}(r) , \tag{91}$$

$$b_1^\alpha = 2\pi\sqrt{\rho_i\rho_s}R_i R_s \frac{1}{\sqrt{3}} \int_0^\infty dr h_{is}^{011,\alpha}(r) , \tag{92}$$

$$b_2 = 2\pi\rho_s R_s^3 \frac{3}{\sqrt{30}} \int_0^\infty \frac{1}{r} dr h_{ss}^{112}(r) , \tag{93}$$

connected with the parameters of the ion and dipole subsystems,

$$\kappa_i^2 R_i^2 = 4\pi e^2 \beta \rho_i R_i^2,$$
$$3y_s = \frac{4}{3}\pi\beta\rho_s p_s^2, \qquad \delta = \frac{R_s}{R_i}, \tag{94}$$

which should be considered together with Eq. (2) for the fraction α.

Only three of the parameters entering (91)–(93) create the system of coupled nonlinear equations which can be represented in the following form,

$$(a_i^0)^2 + 2a_i^0 a_i^1 + a_s^2 = \kappa_i^2 R_i^2, \tag{95}$$

$$a_i^0(K_{si}^{00} + K_{si}^{00}) + a_i^1 K_{si}^{00} + a_s(K_{ss} - 1) = \kappa_i R_i \sqrt{3y_s}, \tag{96}$$

$$(K_{si}^{00})^2 + 2K_{si}^{00}K_{si}^{01} + (K_{ss} - 1)^2 = 3y_s + \beta_6^2/\beta_{12}^4, \tag{97}$$

where

$$K_{si}^{00} = \frac{\delta}{2\Delta}(\Lambda a_i^0 + b_1^0), \tag{98}$$

$$K_{si}^{01} = \frac{\delta}{2\Delta}\left(\Lambda a_i^1 + b_1^1 + \frac{1}{2}(1-\alpha)b_i^0\right), \tag{99}$$

$$K_{ss} - 1 = \frac{\delta}{2\Delta}(\Lambda a_s - \frac{2}{\delta}\beta_3), \tag{100}$$

$$a_i^0 = \frac{1}{D}\left[J_0\beta_6^2 - \frac{1}{12}b_1^0(b_1^0 + b_1^1)(\delta\beta_6 + 3)\right.$$
$$\left. + \frac{1}{4}b_1^0(b_1^1 J_0 - b_1^0 J_1)\right], \tag{101}$$

$$a_i^1 = \frac{1}{D}\left\{\left[\frac{1}{2}(1-\alpha)(1+J_0) + J_1\right]\beta_6^2\right.$$
$$- \frac{1}{12}\delta\left(b_1^0 + b_1^1\right)\frac{1}{2}\left(1-\alpha\right)\left(b_i^0 + b_1^1\right)\beta_6$$
$$\left. - \frac{1}{4}\left(\frac{1+\alpha}{2}b_1^0 + b_1^1\right)\left(b_1^1 + b_1^1 J_0 - b_1^0 J_1\right)\right\}, \tag{102}$$

$$a_s = \frac{1}{2\delta D}\left\{J_0\left(b_1^0 + b_1^1\right)\beta_3 + J_1 b_1^0 \beta_3 + b_1^0 + b_1^1\right.$$
$$\left. \times \left[\frac{1}{6}\delta b_1^0(b_1^0 + 2b_1^1) + \beta_3 + \delta\beta_6\right]\right\}, \tag{103}$$

$$\Delta = \beta_6^2 + \frac{1}{4}b_1^0(b_1^0 + 2b_1^1), \tag{104}$$

$$\Lambda = J_1 b_1^0 + (1 + J_0 + \frac{1}{3}\delta\beta_6)(b_1^0 + b_1^1), \tag{105}$$

$$D = -\frac{1}{2}(1 + J_0)(1 + J_0 + 2J_1)\beta_6^2 + \frac{1}{12}\delta(b_1^0 + b_1^1)$$
$$\times [(b_1^0 + b_1^1)(1 + J_0) + b_1^0 J_1]\beta_6 + \frac{1}{8}[(1 + J_0)b_1^1 - b_1^0 J_1]^2$$
$$-\frac{1}{288}\delta^2 b_1^0(b_1^0 + b_1^1)^2(b_1^0 + 2b_1^1), \tag{106}$$

$$\beta_{3\cdot 2^n} = 1 + \frac{(-1)^n b_2}{3 \cdot 2^n}, \tag{107}$$

$$J_1 = -\frac{1}{2}(1 - \alpha)\left[-\frac{1}{8}\delta^2(b_1^0)^2(1 - \alpha)(1 + J_0 + \frac{1}{3}\delta\beta_6)\right.$$
$$+ \frac{1}{12}\delta J_0(b_1^0)^2(2 - \delta)(\delta\beta_6 + \beta_3) + \frac{1}{6}\delta(b_1^0)^2(\delta\beta_6$$
$$\left. + \beta_3) + (1 + J_0)^2(\delta\beta_6 + \beta_3)^2\right](\delta\beta_6 + \beta_3)^{-1}$$
$$\times \left(\delta\beta_6 + \beta_3 + \frac{1}{24}\delta(1 - \alpha)(b_1^0)^2(2 - \delta)\right)^{-1}, \tag{108}$$

$$b_1^1 = -\left\{\frac{1}{2}b_1^0(1 - \alpha)\left[\frac{1}{12}\delta(b_1^0)^2(2 - \delta) + \delta\beta_6(2 + J_0) + \beta_3(1 + J_0)\right]\right\}$$
$$\times \left[\frac{1}{24}(1 - \alpha)\delta(b_1^0)^2(2 - \delta) + \beta_3 + \delta\beta_6\right]^{-1}. \tag{109}$$

Using the Hoye-Stell scheme [70] for the calculation of thermodynamics in the MSA, extended by Kalyuzhnyi and Holovko [71] for the AMSA case, the thermodynamic properties of the ion-dipole system can be calculated [68, 69]. The excess internal energy of the system is equal to

$$\frac{\beta U^{EL}}{V} = \frac{1}{4\pi}\left[\kappa_i^2 R_i^2(J_0 + J_1) - 2\kappa_i R_i \frac{1}{\delta}\sqrt{3y_s}(b_1^0 + b_1^1) - 6y_s b_2 \frac{1}{\delta^3}\right]. \tag{110}$$

Similarly to the RPM case the pressure, the chemical potentials and free energy contain three different contributions: the hard-sphere contributions (HS), the contributions from the mass action law (MAL) and electrostatic contribution (EL).

For example, the total free energy is equal to

$$\frac{\beta A}{V} = \frac{\beta A^{\text{HS}}}{V} + \frac{\beta A^{\text{MAL}}}{V} + \frac{\beta A^{\text{EL}}}{V}. \tag{111}$$

The chemical potentials can be found according to the relation,

$$\beta \mu_{\text{a}} = \frac{\partial}{\partial \rho_{\text{a}}} \left(\frac{\beta A}{V} \right) = \beta \mu_{\text{a}}^{\text{HS}} + \beta \mu_{\text{a}}^{\text{MAL}} + \beta \mu_{\text{a}}^{\text{EL}}, \tag{112}$$

where $a = i$ or $a = s$.

The expression for the pressure can be found according to the relation,

$$\beta p = \beta \sum_{\text{a}} \rho_{\text{a}} \mu_{\text{a}} - \frac{\beta A}{V} = \beta p^{\text{HS}} + \beta p^{\text{MAL}} + \beta p^{\text{EL}}. \tag{113}$$

For the hard-sphere contribution we can use the thermodynamics of a hard-sphere mixture [72] obtained within the PY theory.

The MAL contribution for the free energy can be presented in the form [9],

$$\frac{\beta A^{\text{MAL}}}{V} = \left[\rho_i \ln \alpha + \frac{1}{2} \rho_i (1 - \alpha) \right]. \tag{114}$$

The electrostatic part of the chemical potentials for ions and dipoles is expressed as,

$$\beta \mu_i^{EL} = \frac{1}{4\pi \rho_i} \left[\kappa_i^2 R_i^2 J_0 - \kappa_i R_i \frac{1}{\delta} \sqrt{3y_s} \, b_1^0 \right], \tag{115}$$

$$\beta \mu_s^{EL} = -\frac{1}{4\pi \rho_s} \left[\frac{\kappa_i R_i}{\delta} \sqrt{3y_s} \, (b_1^0 + b_1^1) + \frac{1}{\delta^3} 6 y_s b_2 \right]. \tag{116}$$

The electrostatic contribution for the pressure is given by,

$$\beta p^{\text{EL}} = \tilde{J} + \frac{\pi}{3} \sum_{a,b} \rho_a \rho_b R_{ab}^3 Sp \left[\mathbf{h}_{ab}(R_{ab}) \mathbf{x}_b \mathbf{h}_{ba}(R_{ab}) \mathbf{x}_a \right]$$

$$+ \frac{1}{6} \rho_i R_i (1 - \alpha) \left(\frac{\partial h_{ii}^{(\text{d})00}(r)}{\partial r} \right)_{r=R_i}, \tag{117}$$

where $\mathbf{x}_i$ has the form (10), $\mathbf{x}_s = 1$,

$$\tilde{J} = \frac{1}{12\pi} \left[\kappa_i^2 R_i^2 (J_0 + J_1) - 4\kappa_i R_i \frac{1}{\delta} \sqrt{3y_s} \, (b_1^0 + b_1^1) - 18 y_s b_2 \frac{1}{\delta^3} \right]. \tag{118}$$

The coefficients $h_{ab}(R_{ab})$ are the contact values of the corresponding harmonics of the electrostatic part of correlation functions,

$$
h_{ii}^{(d)\alpha\beta}(R_i) = \frac{1}{2\pi\rho_i R_i}\left[q_{ii}^{\alpha\beta} + (q_{ii}^{0\beta} + a_i^{\beta})\delta_{\alpha 1}\frac{1}{2}(1-\alpha)\right],
$$

$$
h_{is}^{011\alpha}(R_{is}) = \frac{\sqrt{3}}{2\pi\sqrt{\rho_i\rho_s}R_{is}}\left[q_{is}^{\alpha} + (q_{is}^0 + a_s)\delta_{\alpha 1}\frac{1}{2}(1-\alpha)\right], \quad (119)
$$

$$
h_{ss}^{110}(R_s) = \frac{1}{2\pi\rho_s R_s}\left(q_{ss} - \frac{2b_2}{R_s^2}\frac{\beta_{24}}{\beta_{12}^2}\right),
$$

$$
h_{ss}^{112}(R_s) = \frac{1}{2\pi\rho_s R_s}\left(q_{ss} + \frac{b_2}{R_s^2}\frac{\beta_{24}}{\beta_{12}^2}\right), \quad (120)
$$

where

$$
q_{ii}^{\alpha\beta} = \frac{1}{\Delta}\left[\left(\Delta J_\alpha - \frac{1}{4}\Lambda b_1^\alpha\right)a_i^\beta - \frac{1}{2}b_1^\alpha b_1^\beta + \frac{1}{2}(1-\alpha)\delta_{1\beta}b_1^0\right],
$$

$$
q_{is}^{\alpha} = \frac{1}{\Delta}\left[\left(\Delta J_\alpha - \frac{1}{4}\Lambda b_1^\alpha\right)a_s + b_1^\alpha\frac{1}{\delta}\beta_3\right],
$$

$$
q_{ss} = -\frac{1}{3}\left(b_1^0 + b_1^1\right)a_s - \frac{\Lambda}{2\delta}\beta_3 a_s - \frac{2}{\delta^2}\left(1-\beta_3^2\right), \quad (121)
$$

$$
R_i\left(\frac{\partial h_{ii}^{(d)00}(r)}{\partial r}\right)_{r=R_i} = -\frac{1}{2\pi\rho_i R_i}\Big[\,q_{ii}^{00} + (q_{ii}^{00} + q_{ii}^{01})(q_{ii}^{00} + a_i^0)\,q_{ii}^{00}q_{ii}^{10}
$$
$$
+ q_{is}^0\left(R_s q_{is}^0 + \frac{1}{2\delta}R_s^2(b_1^0 + b_1^1)a_i^0 + \frac{\Lambda}{2\Delta\delta}b_2 a_i^0\right.
$$
$$
+ \left.\frac{1}{\Delta\delta}b_2 b_i^0\right)\Big].
$$

$$(122)$$

For $g_{+-}^{00}(R_i)$ we use the exponential form (16), where $h_{\mathrm{d}}^{00}(R_i)$ is substituted by $h_{ii}^{(d)00}(R_i)$.

Expressions obtained for the free energy, pressure and chemical potentials can be used to study the thermodynamic properties of the electrolyte solutions, in particular, to describe the phase diagram of ionic fluids. Such a possibility is illustrated in Fig. 7, which shows the effect of ion pairing on liquid-liquid coexistence curve in the ion-dipole model as a function of the ion concentration $c_i = \rho_i/(\rho_i + \rho_s)$ and reduced temperature $T^\star = (b_i b_s)^{-1/2}$, $b_s = \beta p_s^2/R_s^3$. The solid line corresponds to the ion-dipole model with the parameter of ion association, $B = 10$. The dashed line corresponds to the ion-dipole model without association, i.e. $(B = 0)$ [64]. The ratio $b_i/b_s = 40$. As we can see from Fig. 7, due to the ion association, the critical temperature decreases and the critical concentration increases similarly to the pure ion model [12, 73]. The explicit consideration of the solvent subsystem leads to a different solvent concentration in two different phases.

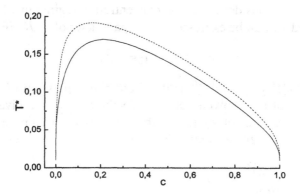

Figure 7. Liquid-liquid coexistence curve for the ion-dipole model in concentration-temperature coordinates. $b_i/b_s = 40$; dashed line – the MSA result, solid line – the AMSA result with $B = 10$.

3.2 Effect of ion association on the dielectric properties of electrolyte solution

The free ions and ion pairs play a distinct role in the dielectric properties of electrolyte solutions. Due to the saturation of the dipole orientation near free ions, the dielectric constant of the system decreases with the increase of free ion concentration. Ion pairs possess the dipole moments and produce an additional contribution to the dielectric properties. Due to the new polarization effect, the dielectric constant of the entire system increases with the increase of the ion pair concentration. It is generally accepted to distinguish the solvent dielectric constant, ϵ_s, and the solution dielectric constant, ϵ [3]. The dielectric constant of the solvent describes the polarization effect of the solvent molecules in the presence of ions; ϵ_s decreases with the increase of ion concentration. The dielectric constant of the solution also includes the polarization effect from the ion pairs that can increase or decrease with the increase of ion concentration. Due to this, $\epsilon > \epsilon_s$, and only for a completely dissociated electrolyte, $(\alpha = 1)$ $\epsilon = \epsilon_s$.

The dielectric constant of the solvent does not include the polarization effect from the ion pairs and can be found from Adelman expression [74]. Thus, in the framework of the AMSA theory [15, 16],

$$\epsilon_s = 1 + 3y_s \frac{\beta_{12}^4}{\beta_6^2} , \tag{123}$$

where β_{12} and β_6 are expressed by the parameter b_2 according to (107).

To analyze the dependence of ϵ_s on the size parameter δ and the degree of association α, we consider a low ion concentration regime in which parameters J_0, b_1^0 and b_2 can be expressed in the parameter of $\kappa_i R_i$ [67]. Hence,

$$\epsilon_s = \epsilon_s^w - a_2 \kappa_i^2 R_i^2 , \tag{124}$$

where $\epsilon_s^w = (\beta_3^0)^2 (\beta_{12}^0)^4 (\beta_6^0)^{-6}$ is the Wertheim expression [74] for the dielectric constant of pure solvent, the set of values $\beta_{3.2n}^0$ are given according to (107) with the change of b_2 to b_2^0, that corresponds to a pure hard-sphere dipolar system and is given by equation [75],

$$\frac{(\beta_3^0)^2}{(\beta_6^0)^4} - \frac{(\beta_6^0)^2}{(\beta_{12}^0)^4} = 3y_s , \tag{125}$$

$$a_2 = 6y_s^2 \delta (\beta_{12}^0)^8 (\beta_6^0)^4 \beta_{24}^0 \left\{ \alpha \left[\delta^2 \frac{(\beta_6^0)^2}{\beta_3^0} \left(\beta_{12}^0 + \frac{1}{4} \beta_3^0 \right) \right. \right.$$
$$\left. \left. + \left(3\delta \frac{\beta_6^0}{\beta_3^0} + 1 \right) \beta_{12}^0 \right] + \frac{1}{4} \frac{\beta_6^0}{\beta_3^0} \delta \left(1 + b_2^0 \right) + \beta_{12}^0 \right\} . \tag{126}$$

The coefficient a_2 describes the character of the decrease of ϵ_s as a function of ion concentration. Since the value of a_2 increases with the increase of parameter δ and decreases when $\alpha \to 0$, the dielectric constant ϵ_s as a function of ion concentration decreases faster with the decrease of ion size while ϵ_s decreases slower with the increase of the degree of ion association $1 - \alpha$.

Figure 8 illustrates the application of proposed theory to interpret the experimental dependence of the solvent dielectric constant, ϵ_s, of the aqueous solutions of nitrate and formate salts on ionic strength $I = 1/2 \cdot \sum_i c_i Z_i^2$ [76]. In the series Ba^{2+}, Cu^{2+}, Y^{3+} the cation size increases. These cations do not create the ion pairs with anion NO_3^- and the dependence of ϵ_s on the ion strength for nitrate salts is in agrement with our previous conclusions about the dependence of the solvent dielectric constant on the ion concentration with change of δ. For the formate salts, these cations create ion pairs with anions $CHOO^-$ due to hydrogen bonds between anion and water molecules from the hydration shell of cation; thus, the degree of association increases with the decrease of ion size. As a result, we obtain the concentration dependence for ϵ_s that is opposite to nitrate salts. Since the considered theory is elaborated for the symmetrical electrolytes and experimental data concern the asymmetri-

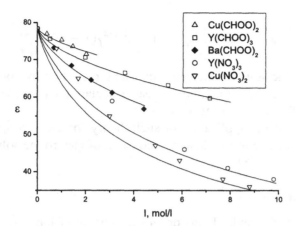

Figure 8. Comparison of the solvent dielectric constant as the function of ionic strength as predicted from the AMSA theory [77] against experimental data [76] for aqueous solutions of nitrate and formate salts.

cal electrolyte, we can consider the agreement of theoretical and experimental results to be more or less satisfactory.

The polarization effect on the ion pair is important in describing the dielectric constant ϵ of the solution. For this purpose it is possible to use the modified version of the MSA-MAL theory in which ion pairs are included as new polar entities with the density $1/2 \cdot \rho_i (1 - \alpha)$ and with the dipole moment $p_i = eR_i$. Correspondingly, the ion density should be changed to the density of free ions $\alpha \rho_i$. Such ion-dipole mixture being taken into consideration, again leads to the form (123) for the solution dielectric constant. However, in the considered case y_s should be changed to

$$y = y_\mathrm{s} + \frac{1}{6}\kappa_i^2 R_\mathrm{ef}^2 (1 - \alpha). \tag{127}$$

The system of equations for the parameter b_2 should be also slightly modified. If we assume for the simplification that ion pairs have the same sizes as the solvent molecules, the system of equations for the parameter b_2 reduces to the system that is similar to that obtained in the MSA theory for ion- dipole mixture [65, 66]. The proposed theory includes two opposite tendencies. Due to the change of κ_i to κ_α, the solution dielectric constant, ϵ, decreases slower with the increase of ion concentration. Due to the change of y_s to y, the dielectric constant ϵ can increase with the increase of ion concentration. For the low ion concentration,

$$\epsilon_s = \epsilon_s^w - a_2(\alpha = 1)\kappa_i^2 R_i^2 \cdot \alpha + \frac{1}{2}\kappa_i^2 R_{ef}^2 (1 - \alpha)\frac{(\beta_{12}^0)^4}{(\beta_6^0)^2} . \tag{128}$$

As we can see, the solution dielectric constant ϵ as the function of ion concentration decreases in the regime of weak association ($\alpha \to 1$)and increases in the regime of strong association ($\alpha \to 0$) .

Some results of the application of such theory for the description of concentration dependence of the dielectric constant of electrolyte solution will be published elsewhere [78].

4.   Summary

In this chapter the revised version of the concept of ion association in the theory of electrolyte solutions is reviewed. The developed approaches are based on the modern theory of associating fluids [9–12]. The background of the revised version of ion association concept is the analytical solution of the AMSA and MSA-MAL approaches for the ion and ion-dipole models. It is shown that the AMSA theory in framework of ion approach is very promising for the description of the thermodynamic properties such as osmotic and activity coefficient of electrolyte solutions in a weakly polar solvent or for the interpretation of electrolyte effect on the intramolecular electron transfer reactions. The MSA-MAL theory in the framework of ion approach is more useful for the description of electrical properties such as electroconductivity of electrolyte solution of the weakly polar solvent. Both theories in the framework of ion approach are useful in the explaining of the specific properties of electrical double layer in the regime of strong ion-ion interaction. For the description of thermodynamic properties such as adsorption coefficient, the AMSA theory appears to be more preferable and for the description of electrical properties such as the capacitance of double layer, the MSA-MAL theory is more acceptable. However, for the description of the properties of double layer for the real electrolytes, the development of the associative concept in the framework of the ion-dipole model is needed. The AMSA theory in the framework of the ion-dipole model is useful for the description of ion concentration dependence of the solvent dielectric constant while the MSA-MAL theory in the framework of ion-dipole model is more preferable for the description of electrolyte solution dielectric constant. The extension of the AMSA and MSA-MAL theories to the ion-dipole model by taking into account solvent-solvent and ion-solvent association effects has been performed [16]. Hence, the possibility to describe the ion association together with the ion solvation effect is very promising. Among them, for example, the treatment of the solvent sharing and solvent separated ion pairs as well as the contact ion pairs are very important for the explanation of the dielectric properties of real electrolyte solutions [79].

Acknowledgements

The majority of different aspects of the results presented in this review were carried out in cooperation with J. Barthel, D. Boda, D. Henderson, Yu. Kalyuzhnyi, V. Kapko, H. Krienke, A. Lyashchenko, P. Piotrowiak, I. Protsykevich, E. Vakarin. The author expresses gratitude to all of them for fruitful and permanent collaboration. The author thanks STCU for partial support of this work under grant No. 1930.

References

[1] Bjerrum, N. *Mat.-Fys. Medd. K. Dan. Vidensk, Selsk*, 1926, **7**, p. 1.

[2] Debye, P., and Hückel, E. *Physik. Z.*, 1923, **24**, p. 185.

[3] Barthel, J., Krienke, H., and Kunz, W. (1998). *Physical Chemistry of Electrolyte Solutions – Modern Aspects*. New York: Springer.

[4] Waisman, E., and Lebowitz, J.L. *J. Chem. Phys.*, 1972, **56**, p. 3086, 3093.

[5] Blum, L. *Mol. Phys.*, 1975, **30**, p. 1529.

[6] Krienke, H., Barthel, J., Holovko, M., Protsykevich, J.A., and Kalyuzhnyi, Yu. *J. Mol. Liq.*, 2000, **87**, p. 191.

[7] Holovko, M.F., and Kalyuzhnyi, Yu.V. *Mol. Phys.*, 1991, **73**, p. 1145.

[8] Blum, L., and Bernard, O. *J. Stat. Phys.*, 1995, **79**, p. 569.

[9] Wertheim, M.S. *J. Stat. Phys.*, **35**, p. 19, 35.

[10] Wertheim, M.S. *J. Stat. Phys.*, 1986, **42**, p. 459, 477.

[11] Holovko, M.F. *Cond. Matter Phys.*, 1999, **2**, No. 2(18), p. 205.

[12] Holovko, M.F. *J. Mol. Liq.*, 2002, **96–97**, p. 65.

[13] Barthel, J., Krienke, H., Holovko, M.F., Kapko, V.J., and Protsykevich, J.A. *Cond. Matter Phys.*, 2000, **3**, No. 3(23), p. 657.

[14] Barthel, J., Krienke, H., Neuder, R., and Holovko, M.F. *Fluid Phase Eq.*, 2002, **194–197**, p. 107.

[15] Holovko, M.F., and Kapko, V.J. *Cond. Matter Phys.*, 1998, **1**, No. 2(14), p. 239.

[16] Holovko, M.F., and Kapko, V.J. *J. Mol. Liq.*, 2000, **87**, p. 109.

[17] Kalyuzhnyi, Yu., Holovko, M.F., and Haymet, A.D.J. *J. Chem. Phys.*, 1991, **95**, p. 1951.

[18] Kalyuzhnyi, Yu., and Holovko, M.F. *J. Mol. Phys.*, 1993, **80**, p. 1165.

[19] Ebeling, W. *Z. Phys. Chem. (Leipzig)*, 1968, **238**, p. 400.

[20] Falkenhagen, H., Ebeling, W., and Hertz, H.G. (1971). *Theorie der Elektrolyte*. Leipzig: S. Hizzel-Verlag.

[21] Vakarin, E.V., Holovko, M.F., and Piotrowiak, P. *Chem. Phys. Let.*, 2002, **363**, p. 7.

[22] Kapko, V.I., Holovko, M.F., Miller, J., and Piotrowiak, P. *J. Phys. Chem. B*, (in press).

[23] Holovko, M., Kapko, V.I., Henderson, D., and Boda, D. *Chem. Phys. Let.*, 2001, **341**, p. 363.

[24] Holovko, M., Kapko, V., Henderson, D., and Boda, D. (in preparation).

[25] Werthein, M.S. *J. Chem. Phys.*, 1986, **85**, p. 2929.

[26] Bernard, O., and Blum, L. *J. Chem. Phys.*, 1996, **104**, p. 4746.

[27] Carnahan, N.F., and Starling, K.E. *J. Chem. Phys.*, 1969, **51**, p. 635.

[28] Blum, L., and Hoye, J.S. *J. Phys. Chem.*, 1977, **81**, p. 134.

[29] Barthel, J., Neuder, R., Poepke, H., and Wittmann, H. *J. Solution Chem.*, 1998, **27**, p. 1055.

[30] Barthel, J., Neuder, R., Poepke, H., and Wittmann, H. *J. Solution Chem.*, 1999, **28**, p. 489, 1263, 1277.

[31] Fuoss, R.M., and Kraus, C.A. *J. Am. Chem. Soc.*, 1933, **55**, p. 2387.

[32] Chhih, A., Turq, P., Bernard, O., Barthel, J., and Blum, L. *Ber. Bunsenges Phys. Chem.*, 1994, **98**, p. 1516.

[33] Turq, P., Blum, L., Bernard, O., and Kunz, W. *J. Phys. Chem.*, 1995, **99**, p. 822.

[34] Delsignore, M., Farber, H., and Petrucci, S. *J. Phys. Chem.*, 1985, **89**, p. 4968.

[35] Marcus, R.A., and N. Sutin, *Biochim. Biophys. Acta*, 1985, **811**, p. 285.

[36] Marcus, R.A. *J. Phys. Chem.*, 1956, **24**, p. 966.

[37] Perng, B.C., Newton, M.D., Raineri, F.O., and Friedman, H.L. *J. Chem. Phys.*, 1996, **104**, p. 7153,7177.

[38] Rossky, P.J., and Simon, J.D. *Nature*, 1999, **370**, p. 263.

[39] Piotrowiak, P., and Miller, J.R. *J. Phys. Chem.*, 1993, **97**, p. 13052.

[40] Closs, C.L., and Miller, J.R. *Science*, 1988, **240**, p. 440.

[41] Holovko, M.F., Kapko, V.I., and Piotrowiak, P. *J. Phys. Chem B*, (in press).

[42] Nielson, R.M., Mc Manis, G.E., Sofford, L.K., and Weaver, M.J. *J. Phys. Chem*, 1989, **93**, p. 2152.

[43] Kuznetsov, A.M., Pheklps, P.K., and Weaver, M.J. *Int. J. Chem. Kinetic*, 1990, **22**, p. 815.

[44] Holovko, M.F., Kapko, V.I., and Piotrowiak, P., (in preparation).

[45] Schatz, T.R., Kobetic, R., and Piotrowiak, P. *J. Photochem and Photobiol. A.*, 1997, **105**, p. 249.

[46] Piotrowiak, P. In Photochemistry Methods for the Study of Electron Transfer. *Advances in Chemistry*, 1998, No. 254, p. 219.

[47] Henderson, D., Blum, L., and Lebowitz, J.L. *J. Electroanal. Chem.*, 1979, **102**, p. 315.

[48] Landau, L.D., and Lifshitz, E.M. (1960). *Electrodynamic of Continuous Media*. Oxford: Pergamon Press.

[49] Blum, L. *J. Phys. Chem*, 1977, **81**,p. 136.

[50] Crozier, P.S., Rowley, R.L., Henderson, D., and Boda, D. *Chem. Phys. Lett*, 2000, **325**,p. 675.

[51] Henderson, D. *J. Mol. Liq.*, 2001, **92**, p. 29.

[52] Boda, D., Chan, K.Y., Henderson, D., and Wasan, D.T. *Chem. Phys. Lett.*, 1999, **308**, p. 473.

[53] Henderson, D., Abraham, F.F., and Barker, J.A. *Mol. Phys.*, 1976, **31**,p. 1291.

[54] Holovko, M.F., and Vakarin, E.V. *Mol. Phys.*, 1995, **84**, p. 1057.

[55] Golovko, M.F., and Yukhnovsij, I.R. (1985). *In the Chemical Physics of Solvation*, part A, chapter 6, ed. by Dogonadze, R.R., Kalman, E., Kornyshev, A.A. and Ulstrup, J. Amsterdam: Elsevier.

[56] Klimontovich, Yu.L., Wilhelmsson, H., Yakimenko, I.P., and Zagorodny, A.G. (1990). *Statistical Theory of Plasma-Molecular Systems*. Moskov: Publishing of Moskov University.

[57] Jancovici, B. *J. Stat. Phys.*, 1982, **28**, p. 43.

[58] di Caprio, D., Stafiej, J., and Badiali, J.P. *Mol. Phys*, 2003, **101**, p. 3197.

[59] Blum, L., and Henderson, D. *J. Chem. Phys.*, 1981, p. 74.

[60] Holovko, M.F., Kapko, V., Henderson, D., and Boda, D., (in preparation).

[61] Blum, L. *Chem. Phys. Lett*, 1974, **26** p. 200; *J. Chem. Phys.*, 1974, **61** p. 2129.

[62] Adelman, S.A., and Deutch, J.M. *J. Chem. Phys.*, 1974, **60** p. 3935.

[63] Yukhnovsij, I.R., and Holovko, M.F. (1980). *Statistical Theory of Classical Equilibrium Systems*. Kyiv: Naukova dumka.

[64] Hoye, J.S., and Lomba, E. *J. Chem. Phys.*, 1988, **88** p. 5790; Hoye, J.S., Lomba, E., and Stell, G. *J. Chem. Phys.*, 1988, **89**, p. 7642.

[65] Blum, L., and Wei, D.Q. *J. Chem. Phys.*, 1987, **87**, p. 555; Wei, D.Q., and Blum, L. *J. Chem. Phys.*, 1987, **87**, p. 2999.

[66] Golovko, M.F., and Protsykevich, I.A. *Chem. Phys. Lett.*, 1987, **142**, p. 463; *J. Stat. Phys.*, 1989, **54**, p. 707.

[67] Kalyuzhnyi, Yu.V., Protsykevich, J.A., and Holovko, M.F. *Chem. Phys. Lett.*, 1993, **215**, p. 1.

[68] Holovko, M.F., and Kapko, V.I. *J. Chem. Phys.*, (in press).

[69] Kapko, V.J. (2001). *Dissertation*. Lviv.

[70] Hoye, J.S., and Stell, G. *J. Chem. Phys.*, 1977, **67**, p. 439.

[71] Kalyuzhnyi, Yu.V., and Holovko, M.F. *J. Chem. Phys.*, 1998, **108**, p. 3709.

[72] Mansoori, G.A., Carnahan, N.F., Starling, K.E., and Leland, T.W. *J. Chem. Phys.*, 1971, **54**, p. 1523.

[73] Kalyuzhnyi, Yu.V. *Mol. Phys.*, 1998, **94**, p. 735.

[74] Adelman, S.A. *J. Chem. Phys.*, 1976, **64**, p. 724.

[75] Wertheim, M. *J. Chem. Phys.*, 1971, **55**, p. 4291.

[76] Lileev, A.S., Balakaeva, I.V., and Lyashchenko, A.K. *Russian J. inorg. Chem.*, 1998, **43**, p. 960.

[77] Holovko, M.F., Kapko, V.I., and Lyashchenko, A.K. (in preparation).

[78] Holovko, M.F., Kapko, V.I., Krienke, A., and Barthel, J. (in preparation).

[79] Buchner, R., Chen, T., and Hefter, G. *J. Phys. Chem. B.*, 2004, **108**, p. 2365.

TOWARDS THE ROLE OF THE RANGE OF INTERMOLECULAR INTERACTIONS

Systematic computer simulations of the fluids with electrostatic interactions

I. Nezbeda [1,2], J. Kolafa [3]

[1] *E. Hála Laboratory of Thermodynamics, ICPF, Academy of Sciences,*
165 02 Prague 6, Czech Republic

[2] *Department of Physics, J. E. Purkyně University,*
400 96 Ústí n. Lab., Czech Republic

[3] *Institute of Physical Chemistry, Prague Institute of Chemical Technology*
166 28 Praha 6, Czech Republic

Abstract   It has been traditionally believed that, unlike simple fluids whose properties are determined primarily by the short-range repulsions, the properties of complex fluids are strongly affected by the long-range electrostatic interactions. In the course of investigations, extensive and systematic computer simulations have been performed on typical quadrupolar, dipolar and associating fluids using available realistic potential models. The structural characteristics as well as dielectric constant and the thermodynamic properties of both the homogeneous liquid and supercritical fluid phases, and vapor-liquid equilibria have also been considered. The obtained results lead to the conclusion that the structure of pure fluids, both polar and associating, is governed by the same molecular mechanism as for simple fluids, *i.e.* by the short-range interactions, whereas the long-range part of the electrostatic forces, regardless of their strength, plays only a marginal role and may be treated as a perturbation only. However, it turns out that for mixtures of charged particles the situation is much more complex and that the observed behavior is very sensitive to the details of intermolecular interactions.

Keywords: Coulomb interactions, structure of fluids, dipole-dipole interaction, dielectric constant, polar fluids, associating fluids, dilute electrolytes

1.   Introduction

The ultimate goal of the statistical mechanics of matter is to provide methods of explaining and predicting the experimentally measurable quantities of a

D. Henderson et al. (eds.), Ionic Soft Matter: Modern Trends in Theory and Applications, 83–108.
© 2005 *Springer. Printed in the Netherlands.*

given substance, in the considered case fluids, in terms of the properties of its constituent particles. Realizing that the observed macroscopic properties result primarily from the established underlying structure of the fluid, the basic problem which modern theories of fluids have to address is: "What interactions play the decisive role in establishing the structure?" An answer to this question is an indispensable first step towards the development of simple theoretically-based models, and hence molecular-based workable expressions for the thermodynamic properties of fluids and more complete understanding of their behavior. For simple fluids, i.e., fluids of almost spherical molecules with negligible electrostatic interactions, this problem was solved already more than three decades ago when it was firmly established that the observed structure is determined primarily by short-range *repulsive* interactions [1, 2]. However, for polar and associating fluids this problem has remained open with prevailing intuitive conviction that in this case the long-range Coulomb interactions must also play an important role.

Classical statistical mechanics views fluids (*i.e.*, gases and liquids) as a collection of N mutually interacting molecules confined to a volume V at a temperature T and specifies the system by a total intermolecular potential energy U, $U(x_1, x_2, \ldots, x_N) \equiv U(1, \ldots, N)$, where x_i stands for a set of generalized coordinates of molecule i. Not only for convenience and simplicity, but as an utmost necessity if a tractable theory is to be ultimately applied, the assumption of pairwise additivity is made at this stage, and U is simplified to

$$U(1, 2, \ldots, N) = \sum_{i<j} u(i, j), \tag{1}$$

where $u(i, j)$ is a pair potential involving the pair i–j. This is a rather crude approximation and fails in the case of dense fluids if for $u(i, j)$ the true pair potentials (i.e., as determined from the properties of dilute gas) are used [3]. However, if effective pair potentials are used in (1), then this equation represents an acceptable and reasonably accurate approximation to the total configurational energy. In this work we will always use effective potentials.

Traditionally, effective pair potentials have been constructed in a form reflecting the properties of the individual molecules. Thus, accounting for the overall molecular electroneutrality, three main classes of fluids made up of (relatively) small molecules have been recognized [3]:

1 Simple fluids, with (nearly) spherical charge distribution and hence with either negligible or even completely absent electrostatic interactions, *i.e.*,

$$u_{\text{normal}}(1, 2) = u_{\text{non-el}}(1, 2). \tag{2}$$

Typical examples are inert gases and lower hydrocarbons.

2 Polar fluids, whose molecules bear non-negligible permanent multipole accounting for a significant deviation of the charge distribution from sphericity. A multipole-multipole interaction term is thus added to $u_{\text{non-el}}$ to construct an effective pair potential,

$$u_{\text{polar}}(1,2) = u_{\text{non-el}}(1,2) + u_{\text{multipole-multipole}} \cdot \tag{3}$$

Typical examples are acetone and acetonitrile with a quite large dipole moment (dipolar fluids), and carbon dioxide with a large quadrupole moment (quadrupolar fluid).

3 Associating fluids, whose molecules, in addition to being polar, are able to form also hydrogen bonds,

$$u_{\text{assoc}}(1,2) = u_{\text{polar}}(1,2) + u_{\text{H-bonding}} \cdot \tag{4}$$

Water and alcohols are well-known examples.

The above classification tends to explain the properties of a more complex fluid in terms of an excess over a less complex (simpler) fluid pointing to a perturbation treatment as a suitable tool for both theory and applications. The properties of fluids belonging to different classes seem thus to be determined by the different types of predominant interactions. Consequently, to be able to understand and thus to predict the macroscopic properties of fluids, it is natural (and important) to determine the effect of the individual terms contributing to u on the macroscopic behavior. However, this need not be the case when the origins of the potential functions are considered. With the advance of computer technology, quantum chemical computation methods have also made considerable progress in the development of reasonably accurate effective pair potentials, but in a form which differs from that of Eqs. (2)–(4). Consequently, the simple physical picture of intermolecular interactions is lost and decompositions (3)–(4) become of little use.

According to our contemporary understanding of intermolecular interactions, the molecules are viewed as bodies made up of individual atoms or groups of atoms that are the seat of two types of interactions: (1) non-electrostatic (non-el) interaction generating a strong repulsion at short separations and a weak (van der Waals) attraction at medium separations, and (2) the long-range Coulombic charge-charge interaction. A common realistic pair potential has thus the form of a site-site potential,

$$u(1,2) \equiv u(R_{12}, \mathbf{\Omega}_1, \mathbf{\Omega}_2) = u_{\text{non-el}}(1,2) + u_{\text{Coul}}(1,2) =$$

$$= \sum_{i\in\{1\}} \sum_{j\in\{2\}} \left\{ u_{\text{non-el}}\left(\left|\mathbf{r}_1^{(i)} - \mathbf{r}_2^{(j)}\right|\right) + \frac{q_1^{(i)} q_2^{(j)}}{\left|\mathbf{r}_1^{(i)} - \mathbf{r}_2^{(j)}\right|} \right\}, \tag{5}$$

where the Lennard-Jones (LJ) potential,

$$u_{\text{LJ}}(r_{ij}) = 4\epsilon_{ij} \left[\left(\frac{\sigma_{ij}}{r_{ij}} \right)^{12} - \left(\frac{\sigma_{ij}}{r_{ij}} \right)^{6} \right], \tag{6}$$

is usually (but not necessarily) chosen for the non-electrostatic site-site part, $u_{\text{non-el}} = u_{\text{LJ}}$. In Eq. (5), R_{12} denotes the separation between the reference points (further referred to as R-sites) of molecules 1 and 2, Ω_k denotes the orientation of molecule k, $r_k^{(i)}$ is the position vector of site i on molecule k, $r_{ij} = |r_1^{(i)} - r_2^{(j)}|$, and $q_k^{(i)}$ is the partial charge of site i of molecule k. Potentials for different compounds thus differ only in the geometrical arrangement of the interaction sites and in the strengths of the individual site-site interactions; effects which might be identified as, e.g., the dipole-dipole interaction or hydrogen bonding are included *implicitly* and result as a net effect of the complex intermolecular force field. A link between the general model (5) and intuitive physical models (2)–(4) is established by expanding (5) in inverse powers of the separation of reference points $(R - R)$; one has however to keep in mind that multipole-multipole interactions are not real physical entities but that they arise from the above mathematical manipulation attempting to approximate the force field of a spatially distributed set of charges [4]. While being reasonable at large separations (and numerically equivalent to the form of (5)), at short intermolecular separations this approximation is inadequate unless a number of multipole-multipole interactions is incorporated.

Accepting the effective pair potential in the above form of (5), one immediately loses the clear (and simple) physical picture of intermolecular interactions. We should therefore examine the relation between the intermolecular interaction and observed properties more deeply which however need not be necessarily disadvantageous. There is one property which might be useful for the characterization of intermolecular interactions: their rate of decay with increasing intermolecular separation or, equivalently, the *range* over which they operate. This is particularly important from the theoretical point of view because theory is rarely able to handle the complex potential $u(1, 2)$ as a whole but must rely on its various decompositions. In fact, the rate of decay has already been used for simple fluids as a criterion to devise a convergent perturbation expansion [1, 2]. One must realize that the usual statement "... the *structural* properties of normal fluids are determined primarily by *repulsive interactions*" [2, 5] means that it is the *short-range part of the total pair potential* $u(1, 2)$ which is the determining factor (and which happens to be also its repulsive part).

The range of intermolecular interactions was recognized as an important factor already a long time ago and was used for instance, along with other quantities, in a classification of fluids by Andersen [6] but a systematic investigation of its effect on the properties of fluids began only recently in connection with studies of water [7, 8]. In addition to the general and never-ending quest for better and deeper understanding of nature, one might consider as an impetus for these investigations also an apparent conflict between traditional arguments on the strong effect of the long-range dipole-dipole interaction on the properties of polar and associating fluids on one side, and the success of various simple short-range models in the interpretation of a number of phenomena [9–14] and their use as a reference system in expressions of thermodynamic properties for applications [15, 16] on the other side.

The purpose of this chapter is to first summarize and analyze recent results on the effect of the range of intermolecular interactions on various properties of *pure* fluids, and then to examine potential validity of the reached conclusions for aqueous electrolytes. The primary focus is on the structure because it is the structure upon which thermodynamic properties are superimposed. The main finding for pure fluids is that the impact of the long-range electrostatic interactions (or, more accurately, of their long-range parts) on their structural properties is suppressed to a perturbation. Consequently, both the structural and thermodynamic properties of both above-specified classes of non-simple fluids, polar and associating, are driven by the same molecular mechanism as normal fluids, namely by the short-range interactions. These short-range interactions (which may be both repulsive and attractive) capture also a good deal of the thermodynamic behavior. Consequences of these findings for a further development of theory and applications are also briefly discussed. Finally, an attempt to examine to what extent these conclusions may remain valid also for mixtures with charged components is made by considering dilute aqueous electrolytes.

2. Theoretical background and technical details

2.1 *Potential models*

It is assumed that systems with fast decaying non-electrostatic interactions (class 1 of our classification) are very well understood nowadays and that they thus do not pose any problem for theory anymore. It is therefore the range of the Coulomb interactions and its effect on the properties of fluids which is of the main concern. To investigate this effect for fluids of class 2 and 3, trial potentials of a different range, u^T, are constructed from the given parent model u so that,

$$u^T(1, 2) = u_{\text{non-el}}(1, 2), \qquad \text{for} \quad R_{12} > R_{\text{range}} . \tag{7}$$

It is only a technical matter how to actually construct such u^Ts. In the early stage of development of theory of fluids it was common to switch off the

Coulomb part completely, *i.e.,* $R_{\text{range}} = 0$. This crude approach must have lead to an evident (and misleading) conclusion that the electrostatic interactions as a whole were very important. Thus, at least a part of $u_{\text{Coul}}(1, 2)$ must be retained if its influence is to be assessed properly. The simplest possibility is to use brute force and simply cut off u_{Coul} at some $R_{12} = R_{\text{range}}$,

$$u^{\text{T}}(1, 2) = \begin{cases} u(1, 2), & \text{for} R_{12} \leqslant R_{\text{range}}, \\ u_{\text{non-el}}(1, 2), & \text{for} R_{12} > R_{\text{range}}. \end{cases} \tag{8}$$

However, this way gives rise to discontinuities of structural properties and in majority of recent studies the trial potential has therefore been defined by

$$\begin{aligned} u^{\text{T}}(1, 2) &= u(1, 2) - S\left(R', R''; R_{12}\right) u_{\text{Coul}}(1, 2) = \\ &= u_{\text{non-el}}(1, 2) + \left[1 - S\left(R', R''; R_{12}\right)\right] u_{\text{Coul}}(1, 2), \end{aligned} \tag{9}$$

where S is a switch function which gradually switches off the interaction at short separations,

$$S(R', R''; r) = \begin{cases} 0, & \text{for} r \leqslant R', \\ (r - R')^2 (3R'' - R' - 2r)/(R'' - R')^3, & \text{for} R' < r < R'', \\ 1, & \text{for} r \geqslant R'', \end{cases} \tag{10}$$

and R' and $R'' \equiv R_{\text{range}}$ are its parameters.

Another possibility is to decompose the electrostatic field of the set of spatially distributed charges into the field of point multipoles and to subtract the long-range part of a certain number of the leading multipole-multipole interaction terms from $u(1, 2)$ [17],

$$\begin{aligned} u^{\text{T}}(1, 2) &= u(1, 2) - S(R', R''; R_{12}) \left[u_{\text{DD}}(1, 2) + u_{\text{DQ}}(1, 2)\right. \\ &\quad \left. + u_{\text{QQ}}(1, 2) + \ldots \right], \end{aligned} \tag{11}$$

where DD, DQ, and QQ stand for the dipole-dipole, dipole-quadrupole, and quadrupole-quadrupole interactions, respectively. This construction does not make the Coulomb contribution to u^{T} strictly zero beyond any R_{range}, but definitely shortens its range and makes it nearly negligible at large separations. In fact, it was shown [18] that for the appropriate choice of (R', R'') the results for u^{T} given by (9), and (11) with only the dipole-dipole term are practically identical. This choice of the short-range system is particularly appealing from the theoretical point of view because it makes it possible to devise a perturbation expansion in powers of the multipole-multipole interactions.

Finally, it is also possible to reverse the problem and dampen the short-range part of the Coulomb interaction making thus its long-range part responsible for the observed behavior,

$$
\begin{aligned}
u_{\mathrm{lr}}^{\mathrm{T}}(1,2) &= u(1,2) - \left[1 - S(R', R''; R_{12})\right] u_{\mathrm{Coul}}(1,2) = \\
&= u_{\mathrm{non\text{-}el}}(1,2) + S(R', R''; R_{12}) u_{\mathrm{Coul}}(1,2),
\end{aligned} \tag{12}
$$

where the non-electrostatic part (or at least a part thereof) must be retained to prevent collapse at short separations.

2.2   Structure and thermodynamics

The primary quantity of interest in studies of fluids is their structure from which all other properties can be derived [19]. All necessary information is provided by the full pair correlation function $g(1,2)$ whose complete experimental determination is however practically impossible. The usual way is to characterize the structure of fluids by the complete set of site-site correlation functions g_{ij},

$$
4\pi r_{ij}^2 g_{ij}(r_{ij}) = \left(1 - \frac{1}{N}\right) V \left\langle \delta \left(r_{ij} - \left|\mathbf{r}_1^{(i)} - \mathbf{r}_2^{(j)}\right|\right)\right\rangle, \tag{13}
$$

where N is the number of particles, V the volume, δ is the Dirac delta distribution, and $\langle \ldots \rangle$ denotes an ensemble average. Another reduction in the number of variables is achieved when only the mutual orientation of selected molecular axes is considered. These correlations are described by the dipole-dipole correlation functions $G_1(r)$, called also "local g-factors" [20],

$$
4\pi r^2 G_1(r) = \left(1 - \frac{1}{N}\right) V \left\langle \delta \left(r - \left|\mathbf{r}_1^{(R)} - \mathbf{r}_2^{(R)}\right|\right) \mathcal{P}_1(\cos \Theta_{12})\right\rangle, \tag{14}
$$

where $\mathcal{P}_1$ is the normalized Legendre polynomial, and Θ_{12} is the angle formed by the chosen axes of molecules 1 and 2. The most important of these functions is function G_1 for its relation to the Kirkwood g-factor and hence to the dielectric constant [21] (see below).

A quantity which has also attracted a good deal of attention in connection with dipolar and associating fluids is the dielectric constant. In finite samples of volume V (used in computer simulations) the natural quantity to measure is the fluctuation of the total dipole moment of the system, $\langle M^2 \rangle$, $M = \sum_{i=1}^{N} \mu_i$, where μ_i are the dipole moments of the individual molecules. For nonpolarizable molecules in the periodic boundary conditions and with the long-range forces treated by the Ewald summation the dielectric constant, ϵ, is given (in CGSE system of units) by [22],

$$\frac{(\epsilon - 1)(2\epsilon' + 1)}{(2\epsilon' + \epsilon)} = \frac{4\pi}{3} \frac{\langle M^2 \rangle}{V k_B T}, \tag{15}$$

where ϵ'_r is the dielectric constant of the surrounding continuum, k_B is the Boltzmann constant and T is the temperature. The fluctuation $\langle M^2 \rangle$ can also be expressed by the integral over the dipole-dipole correlation function G_1,

$$g_1 = \frac{\langle M^2 \rangle}{N \mu^2} = 1 + (N - 1) \langle \cos \Theta_{12} \rangle = 1 + \rho \int G_1(r) \, dr, \tag{16}$$

where g_1 is the Kirkwood g-factor (sometimes called "finite sample Kirkwood factor" to explicitly express its dependence on boundary conditions) introduced to account for correlations between the dipole moments (it equals unity if the dipoles are uncorrelated) [21].

Concerning the thermodynamic properties, the internal energy, E, and pressure, P, are the typical quantities considered. Into the discussion of thermodynamic properties we also include vapor-liquid coexistence properties for their great industrial importance.

The reported results for equilibrium properties were obtained by means of the standard Monte Carlo (MC), molecular dynamics (MD), and Gibbs ensemble (GE) simulation methods [23, 24]. For the trial systems of a finite range the simple spherical cutoff was used, whereas in simulations of the full systems either the Ewald summation or the reaction field method were used. For further technical details we refer the reader to the original papers.

3.   Results and discussion

Following the strategy and methods outlined in the preceding section, the effect of the range of intermolecular interactions on the properties of fluids has been investigated intensively in last years by computing a number of both the structural [complete sets of the 1D site-site correlation functions, selected 2D site-site functions, radial slices through the full pair correlation function $g(1, 2)$, and dipole-dipole correlation functions of the 1st and 2nd order] and thermodynamic properties of homogeneous phases (internal energy, pressure, and dielectric constant), and the thermodynamic properties of coexisting vapor-liquid phases [7, 8, 18, 25, 26]. In these simulations four typical representatives for polar (quadrupolar carbon dioxide and dipolar acetone) and associating (methanol and water) compounds have been considered [for the used realistic potentials, choice of the reference points, and combining rules we refer the reader to the original papers]. One may however argue that for unambiguous conclusions to be drawn, compounds with extreme molecular

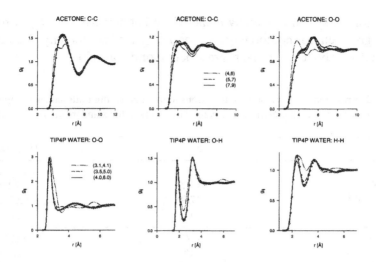

Figure 1. Comparison of the site-site correlation functions of various short-range models (lines) with the full potential models (circles) for acetone and water at ambient conditions. Numbers in the legend denote the switching range.

properties should be included as well. We have therefore extended the above list and performed the same simulations also for acetonitrile (whose dipole moment is nearly twice as large as that of acetone) and hydrogen fluoride with very strong H-bonds. The common realistic potential models have been chosen to describe their interaction (Jorgensen-Briggs model for acetonitrile [27], and the Cournoyer-Jorgensen model for hydrogen fluoride [28]) and selected results are also reported in this paper.

The crucial step affecting the above listed properties is the choice of the switching range (R', R''), i.e., the range over which the long-range part of $u(1,2)$ is gradually turned off: when the potential is switched off at too short separations then too much of the Coulomb interactions may be missing whereas for too large R' and $R'' \equiv R_{\text{range}}$ there may not be any significant difference between u and u^{T}. In Fig. 1 we exemplify therefore how the choice of (R', R'') affects the site-site correlation functions for acetone and water. In general, for all compounds considered we computed the site-site correlation functions for a number of densities both at ambient and supercritical conditions. Since the largest effect is expected (and really observed) at low temperatures (with respect to the existence of the liquid phase) and high densities, for the sake of space the correlation functions only at the ambient conditions are shown ($T = 298\,\text{K}$, $\rho = 790\,\text{kg}\,\text{m}^{-3}$ for acetone; $T = 273.15\,\text{K}$, $\rho = 997\,\text{kg}\,\text{m}^{-3}$ for water). As one could expect, for too short-range potential u^{T} differences in the structure may become even qualitative; nonetheless, the larger separation at which u is switched off, the better agreement between the site-site functions is obtained for the systems given by u and u^T. What however may be surprising is close agreement for not too large R' or R'' (term "not too large" should be interpreted as "about $1.5\times$ the nearest-neighbor distance").

Table 1. Contribution of electrostatic interactions to the total configurational energy in dependence on the switching range. Numbers in parentheses denote the standard error of last digit.

	range	U_{Coul}[kJ/mol]
acetone	(4/6)	$-5.296(22)$
$T = 298\,\text{K}$	(5/7)	$-7.225(16)$
$\rho = 790\,\text{kg m}^{-3}$	(7/9)	$-6.974(21)$
	full	$-6.589(12)$
TIP4P water	(3/4)	$-52.20(12)$
$T = 273.15\,\text{K}$	(4/6)	$-51.77(08)$
$\rho = 1000\,\text{kg m}^{-3}$	(5/7)	$-51.32(09)$
	full	$-50.98(16)$
carbon dioxide	(0/0)	0.000
$T = 228\,\text{K}$	(4/6)	$-2.680(4)$
$\rho = 1120\,\text{kg m}^{-3}$	(5/7)	$-2.952(5)$
	full	$-3.075(7)$

In addition to intuitive arguments, as "screening" of the long-range interactions, we give in Tab. 1 the average Coulomb energies for systems with different range of electrostatic interactions in order to quantify the observed trends. We see that the trends seen for the site-site correlation functions only reflect the similar trends of the Coulomb energies: with increasing R' the total Coulomb energy rapidly approaches that of the full system indicating that the contribution of the long-range part of the electrostatic interactions to the total energy is only marginal. Similar conclusions we may obtain on an even simpler level by analyzing just the energy of dimers. For instance, calculations for the water dimer show that the minimum energy with the O–O distance fixed at 5 Å [i.e., at a rough estimate of the O–O separation at the smooth switching between 4 and 6 Å, $(4,6) = 5 \pm 1$ Å] is within the range of 1/5 to 1/6 (in dependence on the model considered) of the dimer energy (H-bonding energy). This value is small enough (and even changing sign at different orientations) to guarantee only a marginal impact of the neglected interactions on the structure.

Important consequences of this interrelation between the Coulomb energy and structure are demonstrated in Fig. 2. Here we show (i) the site-site functions of carbon dioxide for completely switched off Coulomb interactions (i.e., for $R_{\text{range}} = 0$), and (ii) the correlation functions for the long-range trial system for water defined by u_{lr}^T. As it is seen, even for such an only slightly polar fluid as carbon dioxide the structure of the short-range fluid defined by

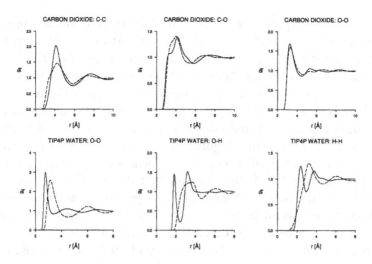

Figure 2. Site-site correlation functions of the short-range model of carbon dioxide at $T = 228K$ and $\rho = 1112 \, \text{kg m}^{-3}$ with completely switched off Coulomb interactions (i.e., $R_{\text{range}} = 0$; dashed line), and the long-range trial system [Eq. (12) with $(R', R'')=(4,6)$] of water at ambient conditions (dashed line), and their comparison with those of the full systems (solid lines).

$R_{\text{range}} = 0$ is different from that of the full system (particularly, locations of extremes are out of phase). It means, in full agreement with energies given in Tab. 1, that the Coulomb interactions cannot be neglected completely which questions the claims used to find in literature that for such systems as carbon dioxide the system with completely switched off electrostatic interactions may be considered as a suitable reference. As for the long-range trial system of water, there is no doubt that the shown results, which are even qualitatively completely off, immediately disqualify any claims on the predominant effect of the long-range electrostatic interactions on the structure of water: If the long-range interaction did play the predominant role in determining the structure of water, then g_{ij}'s would have to be, at least qualitatively, the same.

Closer examination of Fig. 1 shows that if the switching range is about at or slightly beyond the first minimum of g_{ij} of the reference sites, $g_{ij} = g_{\text{RR}}$, no significant difference in the structure between the u and u^{T} systems is found. (An exception is water for which, as a consequence of the tetrahedral arrangement of hydrogen bonds, the second coordination shell is located only 1.6-multiple of the first shell and most of the second shell must therefore be also included.) These results indicate that the influence of the switching range is nearly lost for appropriately defined short-range models u^{T} and it is thus tempting to generalize this finding and formulate the following working hypothesis:

Provided that the trial potential u^T *includes the first coordination shell (plus at least a part of the second shell in the case of water), then the structure of the systems defined by* u *and* u^T *is very similar (nearly identical).*

In other words, the hypothesis claims that the long range part of the Coulomb interactions has only marginal effect on the structure of pure fluids.

Since there is no reason why the above finding for the one-dimensional site-site correlation functions should hold true also for more complex structural properties (and other properties in general), this hypothesis has been carefully examined for a number of other properties, typically along a high density isochore, and along several isotherms covering both the subcritical and supercritical range. Selected results, both those obtained previously and also new ones, are reported and discussed below. As for the switching range, the width of $R'' - R' \approx 2$ Å has been considered as a reasonable compromise between the steepness and smoothness of the tail and has been therefore chosen for all compounds (an exception is acetone for which even slightly narrower range has been used but this is immaterial). With respect to the general goals of these investigations no fine tuning of the switching range has been carried out and the following values for (R', R'')[Å] have been chosen: $(5, 7)$ for carbon dioxide, $(6.6, 8)$ for acetone, $(6.5, 8.5)$ for acetonitrile, $(5.7, 7.7)$ for methanol, and $(4, 6)$ for water and hydrogen fluoride.

3.1   *Pure fluids*

Structural properties.   We begin the discussion on the structural properties by examining various site-site correlation functions g_{ij}. Selected results are shown in Figs. 3–5. These graphs do not seem to require any deeper comments and analysis. In all cases the g_{ij} functions of the trial short-range models with the switching range given above are nearly indistinguishable from those of the full parent models.

Since the site-site correlation functions provide information only on certain averaged correlations, on the basis of their coincidence one cannot make any strong claims concerning details of the orientational arrangement of molecules. To cast more light on such an arrangement we show in Fig. 6 the first dipole-dipole correlation function G_1 (in other words, the ensemble average of $\cos \Theta_{12}$ with fixed distance of the reference points) for acetone and acetonitrile, i.e., for strongly dipolar fluids; the same functions for two strongly associating fluids, water and hydrogen fluoride, are shown in Fig. 7. These figures reveal both different orientational correlations between dipolar and associating fluids, and also a difference in these correlations between the short-range and full systems.

As regards polar fluids, we see that the orientational arrangement in the full- and short-range systems follows the same oscillating pattern and, regardless of the value of the dipole moment, there does not seem to be any significant difference between the full and short-range systems. At short separations, the dipole-dipole correlation functions are negative, which means that the preferred alignment of the dipoles is antiparallel. In the region around the first maximum on the C–C correlation function the dipole-dipole correlation function becomes positive and the parallel alignment is thus preferred. Although with increasing separations the dipole-dipole correlation function gradually dies off, the decay is rather slow and the pronounced oscillating pattern is observed over a large range. To summarize, since the short-range electrostatic field generates the same oscillating behavior as the full field, the observed alternating alignment of dipoles may hardly result from the long-range interactions but are rather propagated via surrounding particles.

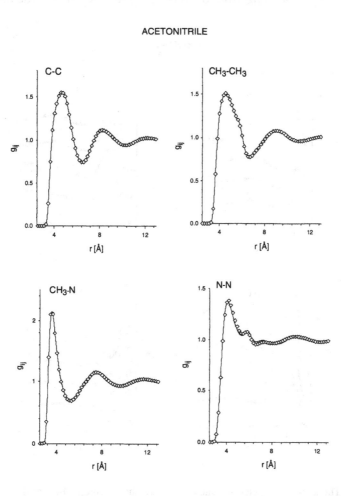

Figure 3. The site-site correlation functions of the short-range (solid lines) and full (symbols) potential models of acetonitrile at $T = 298\,\mathrm{K}$ and $\rho = 765\,\mathrm{kg\,m^{-3}}$.

A different picture of the orientational arrangement we get for associating fluids, see Fig. 7. It is found, maybe surprisingly, that the full realistic models do not exhibit any significant long-range ordering. For both substances, G_1 (as well as higher-order functions) of the long-range systems has a large positive value at short separations but it rapidly decreases with increasing distance and then only fluctuates about zero for medium and large separations. On the other hand, the influence of truncation is much more pronounced than for polar fluids. For the short-range systems function G_1 also begins at a large positive value, but as we move away from close separations it takes on also significant negative values and the oscillations at larger separations are much less damped. Explanation of these observations lies in the competition between the strong short-range hydrogen bonding interaction and the long-range

METHANOL

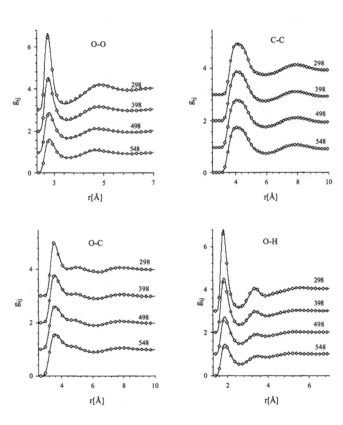

Figure 4. The site-site correlation functions of the short-range (solid lines) and full (symbols) potential models of methanol at $\rho = 762\,\mathrm{kg\,m}^{-3}$ and different temperatures (numbers at curves). The curves are subsequently shifted by 1.

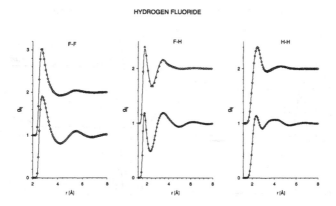

Figure 5. The site-site correlation functions of the short-range (solid lines) and full (symbols) potential models of hydrogen fluoride at $T = 350\,\text{K}$ and $\rho = 1200\,\text{kg}\,\text{m}^{-3}$ (lower set of graphs), and at $T = 500\,\text{K}$ and $\rho = 600\,\text{kg}\,\text{m}^{-3}$ (upper set of graphs). The curves are subsequently shifted by 1.

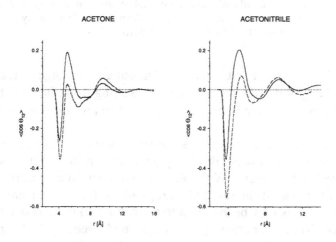

Figure 6. The average values of $\cos\Theta_{12}$ (dipole-dipole correlation functions G_1) of the short-range (dashed lines) and full (solid lines) potentials for strongly polar fluids at ambient conditions.

electrostatic interactions. As a consequence of strong and strongly directional hydrogen bonding, the preferred mutual arrangement evidently exists between the particles in the first coordination shell regardless of the range of the potential. With increasing distance the long-range electrostatic (dipole-dipole) interaction starts playing its role and the competition between the local hydrogen bonding and long-range dipole-dipole interactions in the full-range systems causes the correlations to diminish and no preferred arrangement is observed

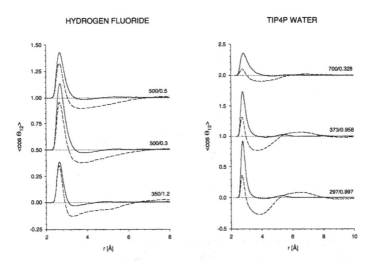

Figure 7.   The average values of $\cos \Theta_{12}$ (dipole-dipole correlation functions G_1) of the short-range (dashed lines) and full (solid lines) potentials of associating fluids at different thermodynamic conditions denoted by numbers at curves (temperature [K]/density [g/cm^3]). The curves are subsequently shifted by 0.5 (for HF) and 1.0 (for H$_2$O).

any more. On the other hand, because of missing long-range interactions in the short-range systems, the regular pattern resulting from the local hydrogen bonding remains unaffected in this case and its effect prevails over a wide range of separations.

Another aspect of the above rather surprising observation can be revealed by viewing water as a fixed network of oxygen atoms (resulting from strong H-bonds) bridged by hydrogens. With this idea in mind we may repeat the classic Pauling's explanation of the residual entropy of ice [29]. There are two rules how hydrogen atoms may be assigned to the bonds: (i) there is one hydrogen per bond, and (ii) there are two hydrogens per node, see Fig. 8. Despite these constraints, there are many different ways to assign hydrogens to the network, giving rise to the residual entropy. Contributions of these different assignments to the internal energy, which is given primarily by the H-bond network, are slightly different due to different dipole-dipole alignments (expressed by G_1) of the *second* neighbors. Reversing this reasoning we may argue that a small change in the water-water energy due to cutting off can cause an easy rearrangement of water molecules over the same network.

A macroscopic measurable quantity which should reflect the above discussed differences in the orientational correlations in different systems is the dielectric constant, see Eqs. (15) and (16). At first sight, one would expect considerable difference in ϵ, particularly between the short- and full-range models of water at low temperatures (differences in the alignment are gradually

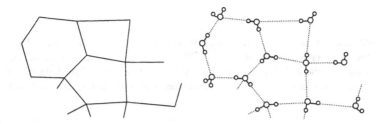

Figure 8. The H-bond network in water (left) with a configuration of hydrogens attached (right).

smeared with increasing temperature). However, as it is seen from Tab. 2, this is not the case. The dielectric constant is an integral quantity depending on the alignment of molecules over the entire range. Thus, the pronounced oscillations of G_1 found in the short-range model of water at ambient conditions tend to cancel the contribution from large separations and the resulting dielectric constant of this model agrees, within the combined pseudoexperimental errors, with that of full realistic water. For acetone, as a representative of polar fluids, there is no qualitative difference in G_1 between the short- and full-range models and the agreement of the respective dielectric constant is therefore not surprising.

Thermodynamic properties. The results presented in the preceding subsection seem to confirm the working hypothesis. Besides its importance for better understanding of the molecular mechanisms determining the properties of fluids, the marginal effect of the long-range part of the Coulomb interactions on the structure is also the necessary condition for developing workable equations for various properties of fluids in a perturbed form. However, it does not say anything on complexity of such equations, i.e., how much of the thermodynamic behavior may be captured by the short-range reference system and hence how many correction terms may be required. Although it is not necessary that the reference fluid provides also a good estimate of the thermodynamics of the investigated fluid (cf., for instance, the properties of soft-repulsive spheres or even just hard spheres serving as a reference for the LJ fluid with those of the LJ fluid itself) it is evident that the more of the behavior of the investigated fluid is captured by its short-range descendant, the better. In this subsection we are going therefore to compare the thermodynamic properties of the short-range models with those of the parent models.

In Tab. 3 we show selected results for the homogeneous phase at typical liquid state points and also at supercritical temperatures. We see that the internal energies for the short-range and parent models agree surprisingly well, particularly for associating fluids. Somewhat larger discrepancy, about 10%, is

Table 2. Dielectric constants of the full and short-range models of TIP4P water and acetone at a number of thermodynamic conditions. $(\epsilon^{min}, \epsilon^{max})$ is the range at the 95% confidence level, and (R'/R'') denotes the switching range of the short-range models. The dielectric constant of the surrounding continuum in the Ewald summation was set to 80 for water and infinity for acetone.

system		T[K]	ρ[kg m^{-3}]	ϵ	$(\epsilon^{min}, \epsilon^{max})$
water	full	298	997	52	(42, 64)
	(4/6)			36	(23, 82)
	full	373	958	40.7	(36, 46)
	(4/6)			34	(26, 47)
	full	700	328	4.78	(4.69, 4.88)
	(4/6)			4.69	(4.55, 4.84)
acetone	full	298	790	9.6	(8.0, 11.7)
	(5/7)			10.0	(6.4, 21.3)
	(6.6/8)			9.8	(6.6, 18.6)
	full	550	556	3.92	(3.57, 4.32)
	(5/7)			5.1	(3.97, 7.02)
	full		278	2.42	(2.27, 2.59)
	(5/7)			2.49	(2.28, 2.75)

Table 3. Configurational energy and pressure of the short- and full-range models at selected state points. Numbers in parentheses denote the error of last digits.

	T [K]	ρ [kg m^{-3}]	U[kJ/mol] short	U[kJ/mol] full	P[MPa] short	P[MPa] full
acetonitrile	298	800	−31.53(3)	−34.48(8)	107.1(19)	283.4(4)
	598	200	−9.06(8)	−10.60(9)	12.01(2)	9.87(2)
		800	−26.38(3)	−27.00(3)	577.1(24)	572.8(19)
water	298	1050	−42.19(23)	−41.78(32)	71.10(4148)	127.4(495)
(TIP4P)		1200	−43.10(27)	−42.37(20)	632.2(543)	696.2(551)
	550	600	−24.35(30)	−24.92(23)	21.98(1274)	14.29(1113)
		1200	−33.57(20)	−33.61(16)	1491(49)	1525(49)
	700	500	−18.37(20)	−19.04(19)	97.98(737)	83.79(688)
		1200	−29.71(19)	−30.00(18)	2018(45)	2033(44)
hydrogen	350	1200	−26.26(06)	−25.94(04)	2379(119)	2634(103)
fluoride	500	600	−17.64(14)	−17.95(11)	461(25)	453(33)
		1200	−23.24(06)	−23.11(05)	5556(103)	5770(82)

found for acetonitrile. This finding cannot be surprising because it only reflects the similar agreement found for the underlying site-site correlation functions: the internal energy is namely determined directly by an integral over $g_{ij}(r)$'s. Agreement for pressure is not as good as for the internal energy but yet it is not unreasonable. The discrepancy is easily understood if we recall that pressure, unlike the internal energy,

(i) is very sensitive to any inaccuracy in the correlation functions,

(ii) is not determined only by the site-site correlation functions but also by angular correlations [30] (as we have shown in the preceding subsection, the angular correlations are more affected by the range than the radial correlations),

(iii) and may be very small or even negative so that calculating percent deviation does not make sense.

From the practical point of view it is very important to know how the long-range interactions affect the coexistence properties, including the location of the critical point. Typical results for densities of the coexisting vapor and liquid phases are shown in Fig. 9. For polar fluids there is only a marginal difference in the behavior of the short-range and full models; slightly larger discrepancies, in correspondence with findings for the homogeneous phase, are found for associating fluids. In [26] the effect of interactions on the location of the

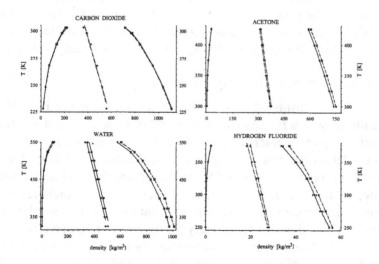

Figure 9. The coexistence densities of the full (solid lines and filled symbols) and short-range models of polar and associating fluids. The straight lines correspond to the rectilinear diameter rule.

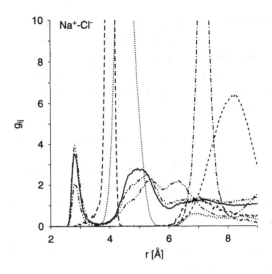

Figure 10. The Na^+–Cl^- correlation function in a molar solution of NaCl. Solid line: Ewald (reference); dashes: truncation $(4, 6)$; dots: truncation $(4, 6)$, ion-ion cut-and-shift at 6; dash-dot: truncation $(7, 10)$; short dashes: truncation $(7, 10)$, ion-ion cut-and-shift at 10; dot-dot-dash and dot-dot-dot-dash: smoothed cut-and-shift at 9 and 13, respectively.

critical point was examined and discussed in detail. It is known that the critical properties are sensitive to the range of the neutral, LJ-type interactions which indicates their sensitivity to the short-range interactions. Since these interactions remain in the process of shortening $u(1, 2)$ intact, and the short-range part of the Coulomb interactions is also completely included, the location of the critical point in the short-range and full systems agree extremely well, see [26].

3.2   Dilute aqueous electrolyte solutions

So far we have considered only pure fluids of neutral molecules. The obtained results show that the spatial arrangement of molecules in these fluids is affected primarily by interactions with nearest neighbors as witnessed by the site-site correlation functions. In other words, the primary driving force determining the properties of pure fluids, no matter whether non-polar, polar, or associating, are short-range interactions (but not necessarily purely repulsive!). A question which immediately arises is, to what extent these conclusions remain valid also for mixtures, particularly for mixtures of charged particles. In the following we therefore use the same strategy as for pure fluids and consider a dilute aqueous electrolyte to examine the effect of the shortening of the interactions.

From the molecular point of view, there are three types of interactions acting simultaneously in aqueous electrolytes: ion-ion (strongest), ion-water, and water-water. All these interactions are, in general, long-ranged. In the simplest case of (infinitely) dilute solutions and large separations of ions, the solvent can be replaced by a homogeneous dielectric (primitive model of electrolytes). However, with decreasing ion-ion separation, the continuum approach is no longer valid and the structure of water (orientation and position of water molecules between ions) has to be considered.

As a test case we consider an aqueous solution of the rock salt of concentration $1\,mol\,dm^{-3}$ (i.e., in simulations we use 10 molecules of NaCl in 555 SPC/E water molecules) described by the potential of Brodholt [31] and consider several possibilities of how to smoothly shorten (switch off) electrostatic interactions for both ions and water.

Figure 10 shows the effect of different truncation schemes on the Na–Cl correlation function which is most sensitive to details of the potential. [The results for the full model treated by means of the Ewald summation (solid line) serve as a benchmark.] It is seen that truncation (9) used for all charge-charge interactions causes serious artifacts – very high peaks [up to 60 for the $(4, 6)$ switching range] at the onset of truncation. To understand this phenomenon let us first mention that in the full-potential system the large cation-anion attraction is effectively screened by intervening water molecules. The $2\,Å$ width of the switching range is too narrow with respect to the depth of the cation-anion potential so that the turning the potential off causes the effective (screened) potential to become positive in the truncated region, while in the region just before the truncation it is comparatively very negative. Similar artifacts are observed for the Na–Na and Cl–Cl correlation functions as well.

To avoid this phenomenon, as a next truncation scheme we may try a milder truncation for the strong ion-ion interactions: the ion-ion potential we cut and shift (to maintain its continuity) while leaving the water-ion and water-water potentials intact. Although the above mentioned artifact becomes less pronounced, it is still large enough to produce wrongly behaving solution.

In the light of the above results it is worth mentioning that popular molecular modelling packages also often implement electrostatic truncation algorithm ("cheap" truncation) which essentially replaces the $1/r$ term by something behaving approximately as $1/r$ at short separations but going *smoothly* to zero at certain r_c. In our version we cut $1/r$ at $0.7r_c$, shift it for $r < 0.7r_c$, and replace it by a quadratic function to obtain continuous potential and forces in interval $[0.7r_c, r_c]$; the potential is zero for $r > r_c$. As it is seen from Fig. 10 (dot-dot(-dot)-dash lines), this method leads to some imbalance in the Na–Cl (and also Na–Na and Cl–Cl) distribution so that it cannot be recommended as a truncation scheme in simulations; nonetheless, the results are at least reasonable. One may expect good results only if r_c is larger than the Debye screening length.

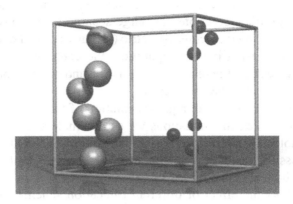

Figure 11.   Snapshot of 6 Cl anions (large spheres) and 6 Na cations (small spheres) dissolved in 244 (4, 6)-truncated SPC/E water molecules (not shown for clarity) in a cubic periodic simulation box and ambient conditions.

Finally, we consider a truncation scheme which may be viewed as a step beyond the primitive model of electrolytes: maintaining the full interaction between ions, and ions and water molecules, we replace realistic water by a short-range model justified by the findings for pure water. From the computational point of view, our implementation within the framework of the Ewald summation (with conducting boundary conditions, $\epsilon' = \infty$) consists of the Ewald calculation for the whole system (water and ions), subtraction of the Ewald terms for the water subsystems, and eventual addition of the truncated terms for the water-ion subsystem. The long-range water-water electrostatic interactions are thus suppressed while the water molecules still fully feel the ions. It turns out that this short-range water gives rise to an unphysical separation of ions of opposite signs and clustering of the like ions, see Fig. 11; we show a snapshot instead of the usual Na–Cl correlation function because it is not well-defined for ions more than half the box size apart. The short-range water appears to act thus as a "much better solvent" than real water. This startling observation can be explained by the continuous approach. In real water, an ion is screened by surrounding dielectric which is polarized. The total energy is a sum of the dielectric-ion interaction and the positive polarization energy of the dielectric. And this polarization energy (or a portion of it beyond the cutoff) is missing in short-range water.

4.   Summary and concluding remarks

Traditionally, common starting point towards understanding the properties of fluids has been an analysis of the effective pair interaction or, more accurately, of its individual parts, and their relation to the observed thermodynamic properties. However, such relations may be misleading because the thermodynamic properties are superimposed on the established structure which is thus the decisive factor and should therefore be the primary focus of such an analysis. Moreover, it has been known from studies of simple fluids that it need not be just an individual part of the pair potential which affects the properties of fluids but more likely some aspects of the *total* potential (cf. the difference in the effect of the repulsive term $(\sigma/r)^{12}$ and the repulsive part of the entire LJ potential) [32]. We have therefore attempted to analyze in detail the connection between the structure and the range of the interaction potential for non-simple fluids.

Starting from realistic site-site pair potential models for polar and associating fluids, we have constructed potentials of different range by smoothly switching off their Coulomb part and analyzed associated structural changes. It has been found that the spatial arrangement of molecules is affected primarily by interactions with nearest neighbors (an exception is water with some impact of the second neighbors as well) as witnessed by the site-site correlation functions. This is not hundred percent true for the angular correlations for which some difference between the short-range and full systems has been found, particularly for associating fluids. Nonetheless, global properties do not seem to be affected by these discrepancies to a larger extent. One can thus conclude that the primary driving force determining the properties of pure fluids, no matter whether non-polar, polar, or associating, are short-range interactions (but not necessarily purely repulsive!) and that the long-range part of electrostatic interactions plays the role of a mere perturbation only.

Once the long-range part of the electrostatic interactions may be ignored at the zeroth level of approximation, one can immediately devise a perturbation expansion about a suitably chosen short-range reference that reproduces the structure of the realistic system at hand. Furthermore, from the computational point of view, this may considerably simplify and hence accelerate molecular simulations which also opens the way for theoretically-based modelling of complex problems. As regards the perturbation expansion, its feasibility has already been exemplified by deriving a theoretically-based equation of state of water whose accuracy may compete even with empirical Eqs. [16]. For accomplishing the expansion, one must be able to describe theoretically the properties of the reference system. This goal is achieved by employing the so called primitive models [33–35] which may be viewed as counterparts of hard spheres for non-simple fluids. Recently, a theoretical method has been developed enabling one to derive a primitive model as a direct descendant of a realistic parent model [34, 35]. These models reproduce, even semi-quantitatively, the structure of realistic fluids and they need not thus serve only for the purpose of the perturbation theory but their main field of applications is in modeling of complex problems, and in both theoretical and computer simulation studies of details of molecular mechanisms (see e.g. [36]) governing the behavior of fluids.

Apart from their theoretical significance, the above findings may also have indirect impact on the technical aspects of molecular simulations. When simulating systems with the Coulomb interactions, one should evidently use, as a rule, some method accounting for the long-range character of interactions (e.g., Ewald summation or the reaction field, etc., [24]). Nonetheless, quite often these techniques are replaced by a simple spherical cutoff and arguments that the differences are only marginal. We have shown that to accept/reject this approach depends on quantities we are interested in. For average structural properties, as the site-site correlation functions, the use of the spherical cutoff seems justified but other properties may be subject to non-negligible errors, see e.g. [37]. One must be therefore very cautious when using the spherical cutoff and drawing general conclusions from results based on its use.

The above conclusions have resulted from an analysis of computer simulation data carried out on *pure* liquids and supercritical fluids, and on liquids in equilibrium with their vapor. One immediate question one should ask concerns thus a more general validity of the reached conclusions. Particularly important problem is to what extent they may remain valid for mixtures. Due to polarizability and other possible effects brought about by electrostatic interactions between unlike species, the pair interaction, and hence the local and, particularly, orientational arrangement may be changed considerably. With respect to a wide variety of mixtures this problem will require rather an extensive investigation. The most difficult mixtures will evidently be solutions of charged objects as e.g. electrolytes.

In electrolyte solutions we have not only moderately long-range effective dipole-dipole interaction but much stronger ion-ion and ion-solvent ones. The effective ion-ion interaction in such a complex solvent as water results from cooperation and cancellation of many terms. Water should behave as a dielectric at larger ion-ion separations while at shorter separations the ion-water interaction becomes important which may cause problems already for developing a realistic potential model: potential parameters of ions and water must be well-balanced to get a realistic description of real solution (for instance, the recent potential of [38] leads to clustering of ions and as result, NaCl is not soluble). Any truncation tends to violate this fine balance and leads to unphysical results; for the purpose of modelling of electrolytes it seems thus inevitable to adopt another approach. The only exception is smooth enough truncation longer than the Debye length, but this is rather a problem of simulation methodology.

Acknowledgements

We would like to acknowledge valuable discussions with, and comments and support of those who have contributed to this project over the last decade, particularly M. Lísal, and M. Předota, Academy of Sciences, Prague, A. Chialvo, Oak Ridge National Laboratory, Oak Ridge, and M. Kettler, University of Leipzig, Leipzig. This work was supported by the Grand Agency of the Czech Republic (Grant No. 203/02/0764), and by the Center for Complex Molecular Systems and Biomolecules (Ministry of Education, Youth and Sports, project LN00A032).

References

[1] Barker, J.A., and Henderson, D. *Rev. mod. Phys.*, 1976, **48**, p. 587.

[2] Hansen, J.P., and McDonald, I.R. (1986). *Theory of Simple Liquids*. London: Academic Press.

[3] Rowlinson, J.S., and Swinton, F.L. (1982). *Liquid and Liquid Mixtures*, 3rd ed. London: Butterworths.

[4] Stratton, J.A. (1941). *Electromagnetic Theory*. New York: Academic Press.

[5] Boublík, T., Nezbeda, I., and Hlavatý, K. (1983). *Statistical Thermodynamics of Simple Liquids and Their Mixtures*. Amsterdam: Elsevier.

[6] Andersen, H.C. *Ann. Rev. Phys. Chem.*, 1975, **26**, p. 145.

[7] Nezbeda, I., and Kolafa, J. *Molec. Phys.*, 1999, **97**, p. 1105.

[8] Kolafa, J., and Nezbeda, I. *Molec. Phys.*, 2000, **98**, p. 1505.

[9] Dahl, L.W., and Andersen, H.C. *J. Chem. Phys.*, 1983, **78**, p. 1980.

[10] Nezbeda, I. *Fluid Phase Equil.*, 2000, **170**, p. 13.

[11] Silverstein, K.A.T., Haymet, A.D.J., and Dill, K.A. *J. Am. Chem. Soc.*, 1998, **120**, p. 3166.

[12] Předota, M., and Nezbeda, I. *Molec. Phys.*, 1999, **96**, p. 1237.

[13] Předota, M., Nezbeda, I., and Cummings, P.T. *Molec. Phys.*, 2002, **100**, p. 2189.

[14] Nezbeda, I. *Molec. Phys.*, 2001, **99**, p. 1631.

[15] Muller, E.A., and Gubbins, K.E. *Ind. Eng. Chem. Res.*, 2001, **40**, p. 2193.

[16] Nezbeda, I., and Weingerl, U. *Molec. Phys.*, 2001, **99**, p. 1595.

[17] Nezbeda, I., and Kolafa, J. *Czech. J. Phys.*, 1990, **B40**, p. 138.

[18] Nezbeda, I., and Lísal, M. *Molec. Phys.*, 2001, **99**, p. 291.

[19] Gray, C.G., and Gubbins, K.E. (1984). *Theory of Molecular Fluids*, vol. 1. Oxford: Clarendon Press.

[20] Steinhauser, O. *Molec. Phys.*, 1982, **45**, p. 335.

[21] Kirkwood, J.G. *J. Chem. Phys.*, 1935, **3**, p. 911.

[22] de Leeuw, S.W., Perram, J.W., and Smith, E.B. *Proc. Roy. Soc. London Ser. A*, 1983, **388**, p. 177.

[23] Allen, M.P., and Tildesley, D.J. (1987). *The Computer Simulation of Liquids*. Oxford: Clarendon Press.

[24] Frenkel, D., and Smit, B. (2003). *Understanding Molecular Simulation: From Algorithms to Applications.* San Diego: Academic Press.

[25] Kolafa, J., Lísal, M., and Nezbeda, I. *Molec. Phys.*, 2001, **99**, p. 1751.

[26] Kettler, M., Nezbeda, I., Chialvo, A.A., and Cummings, P.T. *J. Phys. Chem. B*, 2002, **106**, p. 7537.

[27] Jorgensen, W.L., and Briggs, J.M. *Molec. Phys.*, 1988, **63**, p. 547.

[28] Cournoyer, M.E., and Jogensen, W.L. *Molec. Phys.*, 1984, **51**, p. 119.

[29] Pauling, L. *J. Am. Chem. Soc.*, 1935, **57**, p. 2680.

[30] Labik, S., and Nezbeda, I. *Molec. Phys.*, 1983, **48**, p. 109.

[31] Brodholt, J.P. *Chem. Geol.*, 1998, **151**, p. 11.

[32] Andersen, H.C., Chandler, D., and Weeks, J.D. *Adv. chem. Phys.*, 1976, **14**, p. 105.

[33] Nezbeda, I. *J. Molec. Liquids*, 1997, **73–74**, p. 317.

[34] Vlček, L., and Nezbeda, I. *Molec. Phys.*, 2003, **101**, p. 2987.

[35] Vlček, L., and Nezbeda, I. *Molec. Phys.*, 2004, **102**, p. 485.

[36] Předota, M., Ben-Naim, A., and Nezbeda, I. *J. Chem. Phys.*, 2003, **118**, p. 6446.

[37] Lísal, M., Kolafa, J., and Nezbeda, I. *J. Chem. Phys.*, 2002, **117**, p. 8892.

[38] Koneshan, S., and Jayendran, J.C. *J. Chem. Phys.*, 2000, **113**, p. 8125.

COLLECTIVE DYNAMICS IN IONIC FLUIDS

A comparative study with nonionic fluids

I. Mryglod , T. Bryk, V. Kuporov

Institute for Condensed Matter Physics
of the National Academy of Sciences of Ukraine,
1 Svientsitskii Str., 79011 Lviv, Ukraine

> *One thing I have learned in a long life: that all our science, measured against reality, is primitive and childlike – and yet it is the most precious thing we have.*
>
> —Albert Einstein

Abstract   A solution of strong electrolytes can be considered as a multicomponent system composed of a neutral solvent and a number of ionic species, the concentration of which is constrained by the requirement of overall electroneutrality. The general theoretical framework is developed for the study of collective dynamics in multi-component, chemically nonreactive fluids. On this basis we consider a few simplified dynamical models for ionic fluids and formulate the most adequate one. This allows us to obtain within a unique scheme the results for collective mode spectrum, time correlation functions and generalized transport coefficients, having rather simple physical interpretation in terms of collective excitations. These features are discussed in relation to dynamical properties of a binary liquid of neutral particles, enabling us to study in detail the specific properties caused by the long-range Coulomb interactions between charged particles. The developed theory can be directly extended for the description of the dynamics in more complicated models of ionic solutions.

Keywords: Molten salts, ionic solutions, collective dynamics, time correlation function, collective excitation, transport coefficient, molecular dynamics

1.   Introduction

Charge complex fluids form a wide class of fundamental many-particle systems, which possesses a lot of very specific features due to the role of long-range Coulomb interactions between charged particles. One of the main con-

D. Henderson et al. (eds.), Ionic Soft Matter: Modern Trends in Theory and Applications, 109–141.
© 2005 *Springer. Printed in the Netherlands.*

sequences of this type of interactions is a rather strong coupling of many particles displayed in large space volume, which confers, on the one hand, a typical many-particle character of the problem and, on the other hand, creates a crucial complication for the theory. One of the paradigms of charged complex fluids, which has remained a central area of research in physics and physical chemistry for a long time, is the dynamics of ionic liquids. This category of charged complex fluids comprises molten salts, liquid metals and ionic solutions [1–3].

In this study we restrict our consideration by a class of ionic liquids that can be properly described based on the classical multicomponent models of charged and neutral particles. The simplest nontrivial example is a binary mixture of positive and negative particles disposed in a medium with dielectric constant ϵ that is widely used for the description of molten salts [4–6]. More complicated cases can be related to ionic solutions being neutral multicomponent systems formed by a solute of positive and negative ions immersed in a neutral solvent. This kind of systems widely varies in complexity [7], ranging from electrolyte solutions where cations and anions have a comparable size and charge, to highly asymmetric macromolecular ionic liquids in which macroions (polymers, micelles, proteins, etc) and microscopic counterions coexist. Thus, the importance of this system in many theoretical and applied fields is out of any doubt.

Another restriction is connected with the study of the collective character of the dynamics of multicomponent systems of charged and neutral particles. This means that we concentrate our attention mainly on the dynamical properties caused by processes in which collectives of many particles are involved. The simplest example of such properties is the conductivity in molten salts being closely related to mutual diffusion coefficient via the known Nernst-Einstein relation [1–3]. Collective behavior can be observed in many experimental situations (for instance, scattering experiments) and forms the basis for understanding of many physical properties being important for practical needs. One of the crucial problems in such phenomena is to establish the effect of Coulombic long-range interactions between charged particles on the collective behavior of the multicomponent systems. This last item enforced us to perform our study comparing it with the related many-particle system of neutral particles. We note that the multicomponent nature and the same set of conserved laws make the dynamics of such mixtures of charge and neutral particles rather similar. However, on the other hand, a lot of very specific features can be caused by the main dissimilarities between them, which in addition to the long-range character of interactions, also include its specific "antiferromagetic" properties (in contrast with the repulsive "A–A" and "B–B" interactions the "A–B" interactions are attractive) and an electroneutrality constraint. Thus, one of our goals in this study will be to specify the effect of these different factors on the collective dynamics of ionic liquids.

The collective dynamics in a classical system of charged and neutral particles was the subject of intensive investigations during many years. The activity in this field especially intensified about thirty years ago when the computer experiments provided new significant information and became an important tool for the verification of theory predictions. However, it should be mentioned that indeed *little* progress has been achieved in many aspects being of great interest to the researchers. Among the main achievements important for attaining the goal of this study one should mention the results obtained in the hydrodynamic limits for molten salts, and their link with the non-equilibrium thermodynamics [2, 8]; analytical expressions derived for the hydrodynamic transport coefficients [9–11]; optic-like excitations, observed experimentally in charge density dynamic structural factor, as well as several theories, proposed for their description [4–6]. Nowadays, huge possibilities and powerful methods of computer simulations allow us to study different correlation functions as well as transport coefficients in a wide range of space and temporal scales. However, one has to note once again that the real progress in understanding the physics beyond is still very moderate, even in the cases when rather sophisticated and purely computer-orientated theories are applied. In particular, there is still open the problem of crossover from the hydrodynamic regime to the molecular one in the collective dynamics; there is a real need of developing of the theories for a description of collective excitations, e.g. optic-like collective modes, and of correct determination its contributions to time correlation functions; practically unsolved remains the problem of the calculation of generalized (wavevector- and frequency-dependent) transport coefficients. Some of these issues are considered in this contribution.

2. Theoretical framework

We start with some introductory remarks and the consideration of general framework that can be used for the description of both kinds of multi-component mixtures, i. e., mixtures of neutral particles and mixtures containing charged particles. Let us consider a multi-component fluid in the volume V, containing N_α particles in the α-th species. Some of them can be associated with charged particles, and this point will be specified hereinafter.

2.1 *Hydrodynamic variables*

In order to derive hydrodynamic equations one has to define the microscopic basic set of the slowest (hydrodynamic) variables, which for a mixture may be introduced as follows $\hat{P}_{\mathbf{k}}^{\mathrm{hyd}} = \{\hat{\mathcal{N}}_{\mathbf{k}}, \hat{\mathbf{J}}_{\mathbf{k}}, \hat{E}_{\mathbf{k}}\}$, where $\hat{\mathcal{N}}_{\mathbf{k}} = \{\hat{n}_{\mathbf{k},\alpha}\}$ is a column-vector with the components,

$$\hat{n}_{\mathbf{k},\alpha} = \sum_{i=1}^{N_\alpha} \exp\{i\mathbf{k}\mathbf{R}_i^\alpha\}, \tag{1}$$

being the number density of particles in the α-th species; $\hat{\mathbf{J}}_\mathbf{k}$ is the density of the *total* current,

$$\hat{\mathbf{J}}_\mathbf{k} = \sum_\alpha \hat{\mathbf{J}}_{\mathbf{k},\alpha}, \qquad \hat{\mathbf{J}}_{\mathbf{k},\alpha} = \sum_{i=1}^{N_\alpha} \mathbf{p}_i^\alpha \exp\{i\mathbf{k}\mathbf{R}_i^\alpha\}, \tag{2}$$

with $\hat{\mathbf{J}}_{\mathbf{k},\alpha}$ being the current densities of particles in the α-th species; and

$$\hat{E}_\mathbf{k} = \sum_\alpha \hat{E}_{\mathbf{k},\alpha} = \sum_\alpha \sum_{i=1}^{N_\alpha} e_i^\alpha \exp\{i\mathbf{k}\mathbf{R}_i^\alpha\}, \tag{3}$$

is the *total* energy density, where

$$e_i^\alpha = \frac{(\mathbf{p}_i^\alpha)^2}{2m_\alpha} + \frac{1}{2} \sum_\gamma \left(\sum_{j=1}^{N_\gamma}\right)' V_{\alpha\gamma}\left(|\mathbf{R}_i^\alpha - \mathbf{R}_j^\gamma|\right), \tag{4}$$

$\mathbf{R}_i^\alpha$ and $\mathbf{p}_i^\alpha$ denote the positions and momenta of particles, $V_{\alpha\gamma}(|\mathbf{r} - \mathbf{r}'|)$ is a potential of interparticle interactions, and the symbol $(')$ in (4) indicates that $i \neq j$ when $\gamma = \alpha$.

All the above introduced dynamic variables satisfy the conservation law equation in the local form, which can be written as follows,

$$\frac{\partial}{\partial t} P_{\mathbf{k},l}^{\text{hyd}} - i k \mathcal{I}_{\mathbf{k},l}^{\text{hyd}} = 0. \tag{5}$$

In fact the Eq. (5) can be considered as the definition for the generalized microscopic hydrodynamic fluxes $\mathcal{I}_{\mathbf{k},l}^{\text{hyd}}$.

In general for a ν-component fluid mixture one has a $(\nu+3+1)$-component set $\hat{P}_\mathbf{k}^{\text{hyd}}$ of dynamic variables, containing ν-component of the number densities $\hat{n}_{\mathbf{k},\alpha}$, the three components of the *total* current density $\hat{\mathbf{J}}_\mathbf{k}$, and the *total* energy density $\hat{E}_\mathbf{k}$. However, as follows from the symmetric properties, the number $\hat{n}_{\mathbf{k},\alpha}$ and energy $\hat{E}_\mathbf{k}$ densities are coupled only with the longitudinal component of $\hat{\mathbf{J}}_\mathbf{k}$, directed along k. This is due to the space isotropy of the system. As a result, one may split the set of the hydrodynamic variables into two separate subsets:

(i) $\hat{P}_\mathbf{k}^\mathrm{L} = \{\hat{N}_\mathbf{k}, \hat{J}_\mathbf{k}^l, \hat{E}_\mathbf{k}\}$ for the description of longitudinal dynamics[1],

(ii) and $\hat{P}_\mathbf{k}^\mathrm{T}$ with two transverse components of the total current (perpendicular to $\mathbf{k}$) for the description of transverse dynamics.

It is evident that the final equations for two transverse components should be the same due to the isotropy of the system. Therefore, one has finally the $(\nu + 2)$-component longitudinal set $\hat{P}_\mathbf{k}^\mathrm{L}$ and the two component transverse set $\hat{P}_\mathbf{k}^\mathrm{T}$ of dynamic variables describing a multi-component fluid.

2.2 Macroscopic dynamics

It has been shown [12, 13] that the macroscopic equations of motion for an arbitrary set of the dynamic variables,

$$\hat{P}_\mathbf{k} = \left\{ \hat{P}_\mathbf{k}^1, \hat{P}_\mathbf{k}^2, \dots, \hat{P}_\mathbf{k}^l \right\}, \tag{6}$$

where $\hat{P}_\mathbf{k}$ is a column-vector, could be written in a matrix form as follows,

$$\left\{ i\omega I^{l\times l} - i\Omega_\mathbf{k}^{l\times l} + \tilde{\varphi}_\epsilon^{l\times l}(\mathbf{k}, \omega) \right\} \langle \Delta \hat{P}_\mathbf{k} \rangle^\omega = 0, \tag{7}$$

where $\Delta \hat{P}_\mathbf{k}^i = \hat{P}_\mathbf{k}^i - \langle \hat{P}_\mathbf{k}^i \rangle$, $I^{l\times l}$ is the $(l \times l)$ unit matrix,

$$i\Omega_\mathbf{k}^{l\times l} = (iL_N P_\mathbf{k}, \hat{P}_\mathbf{k}^+) \, (\hat{P}_\mathbf{k}, \hat{P}_\mathbf{k}^+)^{-1}, \tag{8}$$

is the so-called frequency matrix with iL_N being a Liouville operator, and

$$\tilde{\varphi}_\epsilon^{l\times l}(\mathbf{k}, \omega) \equiv \tilde{\varphi}^{l\times l}(\mathbf{k}, \epsilon + i\omega) =$$
$$= \left(\mathcal{R}iL_N P_\mathbf{k}, \frac{1}{i\omega + \epsilon + \mathcal{R}i\hat{L}_N} \mathcal{R}iL_N P_\mathbf{k}^+ \right) \left(\hat{P}_\mathbf{k}, \hat{P}_\mathbf{k}^+ \right)^{-1} \tag{9}$$

is a matrix of the memory functions. Here $\hat{P}_\mathbf{k}^+$ denotes a transposed vector with the elements $\{\hat{P}_{-\mathbf{k}}^i\}$, $\mathcal{R} = 1 - \mathcal{P}$, and the Mori-like projection operator $\mathcal{P}$ is defined by the relation,

$$\mathcal{P} \dots = (\dots, \hat{P}_\mathbf{k}^+)(\hat{P}_\mathbf{k}, \hat{P}_\mathbf{k}^+)^{-1} \hat{P}_\mathbf{k}. \tag{10}$$

In the above equations static correlation functions $(\hat{A}, \hat{B})$ are defined as follows $(\hat{A}, \hat{B}) = \langle \Delta \hat{A} \Delta \hat{B} \rangle$, where $\langle \dots \rangle$ denotes the expectation value.

---

[1]In general, one may choose as the set of longitudinal hydrodynamic variables any other $(\nu + 2)$ variables, constructed as linearly independent combinations of the variables from $\hat{P}_\mathbf{k}^\mathrm{hyd}$.

The matrix equation for the Laplace transform $\tilde{\mathcal{F}}^{l \times l}(k, z)$ of the equilibrium time correlation functions $\mathcal{F}^{l \times l}(k, t)$, where

$$\mathcal{F}^{l \times l}(k, t) = \left\langle \Delta \hat{P}_{\mathbf{k}} \exp\{-\mathrm{i}L_N t\} \Delta \hat{P}_{\mathbf{k}}^{+} \right\rangle, \tag{11}$$

$$\tilde{\mathcal{F}}^{l \times l}(k, z) = \int_0^{\infty} \mathrm{d}t \exp\{-zt\} \mathcal{F}^{l \times l}(k, t) \tag{12}$$

has the structure similar to the Eq. (7), namely,

$$\left\{ zI^{l \times l} - \mathrm{i}\Omega_{\mathbf{k}}^{l \times l} + \tilde{\varphi}^{l \times l}(\mathbf{k}, z) \right\} \tilde{\mathcal{F}}^{l \times l}(k, z) = \mathcal{F}^{l \times l}(k, 0). \tag{13}$$

Hence, the spectrum of collective excitations can be found from the equation,

$$\mathrm{Det} \left| zI^{l \times l} - \mathrm{i}\Omega_{\mathbf{k}}^{l \times l} + \tilde{\varphi}^{l \times l}(\mathbf{k}, z) \right| = 0, \tag{14}$$

that gives in fact [13, 14] the poles of the retarded correlation Green functions, constructed on the set of dynamic variables $\{\hat{P}_{\mathbf{k}}^i\}$. It should be also stressed that the matrix equation for the equilibrium time correlation functions (13) (quite similarly to the Ornstein-Zernike equation in the equilibrium theory) is in fact the *equality* unless the explicit expressions for the frequency and memory function matrixes are used.

As follows from the matrix expressions (8), (8), and (10), in concrete applications it may be very convenient to use the set of orthogonalized dynamic variables possessing the following properties,

$$(\hat{P}_{\mathbf{k}}^i, \hat{P}_{-\mathbf{k}}^j) = \delta_{ij} (\hat{P}_{\mathbf{k}}^i, \hat{P}_{-\mathbf{k}}^i). \tag{15}$$

This expression can be considered as the orthogonality condition and for the orthogonalized dynamical variables one has $(\hat{P}_{\mathbf{k}}, \hat{P}_{\mathbf{k}}^+)_{ji}^{-1} = \delta_{ji}(\hat{P}_{\mathbf{k}}^i, \hat{P}_{-\mathbf{k}}^i)^{-1}$, which allows us to simplify the analytical calculations.

The linearized transport equations (7), the equations for the equilibrium time correlation functions (13), and the equation for collective mode spectrum (14) form a general basis for the study of the dynamic behavior of a multicomponent fluid in the memory function formalism.

2.3 Generalized hydrodynamic equations for a multicomponent fluid

Let us apply the results presented above to the set of dynamic variables which contains the conserved dynamic variables of a multi-component fluid

$\hat{P}_{\mathbf{k}}^{\text{hyd}}$, introduced above. For the longitudinal dynamics one has,

$$\hat{P}_{\mathbf{k}}^{\text{L}} = \left\{ \hat{\mathcal{N}}_{\mathbf{k}}, \hat{J}_{\mathbf{k}}^{\text{L}}, \hat{H}_{\mathbf{k}} \right\}, \qquad (16)$$

where

$$\hat{H}_{\mathbf{k}} = \hat{E}_{\mathbf{k}} - \left(\hat{E}_{\mathbf{k}}, \hat{\mathcal{N}}_{\mathbf{k}}^{+} \right) \left(\hat{\mathcal{N}}_{\mathbf{k}}, \hat{\mathcal{N}}_{\mathbf{k}}^{+} \right)^{-1} \hat{\mathcal{N}}_{\mathbf{k}} = \left(1 - \mathcal{P}_{\mathcal{N}} \right) \hat{E}_{\mathbf{k}}, \qquad (17)$$

is the so-called enthalpy density, and

$$\mathcal{P}_{\mathcal{N}} \ldots = \left(\ldots, \hat{\mathcal{N}}_{\mathbf{k}}^{+} \right) \left(\hat{\mathcal{N}}_{\mathbf{k}}, \hat{\mathcal{N}}_{\mathbf{k}}^{+} \right)^{-1} \hat{\mathcal{N}} =$$

$$= \sum_{\alpha\gamma} \left(\ldots, \hat{n}_{-\mathbf{k},\alpha} \right) \left(\hat{\mathcal{N}}_{\mathbf{k}}, \hat{\mathcal{N}}_{\mathbf{k}}^{+} \right)^{-1}_{\alpha\gamma} \hat{n}_{\mathbf{k},\gamma}, \qquad (18)$$

is Mori-like projection operator, defined on the number densities $\{\hat{n}_{\mathbf{k},\alpha}\}$. We note that contrary to the case of simple fluids the hydrodynamic set (16) is not completely orthogonal in the sense of (15) due to of $(\hat{n}_{\mathbf{k},\alpha}, \hat{n}_{-\mathbf{k},\beta}) \neq 0$ for $\alpha \neq \beta$. However, such a choice is convenient for the further comparison of the results obtained in the case of simple liquids [15, 16].

Using the properties of correlation functions under time inversion and spatial symmetry operations, one can show [17] that in general the structure of the longitudinal $(l+2) \times (l+2)$ hydrodynamic frequency matrix is as follows,

$$i\Omega^{\text{L}}(k) = \begin{pmatrix} 0^{l\times l} & i\Omega_{nj}^{l\times 1}(k) & 0^{l\times 1} \\ i\Omega_{jn}^{1\times l}(k) & 0 & i\Omega_{jh}(k) \\ 0^{1\times l} & i\Omega_{hj}(k) & 0 \end{pmatrix}, \qquad (19)$$

where $0^{\alpha\times\beta}$ denotes the $\alpha \times \beta$ matrix with all zero elements.

For arbitrary finite k all the elements of the memory function matrix are nonzero, so that this matrix $\varphi^{\text{L}}(k,t)$ in the longitudinal case has the structure,

$$\varphi^{\text{L}}(k,t) = k^2 \begin{pmatrix} \phi_{nn}^{l\times l}(k,t) & \phi_{nj}^{l\times 1}(k,t) & \phi_{nh}^{l\times 1}(k,t) \\ \phi_{jn}^{1\times l}(k,t) & \phi_{jj}(k,t) & \phi_{jh}(k,t) \\ \phi_{hn}^{1\times l}(k,t) & \phi_{hj}(k,t) & \phi_{hh}(k,t) \end{pmatrix}. \qquad (20)$$

Using the Eqs. (7), (13) and (14) for the hydrodynamic set of dynamic variables $\hat{P}_{\mathbf{k}}^{\text{L}}$ we obtain the basis equations of generalized hydrodynamics for longitudinal components of multi-component fluids [14]. The generalized transport coefficients of the system can be defined by standard manner via the elements of matrix $\tilde{\varphi}^{\mathrm{L}}(k, z)$.

For the transverse fluctuations with one dynamic variable $\hat{P}_{\mathbf{k}}^{\mathrm{T}} = \hat{J}_{\mathbf{k}}^{\mathrm{T}}$, one has the generalized transport equation in the form,

$$\left\{ i\omega + \tilde{\varphi}_{jj}^{\mathrm{T}}\left(k, \omega + i\varepsilon\right) \right\} \left\langle \Delta \hat{J}_{\mathbf{k}}^{\mathrm{T}} \right\rangle^{\omega} = 0, \tag{21}$$

that is quite similar to the case of a simple fluid. The generalized shear viscosity can be defined via the memory function $\tilde{\varphi}_{jj}^{\mathrm{T}}(k, z)$.

Since the elements of the hydrodynamic frequency matrix (20) are expressed via the static correlation functions (see (8)), constructed on the densities of conserved variables and its first time derivatives, this allows us to express them via the so-called generalized k-dependent thermodynamic quantities, namely, one gets [17],

$$i\Omega^{\mathrm{L}}(k) = \begin{pmatrix} 0^{l \times l} & \dfrac{ik}{\bar{m}}\mathbf{c} & 0^{l \times 1} \\[2ex] ik\dfrac{\mathbf{v}^{+}(k)}{\kappa_{\mathrm{T}}(k)} & 0 & ik\dfrac{\alpha_{\mathrm{P}}(k)V}{\kappa_{\mathrm{T}}(k)C_{\mathrm{V}}(k)} \\[2ex] 0^{1 \times l} & ik\dfrac{\alpha_{\mathrm{P}}(k)T}{\rho\kappa_{\mathrm{T}}(k)} & 0 \end{pmatrix}, \tag{22}$$

where $\mathbf{c}$ and $\mathbf{v}$ are column-vectors with the components c_{α} and $v_{\alpha}(k)$, respectively. The generalized k-dependent thermodynamic quantities in (22) are defined by the expressions [17],

$$\frac{1}{\kappa_{\mathrm{T}}(k)} = Nnk_{\mathrm{B}}T \sum_{\alpha,\gamma=1}^{\nu} c_{\alpha} \, (\hat{\mathcal{N}}_{\mathbf{k}}, \hat{\mathcal{N}}_{\mathbf{k}}^{+})_{\alpha\gamma}^{-1} \, c_{\gamma}|_{k \to 0} \to -\frac{1}{V}\left(\frac{\partial V}{\partial P}\right)_{T,N_{\gamma}}, \tag{23}$$

for *the generalized compressibility* $\kappa_{\mathrm{T}}(k)$;

$$\frac{v_{\gamma}(k)}{\kappa_{\mathrm{T}}(k)} = \frac{N}{\beta} \sum_{\alpha=1}^{\nu} c_{\alpha}(\hat{\mathcal{N}}\hat{\mathcal{N}}^{+})_{\alpha\gamma}^{-1}\Big|_{k \to 0} \to N \sum_{\alpha=1}^{l} c_{\alpha}\left(\frac{\partial \mu_{\alpha}}{\partial N_{\gamma}}\right)_{T,V,N_{\bar{\gamma}}}, \tag{24}$$

for *the k-dependent partial molar volume*, per molecule, of species γ;

$$k_{\mathrm{B}}T^{2}C_{\mathrm{V}}(k) = (\hat{H}_{\mathbf{k}}, \hat{H}_{-\mathbf{k}}), \tag{25}$$

for *the generalized specific heat at constant volume*; and

$$\alpha_P(k) = \frac{1}{ik}(iL_N\hat{J}_{\mathbf{k}}, \hat{H}_{-\mathbf{k}})\frac{\kappa_T(k)}{k_B T^2 V}\bigg|_{\mathbf{k}\to 0} \to \frac{1}{V}\left(\frac{\partial V}{\partial T}\right)_{P,N_\alpha}, \qquad (26)$$

for *the generalized linear thermal expansion coefficient*.

Note that as it follows from (23) and (24), one has the equality,

$$\sum_{\alpha=1}^{\nu} c_\alpha v_\alpha(k) = v = \frac{V}{N}, \qquad (27)$$

being known from the standard thermodynamic treatment. Moreover, it can be proved rigorously that another thermodynamic relation, known for the ratio of specific heats at constant pressure and constant volume, is also valid for the generalized thermodynamic quantity $\gamma(k) = C_P(k)/C_V(k)$, namely,

$$\gamma(k) = 1 + \frac{TV\alpha_P^2(k)}{C_V(k)\kappa_T(k)}. \qquad (28)$$

2.4 *Generalized transport coefficients*

The memory functions [see, e.g., (8)] are defined based on the generalized fluxes,

$$ikI_{k,\alpha}^d = (1-\mathcal{P}_H)iL_N\hat{N}_{k,\alpha} = \frac{ik}{m_\alpha}\left(\hat{J}_{k,\alpha} - \frac{m_\alpha c_\alpha}{\bar{m}}\hat{J}_k\right), \qquad (29)$$

$$ikI_{k,j}^d = (1-\mathcal{P}_H)iL_N\hat{J}_k = ik\sigma^{zz}(k) - \sum_{\alpha=1}^{l} i\Omega_{jn}^\alpha(k)\hat{N}_{k,\alpha} - i\Omega_{jh}(k)\hat{H}_k, \qquad (30)$$

$$ikI_{k,h}^d = (1-\mathcal{P}_H)iL_N\hat{h}_k = ikI_{k,h} - i\Omega_{hj}(k)\hat{J}_k, \qquad (31)$$

where $\mathcal{P}_H$ is the Mori-like projection operator, constructed on the set of all the hydrodynamic variables, and the explicit expression for $I_{k,h}$ follows from the definition,

$$ikI_{k,h} = iL_N\hat{H}_k = ikI_{k,\varepsilon} - \sum_{\alpha\beta}\left(\hat{E}_k, \hat{N}_{-k,\alpha}\right)\left(\hat{N}_k, \hat{N}_k^+\right)_{\alpha\beta}^{-1}\hat{J}_{k,\beta} \qquad (32)$$

with $I_{k,\varepsilon}$ being the microscopic flux of energy. For the sake of simplicity we suppose that wavevector k is orientated along the OZ axis.

Thereafter the Laplace transforms of the hydrodynamic memory functions can be written via the generalized transport coefficients $L_{ij}(k,z)$ as follows,

$$\tilde{\varphi}_{ij}(\mathbf{k},z) = k^2 \tilde{\phi}_{ij}(\mathbf{k},z) = k^2 V k_{\mathrm{B}} T \sum_g \tilde{L}_{ig}\left(k,z\right) \left(\hat{P}_{\mathbf{k}}^{\mathrm{hyd}}, \hat{P}_{-\mathbf{k}}^{\mathrm{hyd}}\right)_{gj}^{-1}, \quad (33)$$

where $i,j,g = \{\alpha, j, h\}$, $\alpha = 1, 2, \ldots, \nu$, and we have introduced the generalized transport coefficients, $\tilde{L}_{ij}(k,z)$, as follows:

$$L_{ij}(k,z) = \frac{\beta}{V} \int\limits_0^\infty dt \exp\{-zt\} \left(I_{k,i}^d, \exp\left\{-(1-\mathcal{P}_{\mathrm{H}})\mathrm{i}L_{\mathrm{N}}t\right\} I_{-k,j}^d\right).$$

Taking into account the properties of the Mori operator, $\mathcal{P}_{\mathrm{H}}$, it is easy to prove that in the limit $k \to 0$, where the equality,

$$\left(I_{k,l}^d, \exp\left\{-(1-\mathcal{P}_H)\mathrm{i}L_{\mathrm{N}}t\right\} I_{-k,j}^d\right)\Big|_{k\to 0} = \left(\Delta I_l^d, \exp\{-\mathrm{i}L_{\mathrm{N}}t\}\Delta I_j^d\right)$$

is performed, the generalized transport coefficients $L_{ij}(k,z)$, defined above, explicitly reproduce the structure of well-known Green-Kubo formulas for the hydrodynamic transport coefficients in the limit $(k, \omega) \to 0$. Note also that the symmetry properties of the transport coefficients immediately follow from the symmetry properties of the generalized I_i^d fluxes under transformations of time and space coordinates.

The generalized transport coefficients $L_{ij}(k,z)$ can be directly related to transport coefficients, well-known from the nonequilibrium thermodynamics treatment, for instance: $L_{jj}(k,z) = \eta_l(k,z)$ gives just *the generalized longitudinal viscosity*; $L_{hh}(k,z) = T\lambda(k,z)$ defines *the generalized thermal conductivity* $\lambda(k,z)$; $L_{nn}^{\alpha\beta}(k,z) = nD_{\alpha\beta}(k,z)/k_{\mathrm{B}}T$ is used in defining *the generalized mutual diffusion coefficients* $D_{\alpha\beta}(k,z)$; and $L_{nh}^\alpha(k,z) = nTD_{\mathrm{T}}^\alpha(k,z)$ describes *the generalized thermal diffusion* in the α-th species.

In a similar way, for the transverse dynamics one can introduce the generalized shear viscosity $L_{jj}^{(T)}(k,z) = \eta(k,z)$, defined on the nondiagonal element of the stress tensor $\sigma^{zx}(k)$. Comparing the results obtained within the proposed rigorous statistical approach with other known theories, two important consequences should be noted:

(i) all the fluxes, used in defining the transport coefficients, are written in the microscopic form being convenient for subsequent application in computer simulations[2];

(ii) in addition to the well-known transport coefficients there appear new ones [20, 21], namely $L_{jh}(k, z) = -i\xi(k, z)$ and $L_{nj}^{\alpha}(k, z) = -i\zeta_{\alpha}(k, z)$, describing the dynamic coupling between the processes with different tensor dimensionality.

In the case of the present interest one has such an example in the matrix elements φ_{jh}, $\varphi_{jn}^{1\times\alpha}$, and φ_{hj}, $\varphi_{nj}^{\alpha\times1}$ in (20). In the hydrodynamic limit the corresponding transport coefficients tend to zero and they contribute in higher order with respect to k in comparison with all other terms in the matrix (20) being proportional to k^2.

Let us summarize the expressions for the elements of the memory functions matrix (see (20)). For the matrix $\tilde{\phi}_{ij}^{L}(\mathbf{k}, z)$ one has:

$$\tilde{\phi}_{nn}^{\alpha\delta}(k, z) = N \sum_{\gamma=1}^{l} D_{\alpha\gamma}(k, z) \left(\hat{\mathcal{N}}\hat{\mathcal{N}}^{+}\right)_{\gamma\delta}^{-1}, \qquad \tilde{\phi}_{jj}(k, z) = \frac{1}{\rho}\eta(k, z),$$

$$\tilde{\phi}_{hh}(k, z) = \frac{V}{C_V(k)}\lambda(k, z),$$

$$\tilde{\phi}_{hn}^{1\times\alpha}(k, z) = \frac{NT}{\beta} \sum_{\gamma=1}^{l} D_{T}^{\gamma}(k, z) \left(\hat{\mathcal{N}}\hat{\mathcal{N}}^{+}\right)_{\gamma\alpha}^{-1},$$

$$\tilde{\phi}_{nh}^{\alpha\times1}(k, z) = \frac{N}{C_V(k)}D_{T}^{\alpha}(k, z), \qquad \tilde{\phi}_{jh}(k, z) = -\frac{iV}{TC_V(k)}\xi(k, z),$$

$$\tilde{\phi}_{hj}(k, z) = -\frac{i}{\rho}\xi(k, z), \qquad \tilde{\phi}_{nj}^{\alpha\times1}(k, z) = -\frac{i}{\rho}\zeta_{\alpha}(k, z),$$

$$\tilde{\phi}_{jn}^{1\times\alpha}(k, z) = -ik_B TV \sum_{\gamma=1}^{l} \zeta_{\gamma}(k, z) \left(\hat{\mathcal{N}}\hat{\mathcal{N}}^{+}\right)_{\alpha\gamma}^{-1}.$$

As was already mentioned, the generalized transport coefficients $\xi(k, z)$ and $\zeta_{\alpha}(k, z)$, describing the dynamic cross-correlation coupling, tend to zero in the hydrodynamic limit. However, if a wavenumber k is finite they can contribute significantly to the dynamic behavior of the system.

---

[2]See, for instance, the discussion of this problem in [10, 18, 19], where Soret coefficient for binary mixtures was studied.

Note that not all the generalized transport coefficients in the matrix (20) are independent. Taking into account that,

$$\sum_{\alpha=1}^{\nu} m_\alpha I_{k,\alpha}^d = \sum_{\alpha=1}^{\nu} \left(\hat{J}_{k,\alpha} - \frac{m_\alpha c_\alpha}{\bar{m}} \hat{J}_k \right) \equiv 0, \qquad (34)$$

a few useful relations for the generalized transport coefficients can be easily derived. In particular, one has,

$$\sum_{\alpha=1}^{\nu} m_\alpha D_{\alpha\beta}(k, z) = \sum_{\beta=1}^{\nu} D_{\alpha\beta}(k, z) m_\beta \equiv 0, \qquad (35)$$

and

$$\sum_{\alpha=1}^{\nu} m_\alpha D_{\mathrm{T}}^\alpha(k, z) \equiv 0, \qquad (36)$$

for the generalized mutual diffusion and thermal diffusion coefficients, respectively. It is evident that a similar relation can be also written for the non-hydrodynamic coefficients $\zeta_\gamma(k, z)$. In fact the relations for (k, ω)-dependent transport coefficients (35) and (36) generalize the expressions previously known for the hydrodynamic transport coefficients.

3. Generalized hydrodynamics of a binary mixture

Let us consider now more in detail the simplest nontrivial case of a multi-component mixture that is a binary fluid. We first simplify some expressions presented above, and let us start from the elements of the memory functions matrix. Taking into account the relations (35) and (36), we can introduce the normalized generalized mutual diffusion coefficient $\bar{D}(k, z)$, the normalized generalized thermal diffusion coefficient $\bar{D}_{\mathrm{T}}(k, z)$, and the cross-correlation coefficient $\bar{\zeta}(k, z)$ as follows,

$$D_{\alpha\beta}(k, z) = (-1)^{\alpha+\beta} \frac{\bar{D}(k, z)}{m_\alpha m_\beta},$$

and

$$D_{\mathrm{T}}^\alpha(k, z) = (-1)^{\alpha+1} \frac{\bar{D}_{\mathrm{T}}(k, z)}{m_\alpha}, \qquad \zeta_\alpha(k, z) = (-1)^{\alpha+1} \frac{\bar{\zeta}(k, z)}{m_\alpha},$$

where $\alpha, \beta = 1, 2$. Thus, for the corresponding elements of the matrix of memory functions one obtains,

$$\phi_{nn}^{11}(k, z) = n\bar{D}(k, z)\frac{g_1(k)}{m_1}, \qquad \phi_{nn}^{12}(k, z) = -n\bar{D}(k, z)\frac{g_2(k)}{m_1},$$

$$\phi_{nn}^{21}(k, z) = -n\bar{D}(k, z)\frac{g_1(k, z)}{m_2}, \qquad \phi_{nn}^{22}(k, z) = n\bar{D}(k, z)\frac{g_2(k)}{m_2},$$

$$\phi_{hn}^{1}(k, z) = \frac{nT}{\beta}\bar{D}_T(k, z)g_1(k), \qquad \phi_{hn}^{2}(k, z) = -\frac{nT}{\beta}\bar{D}_T(k, z)g_2(k),$$

$$\phi_{nh}^{1}(k, z) = \frac{N}{C_V(k)m_1}\bar{D}_T(k, z), \qquad \phi_{nh}^{2}(k, z) = -\frac{N}{C_V(k)m_2}\bar{D}_T(k, z),$$

and

$$\phi_{nj}^{1}(k, z) = -\frac{i}{\rho m_1}\bar{\zeta}(k, z), \qquad \phi_{nj}^{2}(k, z) = \frac{i}{\rho m_2}\bar{\zeta}(k, z),$$

$$\phi_{jn}^{1}(k, z) = -\frac{i}{\beta}\bar{\zeta}(k, z)g_1(k), \qquad \phi_{jn}^{2}(k, z) = \frac{i}{\beta}\bar{\zeta}(k, z)g_2(k),$$

where the k-dependent functions $g_1(k)$ and $g_2(k)$ can be expressed via the elements of the matrix of direct correlation functions $C_{\alpha\gamma}(k)$ (see Appendix), namely,

$$g_1(k) = \frac{1}{\rho_1} - \left[\frac{1}{m_1}C_{11}(k) - \frac{1}{m_2}C_{21}(k)\right], \qquad (37)$$

$$g_2(k) = \frac{1}{\rho_2} - \left[\frac{1}{m_2}C_{22}(k) - \frac{1}{m_1}C_{12}(k)\right]. \qquad (38)$$

Note that for a mixture of neutral particles in the hydrodynamic limit one has:

$$g_1 = \frac{1}{m_1}\left(\frac{\partial\mu_1}{\partial n_1}\right)_{V,T,n_2} - \frac{1}{m_2}\left(\frac{\partial\mu_1}{\partial n_2}\right)_{V,T,n_1} = \frac{1}{\rho}\left(\frac{\partial\mu_1}{\partial x}\right)_{V,T,M},$$

$$g_2 = -\frac{1}{m_1}\left(\frac{\partial\mu_2}{\partial n_1}\right)_{V,T,n_2} + \frac{1}{m_2}\left(\frac{\partial\mu_2}{\partial n_2}\right)_{V,T,n_1} = -\frac{1}{\rho}\left(\frac{\partial\mu_2}{\partial x}\right)_{V,T,M}$$

with $x = x_1$ denoting the mass-concentration.

Having all the matrix elements needed for the description of macrodynamics of the system in terms of generalized thermodynamic quantities and generalized transport coefficients, we can write down the generalized transport equations [see (7)] in the explicit form. In particular, for the averaged partial densities one gets[3]:

$$\left[i\omega + k^2\bar{D}\frac{ng_1}{m_1}\right]\langle n_{\mathbf{k},1}\rangle^\omega - k^2\bar{D}\frac{ng_2}{m_1}\langle n_{\mathbf{k},2}\rangle^\omega - ik\frac{c_1}{\bar{m}}\langle J_{\mathbf{k}}\rangle^\omega$$

$$- k^2\frac{i\bar{\zeta}}{\rho m_1}\langle J_{\mathbf{k}}\rangle^\omega + k^2\bar{D}_{\mathrm{T}}\frac{N}{m_1 C_{\mathrm{V}}}\langle H_{\mathbf{k}}\rangle^\omega = 0\,, \qquad (39)$$

$$\left[i\omega + k^2\bar{D}\frac{ng_2}{m_2}\right]\langle n_{\mathbf{k},2}\rangle^\omega - k^2\bar{D}\frac{ng_1}{m_2}\langle n_{\mathbf{k},1}\rangle^\omega - ik\frac{c_2}{\bar{m}}\langle J_{\mathbf{k}}\rangle^\omega$$

$$+ k^2\frac{i\bar{\zeta}}{\rho m_2}\langle J_{\mathbf{k}}\rangle^\omega - k^2\bar{D}_{\mathrm{T}}\frac{N}{m_2 C_{\mathrm{V}}}\langle H_{\mathbf{k}}\rangle^\omega = 0\,. \qquad (40)$$

Two other generalized transport equations describe the dynamics of viscoelastic and thermal fluctuations and have the following structure,

$$\left[i\omega + \frac{k^2}{\rho}\eta_l\right]\langle J_{\mathbf{k}}\rangle^\omega - \left[ik\frac{v_1}{\kappa_{\mathrm{T}}} + k^2\frac{i\bar{\zeta}g_1}{\beta}\right]\langle n_{\mathbf{k},1}\rangle^\omega$$

$$- \left[ik\frac{v_2}{\kappa_{\mathrm{T}}} - k^2\frac{i\bar{\zeta}g_2}{\beta}\right]\langle n_{\mathbf{k},2}\rangle^\omega - \left[ik\frac{\alpha_{\mathrm{P}}V}{\kappa_{\mathrm{T}}C_{\mathrm{V}}} + k^2\frac{i\xi V}{TC_{\mathrm{V}}}\right]\langle H_{\mathbf{k}}\rangle^\omega = 0, \quad (41)$$

$$\left[i\omega + k^2\frac{\lambda V}{C_{\mathrm{V}}}\right]\langle H_{\mathbf{k}}\rangle^\omega - \left[ik\frac{\alpha_{\mathrm{P}}T}{\rho\kappa_{\mathrm{T}}} + k^2\frac{i\xi}{\rho}\right]\langle J_{\mathbf{k}}\rangle^\omega$$

$$+ k^2\frac{nT\bar{D}_{\mathrm{T}}}{\beta}\left[g_1\langle n_{\mathbf{k},1}\rangle^\omega - g_2\langle n_{\mathbf{k},2}\rangle^\omega\right] = 0\,. \qquad (42)$$

The generalized transport equations presented above give a complete picture of the dynamics close to an equilibrium state for arbitrary temporal and spatial scales.

For practical reasons in case of a binary mixture it is often more convenient to use instead the partial densities $\hat{n}_{\mathbf{k},\alpha}$ with $\alpha = 1, 2$ their linear combinations. For instance, one can define the corresponding dynamic variables in the form,

$$\hat{\rho}_{\mathbf{k}} = \hat{\rho}_{\mathbf{k},1} + \hat{\rho}_{\mathbf{k},2}\,, \qquad \hat{x}_{\mathbf{k}} = \frac{1}{\bar{m}}\left[x_2\hat{\rho}_{\mathbf{k},1} - x_1\hat{\rho}_{\mathbf{k},2}\right], \qquad (43)$$

where $\hat{\rho}_{\mathbf{k},\alpha} = m_\alpha\hat{n}_{\mathbf{k},\alpha}$ are the partial mass densities. In fact the variables $\hat{\rho}_{\mathbf{k}}$ and $\hat{x}_{\mathbf{k}}$ denote the densities of total mass and mass-concentration, respectively. One of the advantages of such a choice is the simple relation connecting

---

[3]For the sake of simplicity here and below we omit the k- and (k, ω)-dependence in generalized thermodynamic quantities and generalized transport coefficients.

the mass-concentration density to the longitudinal mass-concentration current $\hat{J}_{\mathbf{k},x}$,

$$iL_{\mathrm{N}}\hat{x}_{\mathbf{k}} = \frac{ik}{\bar{m}}\hat{J}_{\mathbf{k},x}.$$

The dynamic variable $J_{\mathbf{k},x}$ is orthogonal [22–24] to the longitudinal total mass-current density $\hat{J}_{\mathbf{k}} \equiv \hat{J}_{\mathbf{k},\rho}$. This makes especially convenient the theoretical treatment of dynamical processes in small k region, where the collective type of the dynamics prevails [22–24]. Using the definition (43), the Eqs. (39)–(42) can be easily rewritten for new set of dynamic variables, so that we obtain:

$$i\omega\langle\hat{\rho}_{\mathbf{k}}\rangle^{\omega} - ik\langle\hat{J}_{\mathbf{k}}\rangle^{\omega} = 0, \tag{44}$$

$$\left[i\omega + k^2 n\bar{D}A(k)\right]\langle\hat{x}_{\mathbf{k}}\rangle^{\omega} + k^2\frac{n\bar{D}}{\bar{m}}B(k)\langle\hat{\rho}_{\mathbf{k}}\rangle^{\omega}$$
$$- k^2\frac{i\bar{\zeta}}{\rho\bar{m}}\langle\hat{J}_{\mathbf{k}}\rangle^{\omega} + k^2\bar{D}_{\mathrm{T}}\frac{N}{\bar{m}C_{\mathrm{V}}}\langle\hat{H}_{\mathbf{k}}\rangle^{\omega} = 0, \tag{45}$$

$$\left[i\omega + k^2\frac{\eta l}{\rho}\right]\langle\hat{J}_{\mathbf{k}}\rangle^{\omega} - \left[ik\frac{\bar{m}\rho}{\beta}B(k) + ik^2\frac{\bar{m}\bar{\zeta}}{\beta}A(k)\right]\langle\hat{x}_{\mathbf{k}}\rangle^{\omega}$$
$$- \left[\frac{ik}{\rho\kappa_{\mathrm{T}}} + k^2\frac{i\bar{\zeta}}{\beta}B(k)\right]\langle\hat{\rho}_{\mathbf{k}}\rangle^{\omega} - \left[ik\frac{\alpha_{\mathrm{P}}V}{\kappa_{\mathrm{T}}C_{\mathrm{V}}} + k^2\frac{i\xi V}{TC_{\mathrm{V}}}\right]\langle\hat{H}_{\mathbf{k}}\rangle^{\omega} = 0, \tag{46}$$

$$\left[i\omega + k^2\frac{\lambda V}{C_{\mathrm{V}}}\right]\langle\hat{H}_{\mathbf{k}}\rangle^{\omega} + k^2\frac{nT\bar{D}_{\mathrm{T}}}{\beta}B(k)\langle\hat{\rho}_{\mathbf{k}}\rangle^{\omega}$$
$$+ k^2\frac{\rho T\bar{D}_{\mathrm{T}}}{\beta}A(k)\langle\hat{x}_{\mathbf{k}}\rangle^{\omega} - \left[ik\frac{\alpha_{\mathrm{P}}T}{\rho\kappa_{\mathrm{T}}} + k^2\frac{i\xi}{\rho}\right]\langle\hat{J}_{\mathbf{k}}\rangle^{\omega} = 0, \tag{47}$$

where the k-dependent coefficients $A(k)$ and $B(k)$ are defined as follows,

$$A(k) = \frac{1}{m_1}g_1(k) + \frac{1}{m_2}g_2(k), \tag{48}$$

$$B(k) = \frac{x_1}{m_1}g_1(k) - \frac{x_2}{m_2}g_2(k). \tag{49}$$

The expression for $B(k)$ can be rewritten via the partial direct correlation functions $C_{\alpha\beta}$ (see Appendix) and one gets,

$$B(k) = \frac{1}{\rho}\frac{m_2 - m_1}{m_1 m_2} + \frac{x_1 - x_2}{m_1 m_2}C_{12}(k) + \left[\frac{x_2}{m_2^2}C_{22}(k) - \frac{x_1}{m_1^2}C_{11}(k)\right]. \tag{50}$$

Hence, it is seen in (50) that this factor actually describes the effects of dissimilarities between the pure components forming the mixture. In particular, the first term is nonzero for particles of different masses, the second one is connected with the mass-concentration difference, and the last term can be significant for particles of different sizes. Considering the simplest case of symmetrical binary mixture (with $m_1 = m_2 = m$ and $V_{11}(R) = V_{22}(R) \neq V_{12}(R)$), we find that $B(k) = (c_2 - c_1)[C_{11}(k) - C_{12}(k)]/m^2$, so that this term is nonzero for $c_1 \neq c_2$. Less sensitive in this sense is the factor $A(k)$.

It can be testified that considering the equations of macrodynamics (44)–(47) in the hydrodynamic limit, all the analytical results known in the literature for a binary liquid [25, 26] are reproduced.

4.   Molten salts: model of a binary mixture of charged spheres

In practice, in describing a binary mixture of charged particles, another set of dynamic variables is widely used, namely, instead of partial densities $\hat{n}_{\mathbf{k},\alpha}$ or the set (43), the mass density $\hat{\rho}_{\mathbf{k}}$ and the charge density $\hat{q}_{\mathbf{k}}$ are utilized. However, it should be mentioned that due to the electroneutrality constraint the charge density $\hat{q}_{\mathbf{k}}$ can be simply connected with the mass-concentration density $\hat{x}_{\mathbf{k}}$, introduced above. In particular, one has,

$$\hat{q}_{\mathbf{k}} = q_1 \hat{n}_{\mathbf{k},1} + q_2 \hat{n}_{\mathbf{k},2} = \bar{m} \left[\frac{q_1}{m_1} - \frac{q_2}{m_2} \right] \hat{x}_{\mathbf{k}} = \bar{m} I \hat{x}_{\mathbf{k}}, \qquad (51)$$

where q_α denotes the charge of particles in the α-th species and $I = q_1/m_1 - q_2/m_2$. Thus, all the results, presented in the previous section, can be directly applied to the case considered here.

In order to compare the equations derived above let us consider Eq. (45) in the hydrodynamic region, when k is small. Note that the first Eq. (44) is just the mass-density conservation law, and to analyze the second one (45) one has to know the properties of the factors $A(k)$ and $B(k)$ in the hydrodynamic limit. Note that the cross-correlation coefficients $\bar{\zeta}$ and ξ can be neglected in this case.

To find the long-wavelength limit of $A(k)$ one has to recall the definitions (37), (38), and (48), as well as the properties of direct correlation functions $C_{\alpha\gamma}(k)$ when $k \to 0$. For the mixture of charge particles the long-range contributions to the pair potentials are entirely determined by the Coulomb interactions [2]. Hence, when k is small, one has,

$$C_{\alpha\gamma}(k) \simeq -\beta \frac{4\pi q_\alpha q_\gamma}{\varepsilon k^2},$$

where ε is the dielectric constant.

The factor $B(k)$ can be easily calculated from (50), using the known asymptotic behavior for $C_{\alpha\gamma}(k)$ and the electroneutrality condition. One gets,

$$B(k)\big|_{k\to 0} = B_0 = \text{const}$$

with the value B_0 depending mainly on mass and size differences of the particles. On the other hand, for the factor $A(k)$ one has,

$$A(k)\big|_{k\to 0} \simeq \frac{4\pi I^2 \beta}{\varepsilon k^2}$$

. This means that the corresponding coupling constant in (45) increases in
long-wavelength limit and the relevant term eventually becomes even dominant
in comparison with other ones. Therefore the Eq. (45) can be simplified and in
the limit $k \to 0$ we get,

$$\frac{\partial}{\partial t} \langle q_{\mathbf{k}} \rangle^t + \frac{1}{\tau_D} \langle q_{\mathbf{k}} \rangle^t = 0,$$

with the finite relaxation time given by

$$\tau_D = \frac{\varepsilon}{4\pi n \bar{D} I^2 \beta} = \frac{\varepsilon}{4\pi \sigma}.$$

The expression $\sigma = n\bar{D}I^2\beta$ for the ionic conductivity σ follows directly
from the comparison with the phenomenological theories [2, 8] and indeed
gives the well-known Nernst-Einstein relation connecting the conductivity with
the mutual diffusion coefficient. Note that $\bar{D} = \lim_{k,\omega \to 0} \bar{D}(k, z)$ is the hydrodynamic value of the normalized mutual diffusion coefficient.

Hence, one may conclude that in the limit $k \to 0$ the dynamics of the charge
fluctuations is completely determined by relaxation processes with the finite
(nonzero) relaxation time. In this sense we can speak about the fast kinetic-
like behavior of the charge fluctuations in the model considered. This results in
the effective independence of the other hydrodynamic Eqs. (44), (46), and (47),
from the time evolution of fast charge subsystem, so that the hydrodynamics of
a binary mixture of charge particles becomes rather similar to the case of simple liquids. However, we have to remember that in the hydrodynamic limit the
additional (comparing with simple liquids) well-defined transport coefficients,
namely the mutual $\bar{D}$ and $\bar{D}_T$ thermal diffusion coefficients, exist in the system that play a crucial role in the electric and the thermoelectric properties,
respectively.

Another kind of dissimilarities both from simple liquids and a binary mixtures of neutral particles is caused by the factor $B(k = 0) = B_0$. Let us use
the results obtained above for deriving the last two hydrodynamic equations
[see (41) and (42)], describing the longitudinal dynamics of the model in small
k limit. It can be done directly from the Eqs. (46) and (47), taking into account
the existing relations between two subsets of dynamic variables, describing the
density fluctuations, namely, $\{\hat{\rho}_{\mathbf{k}}, \hat{x}_{\mathbf{k}}\}$ and $\{\hat{\rho}_{\mathbf{k}}, \hat{q}_{\mathbf{k}}\}$ [see Eq. (51)]. In small k

domain one gets,

$$\left[\mathrm{i}\omega + k^2 \frac{\eta_l}{\rho} \right] \langle J_{\mathbf{k}} \rangle^\omega - \frac{\mathrm{i}k}{\rho \kappa_{\mathrm{T}}} \langle \rho_{\mathbf{k}} \rangle^\omega - \mathrm{i}k \frac{\alpha_{\mathrm{P}} V}{\kappa_{\mathrm{T}} C_{\mathrm{V}}} \langle H_{\mathbf{k}} \rangle^\omega = 0, \qquad (52)$$

$$\left[\mathrm{i}\omega + k^2 \frac{\lambda V}{C_{\mathrm{V}}} \right] \langle H_{\mathbf{k}} \rangle^\omega - k^2 \frac{nT\bar{D}_{\mathrm{T}}}{\beta} B_0 \langle \rho_{\mathbf{k}} \rangle^\omega - \mathrm{i}k \frac{\alpha_{\mathrm{P}} T}{\rho \kappa_{\mathrm{T}}} \langle J_{\mathbf{k}} \rangle^\omega$$
$$+ \frac{4\pi n T \bar{D}_{\mathrm{T}} I}{\varepsilon} \langle q_{\mathbf{k}} \rangle^\omega = 0, \qquad (53)$$

where smaller contributions being higher order with respect to k are neglected. It is seen that two terms on the left hand side of (53) make the difference from the case of simple liquid. Namely, the second one is proportional to the value of B_0 that depends on mass and size dissimilarities of particles. Note that for the restricted primitive model with $m_1 = m_2$ one has $B_0 = 0$. And the last term on the left hand side of (53) describes the contribution, caused by the charge fluctuations. It may be shown that such a contribution is important in the hydrodynamic limit and can change the damping coefficients of the hydrodynamic excitations. In order to prove this statement let us consider the Eq. (45) in small ω limit ($\omega \ll 1/\tau_{\mathrm{D}}$). This gives the solution,

$$\frac{4\pi}{\varepsilon} \langle \hat{q}_{\mathbf{k}} \rangle^\omega \simeq -k^2 \frac{1}{\beta I} B_0 \langle \hat{\rho}_{\mathbf{k}} \rangle^\omega - k^2 \frac{V}{I \beta C_{\mathrm{V}}} \frac{\bar{D}_{\mathrm{T}}}{\bar{D}} \langle \hat{H}_{\mathbf{k}} \rangle^\omega \,,$$

that will produce the additional hydrodynamic terms in (53) after the corresponding substitution. Hence, collecting the results of analytical study performed above, one can conclude that the hydrodynamic behavior of a binary mixture of charge particles has several very specific features. On the one hand, when k tends to zero, the charge fluctuations become effectively decoupled from the other hydrodynamic processes and are described by kinetic-like relaxing mode – this makes the hydrodynamics rather similar to the case of simple fluids. On another hand, they can significantly contribute into the hydrodynamics causing the changes of the corresponding damping coefficients. Moreover, the hydrodynamic coefficients, describing the mutual and thermal diffusions, that are well-defined for a binary liquid and absent in simple fluids, play an important role in the electric and the thermoelectric properties of a mixture of charged particles. Thus, even in the hydrodynamic limit the collective dynamical behavior of ionic liquids remains an interesting subject for further investigations.

5.   Generalized collective mode approach

The main problem in implementations of the generalized hydrodynamic equations, considered in the previous sections, is connected with the need to know the (k, ω)-dependent memory functions. Only having these functions one can solve the generalized hydrodynamic equations and obtain the main dynamic characteristics being interesting for the system considered. There are two alternatives in achieving such a goal. First, we may derive additional equations for memory functions and try to solve them in a certain approximation. In particular, this is a way of the mode coupling theory [5, 27]. Another possibility is based on the applications of some approximated formulas for memory functions that permit to explicitly reproduce its certain properties, for instance the sum rules [15, 16, 26].

During the last decade an essential progress in understanding the dynamical properties of fluids was achieved in connection with the method of generalized collective modes (GCM). The concept of generalized collective modes for the study of time correlation functions of a dense monoatomic fluid was initially proposed in [28–30]. It is important to note that this approach is based mainly on two physical ideas which were already formulated in the literature. The first idea is to use for the study of generalized hydrodynamics an extended set of dynamic variables containing, in addition to the conserved variables, their higher-order time derivatives. In such a way the kinetic processes which are important for smaller time scales become incorporated into the consideration. And the second one is, in fact, a physical assumption that the Markovian approximation for *higher order memory functions* can be applied even for non-hydrodynamic values of k. Nevertheless, one has to keep in mind that the Markovian approximation for higher order memory functions enables us to study the processes which are essentially non-Markovian from the viewpoint of standard hydrodynamics. Under these conditions the problem of generalized hydrodynamic description is reduced to the eigenvalues and eigenvectors problem for the generalized hydrodynamic operator, most of the elements of which can be calculated.

In paper [30], where the GCM approach within the five-variable description was applied for the first time to the study of a simple fluid with a continuous potential, some elements of the generalized hydrodynamic matrix were considered as adjustable parameters: namely, three k-dependent transport coefficients were determined from the weighted least-squares-fitting procedure as such that gave the best fit to four MD correlation functions, and, in addition, three static correlation functions were also used as adjustable parameters for increasing the accuracy in the fitting procedure. In this way the generalized collective mode spectrum for the Lennard-Jones fluid was calculated and it was shown that all the time correlation functions, constructed on the extended five-component variable, could be fitted consistently by five exponentials that correspond to the five generalized k-dependent modes. Thereafter, several attempts have been made to extend the GCM approach to the investigation of more complicated fluid-like systems such as molecular liquids, mixtures and colloidal suspensions (e.g., see [31, 32]). Very promising results have been obtained for binary mixtures. In particular, the theoretical and experimental investigations showed that the dynamic structure factors $S(k, \omega)$ of liquid $Li_{0.8}$–$Pb_{0.2}$ [33], dense gas mixtures He–Ne [34] and He–Ar [35] display a behavior which can be explained in terms of more than one pair of propagating generalized collective modes only. However, in all the theoretical works listed above the same procedure based on adjustable parameters was used. Later formulations of the GCM approach (see, e.g., [14, 36–38]) permits to avoid this problem and give the powerful way of constructing the generalized hydrodynamic theory of simple fluids and their mixtures.

In order to draw the main ideas of the GCM approach let us recall a general representation for the Laplace transform of an equilibrium time correlation function $= \langle A_k(t) A_k^*(0) \rangle$, derived by Mori [39],

$$\tilde{F}(k, z) = \cfrac{\Gamma_0}{z + \cfrac{\Gamma_1(k)}{z + \cfrac{\Gamma_2(k)}{z + \dots \cfrac{\Gamma_{s-1}(k)}{z + \tilde{\varphi}_s(k, z)}}}}, \qquad (54)$$

where

$$\Gamma_l(k) = \left\langle \hat{Y}_k^{(l+1)} \hat{Y}_{-k}^{(l+1)} \right\rangle \left\langle \hat{Y}_k^{(l)} \hat{Y}_{-k}^{(l)} \right\rangle^{-1},$$

$$\tilde{\varphi}_s(k, z) = \left\langle \hat{Y}_k^{(s+1)} \frac{1}{z + (1 - \mathcal{P}_s) i \hat{L}_N} \hat{Y}_{-k}^{(s+1)} \right\rangle \left\langle \hat{Y}_k^{(s)} \hat{Y}_{-k}^{(s)} \right\rangle^{-1},$$

with

$$\hat{Y}_k^{(0)} = \hat{A}_k, \qquad \hat{Y}_k^{(1)} = (1 - \mathcal{P}_0) i L_N \hat{A}_k,$$

$$\hat{Y}_k^{(2)} = (1 - \mathcal{P}_1) i L_N \hat{Y}_k^{(1)}, \qquad \dots, \qquad \hat{Y}_k^{(s)} = (1 - \mathcal{P}_{s-1}) i L_N \hat{Y}_k^{(s-1)},$$

$$\mathcal{P}_l = \sum_{j=0}^{l} \Delta \mathcal{P}_j, \qquad \Delta \mathcal{P}_j = \left\langle \dots \hat{Y}_{-k}^{(j)} \right\rangle \left\langle \hat{Y}_k^{(j)} \hat{Y}_{-k}^{(j)} \right\rangle^{-1} \hat{Y}_k^{(j)}.$$

For the memory function $\tilde{\varphi}_s(k, z)$ one can write the expression

$$\tilde{\varphi}_s(k, z) = \frac{\Gamma_s(k)}{z + \tilde{\varphi}_{s+1}(k, z)}, \qquad (55)$$

that can be considered as certain recurrent relation between lower and higher order memory functions. All other quantities in (54) are indeed the static correlation functions that have to be calculated by using the means of an equilibrium theory. In order to obtain the numerical results for $F(k, t)$ we have to calculate $\tilde{\varphi}_s(k, z)$. For example, one can use the Markovian approximation for $\tilde{\varphi}_s(k, z)$, $\tilde{\varphi}_s(k, z) \simeq \tilde{\varphi}_s(k, 0)$, where the function $\tilde{\varphi}_s(k, 0)$ is simply connected with the correlation time

$$\tau_0(k) = \frac{1}{F(k, 0)} \int\limits_0^\infty dt \, F(k, t)$$

that can be either calculated based on some ansatz (see, i.e., [40]) or obtained directly from molecular dynamic (MD) simulations.

Another alternative can be proposed using the equations of macrodynamics (7) for the extended set of dynamic variables

$$\mathbf{P}_k = \left\{ A_k, iL_N A_k, \ldots, (iL_N)^{(s-1)} A_k \right\}.$$

This gives the matrix equation,

$$[z\mathbf{I} + \mathbf{T}(k)] \, \tilde{\mathbf{F}}(k, z) = \mathbf{F}(k),$$ (56)

with the matrix $\mathbf{T}(k)$ defined by

$$\mathbf{T}(k) = -i\bar{\Omega}(k) + \tilde{\varphi}_s(k, 0) = \frac{\mathbf{F}(k, 0)}{\tilde{\mathbf{F}}(k, 0)},$$

where the matrix of time correlation functions $\mathbf{F}(k, t)$ is constructed on the variables $\mathbf{P}_k$. Thus, solving the eigenvalues problem,

$$\sum_{j=0}^s T_{ij}(k)\hat{X}_{j,\alpha} = z_\alpha \hat{X}_{i,\alpha},$$

one has the analytical expression for the Laplace transforms of time correlation functions,

$$\tilde{F}_{ij}(k, z) = \sum_{\alpha=0}^s \frac{G_\alpha^{ij}(k)}{z + z_\alpha(k)},$$ (57)

with weight coefficients,

$$G_\alpha^{ij}(k) = \sum_{l=0}^s \hat{X}_{i,\alpha}\hat{X}_{\alpha,l}^{-1}F_{lj}(k).$$

It should be stressed that in fact the eigenvalues $\{z_\alpha(k)\}$ give us the collective modes spectrum.

Comparing the two ways considered above one may conclude that the later approach – the generalized collective mode approach – has some important advantages. In particular, this method is especially promising in combination with molecular dynamics, because the time correlation time $\tau_0(k)$, appearing in $\mathbf{T}(k)$, can be directly calculated in MD simulations. Moreover, the eigenvalues problem can be formulated for initial set of nonorthogonal dynamic variables $\mathbf{P}_k = \{A_k, iL_N A_k, \ldots, (iL_N)^{(s-1)} A_k\}$, the dynamics of which is directly accessible in the simulations, instead of the variables $\hat{Y}_k^{(p)} = (1 - \mathcal{P}_{p-1}) iL_N \hat{Y}_k^{(p-1)}$ containing the projection operators $\mathcal{P}_{p-1}$. The theory is formulated in matrix form convenient for real calculations and permits within the unique approach to obtain the collective mode spectrum and weight coefficients for different time correlation functions.

In order to study the collective behavior of real fluids for arbitrary temporal and spacial scales one should on equal basis consider all the dynamic processes having the relaxation times having close values. This means that for the needs of generalized hydrodynamic theory, when all the hydrodynamic processes are strongly coupled, the hydrodynamic variables have to be incorporated into the initial subset $\mathbf{A}_k$ forming the extended set of dynamic variables as follows $\mathbf{P}_k = \{\mathbf{A}_k, iL_N \mathbf{A}_k, \ldots, (iL_N)^{(s-1)} \mathbf{A}_k\}$. In this case the Mori approach gives us the representation for time correlation functions in the form of branched continue fractions. Otherwise, the GCM approach can be easily generalized keeping all the advantages discussed above for scalar dynamic variable. Such a program has been performed for simple and binary liquids, and extended later to a case of multicomponent fluids. The details can be found in [14, 17]. The numerical results obtained within the GCM approach for simple and binary liquids are convenient for the interpretation in terms of generalized collective modes, especially in the range of crossover from the hydrodynamic behavior to the molecular regime, when the fast kinetic-like processes are involved in the collective dynamics. Moreover, the theory also allows us to calculate the generalized transport coefficients which was the subject of intensive recent investigations in the literature. In addition, considering the results found for separated subsets of dynamic variables, forming the set $\mathbf{P}_k$, one may obtain useful information about the mode coupling effects and establish the physical mechanisms responsible for mode formation in different spacial scales.

6. Numerical results and discussion

Let us now discuss the results, obtained in MD simulations of two liquid binary systems: a molten salt and a Lennard-Jones binary liquid. The eight-variable GCM approach is applied in the analysis of collective dynamics in these two model systems.

Computer simulations for a molten salt LiF at the temperature 1287 K and density 1729.29 kg/m^3 were performed in the standard microcanonical ensemble on a two model system of 500 and 1000 particles in a cubic box subject to periodic boundary conditions. Potentials in the Fumi-Tosi form for LiF were taken from [41]. The long-range interaction was treated by Ewald method. Fifteen wavenumbers were sampled in MD simulations with the smallest value of $k_{min}^{LiF} = 0.2711\text{Å}^{-1}$. In the case of equimolar Lennard-Jones binary liquid we have simulated liquid KrAr at temperature 116 K and density 1870.28 kg/m^3 using the system of 864 particles in a cubic box. Twenty two k-points were sampled in MD simulations of liquid KrAr, and the smallest wavenumber reached in MD was $k_{min}^{LJ} = 0.1735$ Å^{-1}. Regular production runs took over $3 \cdot 10^5$ time steps, while for three lowest k-values in the case of KrAr, in order to obtain the desired convergence of relevant static averages and time correlation functions, the system has been simulated over $2.1 \cdot 10^6$ time steps.

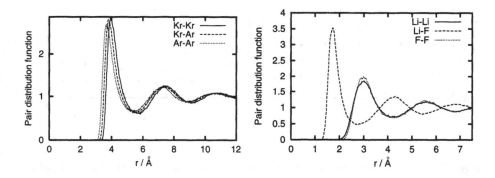

Figure 1. Radial distribution functions calculated for a Lennard-Jones KrAr binary mixture (left) and molten LiF alloy (right).

About the type of local structure in the two simulated binary fluids one may judge from the shape of the partial radial distribution functions, shown in Fig. 1. Attractive and repulsive character of interactions between different and similar particles in LiF creates the situation when the A–B structure becomes dominant. This results in a well-pronounced peak in the Li–F distribution function that is located at a smaller distance than the corresponding peaks in the Li–Li and F–F radial distribution functions. In liquid KrAr mixture the local structure is completely different: the sequence of partial radial distribution functions in the left frame of Fig. 1 reflects the difference in the Lennard-Jones parameter σ in interatomic potentials.

Due to the striking difference in the partial distribution functions between molten salts and Lennard-Jones liquid mixtures the structure factors of KrAr and LiF display opposite tendencies in total density and concentration correlations. Fig. 2 shows the static structure factors, defined on the variables (43) and obtained directly in MD simulations via fluctuation expressions. It is seen that contrary to the case of KrAr the "concentration-concentration" structure factor has in LiF a well observed peak that points out the local "antiferromagnetic"-like ordering of particles in different species. Another specific feature, seen in Fig. 2, is observed in small k domain where the function $S_{xx}(k)$ tends to zero in longwavelength limit. This is caused by the long-range Coulombic interactions being dominant in LiF. Different behavior is observed also for the total density structure factors $S_{tt}(k)$: in the Lennard-Jones mixture it has the shape typical of pure fluids with well pronounced first maximum, while in molten salts the first maximum at $k \approx 2.3 \text{ Å}^{-1}$ is strongly overdamped.

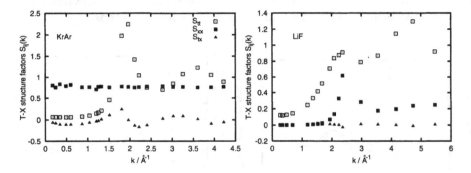

Figure 2.   Static structure factors of Lennard-Jones KrAr binary mixture (left) and molten LiF alloy (right).

Very specific features caused by Coulombic interactions in molten salts can be observed by comparing the results obtained for the generalized thermodynamic quantities of Lennard-Jones and Coulombic binary mixtures (see Fig. 3). Rather strong dissimilarities are seen practically for all the quantities studied. For instance, quite different behavior in small k-range is observed for the generalized specific heat $C_V(k)$ and the ratio of generalized specific heats $\gamma(k)$ at constant pressure and constant volume. Instead of decreasing $C_V(k)$ observed for KrAr mixture, when k increases, we see for LiF the inverse of the behavior. Rather strong static coupling between viscoelastic and thermal fluctuations, described by $\gamma(k)$, can be seen in LiF for k close to the position of the first peaks in structure factors $S_{tt}(k)$ and $S_{xx}(k)$. For the generalized thermal expansion coefficient $\alpha_T(k)$ there exists the range of k where this coefficient becomes negative. Note that for finite k the main reason for the appearance of the observed dissimilarities is connected with "antiferromagnetic" character

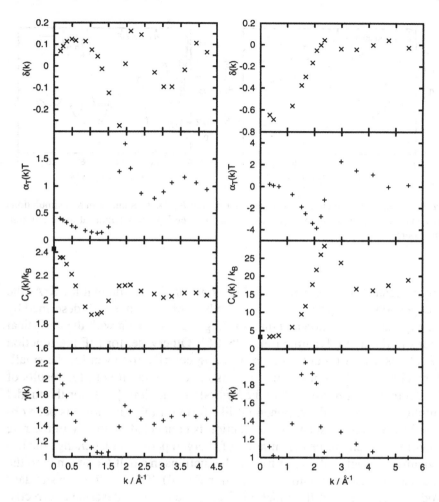

Figure 3. Generalized thermodynamic quantities calculated for a Lennard-Jones KrAr binary mixture (left) and molten LiF alloy (right): the generalized dilatation $\delta(k)$; the generalized linear thermal expansion coefficient $\alpha_T(k)$; the generalized specific heat at constant volume $C_V(k)$ (the filled boxes at $k = 0$ correspond to the values obtained directly in MD simulations); and the generalized ratio of specific heats $\gamma(k)$.

of the pair interactions, so that additional investigations for binary liquids of neutral particles with such type of interactions can be very useful for deeper understanding of the behavior of k-dependent thermodynamic quantities.

Let us now consider some dynamical properties, and we start from the time correlation functions. Fig. 4 plots the MD results obtained for time correlation function of "concentration-concentration" at three values of k. The most interesting feature is seen at smallest wavenumbers, presented in Fig. 4. Com-

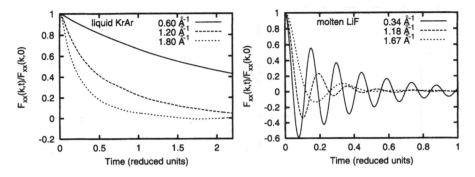

Figure 4.   Autocorrelation mass-concentration function $F_{xx}(k,t)$ obtained in MD simulations for a Lennard-Jones KrAr binary mixture (left) and molten LiF alloy (right) at three different wavenumbers k.

paring the results found for KrAr and LiF one may conclude that for KrAr the time-dependence of $F_{xx}(k,t)$ in the hydrodynamic region is well described by single exponential function and that it is in good agreement with the analytical solutions in the hydrodynamic limit [25, 26]. Otherwise, for LiF the function $F_{xx}(k,t)$ has a well-pronounced propagating contribution even for the smallest k reached in MD simulations. This is in contradiction with the results of analytical treatment performed in the hydrodynamic limit [2, 8] for the model of a molten salt, where the exponential-like form for this function has been obtained. The reason for such a contradiction is connected mainly with the role of Coulombic interactions in a mixture of charged particles that change the hydrodynamic character of concentration fluctuations in small k domain onto the kinetic one (see the discussion, presented in Sec. 4). Thus, within the standard hydrodynamic treatment the function $F_{xx}(k,t)$ cannot be described correctly. This problem was considered more in detail in [42], where appropriate expression for $F_{xx}(k,t)$ has been derived.

Nowadays, generalized collective mode spectra for binary liquids are the subject of intensive studies in the literature. Within the GCM approach such investigations have been performed for several models of real fluids and the main findings herein are as follows:

(i) it was shown that in binary mixtures of neutral particles under some condition one can observe the optic phonon-like propagating modes, describing out-of-phase oscillations of particles in different species;

(ii) the analytical expression for such a condition has been derived and justified for several binary liquids;

(iii) the crossover from the cooperative behavior to partial one was estab-
    lished in collective dynamics of binary fluids when k increases from the
    hydrodynamic values to the wavenumbers typical of molecular regime.

In order to illustrate these statements let us consider the results obtained within
the eight-variable GCM approach for two lowest propagating modes in KrAr
and LiF (see Fig. 5). In both cases we found a pair of generalized sound modes
(see filled boxes in upper parts) with linear dispersion in small k range and a
pair of optic-like propagating modes with finite damping coefficients (lower
parts of Fig. 5) when k goes to zero. This means that the optic-like modes are
in fact kinetic ones and do not contribute to the hydrodynamic time correlation
functions of a binary liquid of neutral particles in small k limit. However this is
not true for a binary mixture of charged particles which will be seen hereinafter.

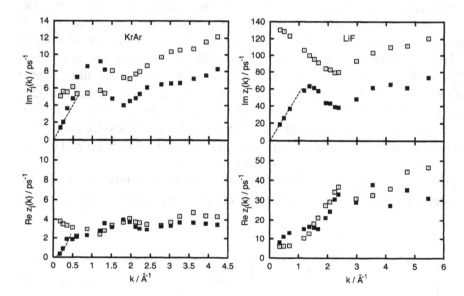

Figure 5. Dispersions (imaginary parts of eigenvalues) and damping coefficients (real parts)
of two lowest propagating modes obtained for a Lennard-Jones KrAr binary mixture (left) and
molten LiF alloy (right) within the eight-variable GCM approach.

Note that the main dissimilarities seen in Fig. 5 are related with the k-
dependence of dispersions and damping coefficients of optic-like mode at small
k domain. For example, for LiF the characteristic frequency of these excitations decreases when k increases. An inverse situation is observed for KrAr.
One may suppose that such a dissimilarity is caused for finite k mainly by
an "antiferromagnetic" type of interactions in a mixture of charged particles,
whereas the long-range characte of Coulombic potential becomes a crucial fac-

tor only in very small k range. Another point, that should be mentioned, concerns the damping coefficients. It is seen that the damping coefficients for both propagating modes in LiF are lying very closely in small k range. Thus, this creates a good condition for an observation of the corresponding well defined propagating contributions in time correlation functions. In fact it is seen in Fig. 4.

Time correlation functions within the GCM approach can be written in the following general form [compare with (57)]:

$$\frac{F_{ij}^{(GCM)}(k,t)}{F_{ij}(k,0)} = \sum_{\alpha}^{l_r} A_{ij}^\alpha(k)e^{-d_\alpha(k)t} + \sum_{\alpha}^{l_p}\{B_{ij}^\alpha(k)\cos[\omega_\alpha(k)t] + C_{ij}^\alpha(k)\sin[\omega_\alpha(k)t]\}e^{-\sigma_\alpha(k)t}, \tag{58}$$

where $d_\alpha(k)$ are damping coefficients for the purely relaxing modes, and $\omega_\alpha(k)$ and $\sigma_\alpha(k)$ denote the dispersions and damping coefficients for collective propagating excitations. It is evident that the sum $l_r + 2l_p$ gives a total number of dynamic variables l considered within the GCM approach. The amplitudes $A_{ij}^\alpha(k)$, $B_{ij}^\alpha(k)$, and $C_{ij}^\alpha(k)$ describe different kinds of contributions related with relaxing and propagating collective modes, respectively. In fact, Eq. (57) generalizes the known hydrodynamic expression [15, 16] onto the case of additional nonhydrodynamic collective excitations in the liquid, and the knowledge of corresponding amplitudes allows us to judge the role of each type of collective excitations at certain range of wavenumbers k.

Fig. 6 presents the results obtained for the main mode contributions to "density-density" $F_{NN}^{(GCM)}(k,t)$ and "charge density-charge density" $F_{QQ}^{(GCM)}(k,t)$

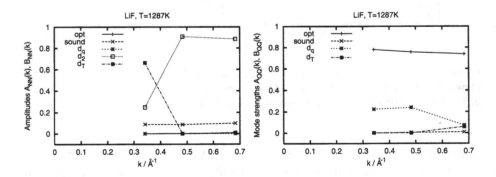

Figure 6. Main amplitudes describing the mode contributions to time correlation functions "density-density" $F_{NN}^{(GCM)}(k,t)$ (right) and "charge density-charge density" $F_{QQ}^{(GCM)}(k,t)$ (left) in LiF at small k domain.

time correlation functions of LiF at small k domain. Beside the amplitudes, describing the propagating modes, in this figure one can also see the terms associated with the contributions from the lowest lieing relaxing collective modes: thermal diffusivity $d_T(k)$, mutual diffusion mode $d_q(k)$, and structural relaxation kinetic mode $d_2(k)$.

Contrary to the case of a binary mixture of neutral particles, studied in our several papers [22–24], one can see in Fig. 6 that the contribution of optic-like propagating modes to the function $F_{QQ}^{(GCM)}(k,t)$ is finite (and even dominant) in small k limit. This explains the results presented in Fig. 4 and supports the conclusion about the dominant role of nonhydrodynamic processes in charge density fluctuations. We have to recall that in mixture of neutral particles this contribution tends to zero as k^2 [24] when k goes to zero. Considering the behavior of the corresponding amplitudes to $F_{NN}^{(GCM)}(k,t)$, one can see in Fig. 6 that the main contributions to this function are caused solely by the hydrodynamic modes when k is small. This means that in small k limit the hydrodynamics of binary mixture of charged particles becomes formally similar to the case of simple fluids. This is in agreement with the results of analytical study, presented in Sec. 5. However, numerous specific features make the difference between these two cases. In particular, quite unusual behavior can be found for the k-dependent transport coefficients.

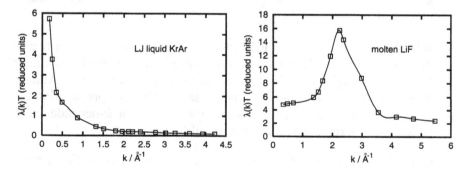

Figure 7. Generalized k-dependent thermal conductivity obtained for a Lennard-Jones KrAr binary mixture (left) and molten LiF alloy (right).

Figure 7 presents the numerical results, obtained for k dependent thermal conductivity $\lambda(k)$ by combining the MD simulations and GCM approach for KrAr and LiF. The difference between these two cases is seen on the qualitative level. For a mixture of neutral particles (right) the expected behavior, described by the Lorentzian-like function, is observed. Otherwise, we found the increase in $\lambda(k)$ when k becomes larger for molten LiF, and a well pronounced peak is seen at wavenumber k_p where the first peak of static structure factor $S_{xx}(k)$ is located (see Fig. 1).

One of the reasons explaining quite different behavior of $\lambda(k)$ has been already mentioned – the "antiferromagnetic" type of interactions in molten salts. Note that the k-dependence of other generalized transport coefficients was found qualitatively quite similar for liquid KrAr and LiF. This allows us to conclude that in fact in the function $\lambda(k)$ the crossover effect is observed caused by the long-ranged Coulombic interactions and discussed in Sec. 5. In order to verify this statement additional study should be performed.

7.   Summary

In this contribution we have reviewed the recent results concerning the collective dynamics of charged liquids. In order to establish the role of long-range Coulombic interactions we have concentrated our attention on the comparison of the results obtained for binary mixtures of neutral and charged particles. Such a comparison has been performed on two levels of consideration – on the level of analytical theories and numerical simulations. The main conclusions from our studies are as follows.

- General theoretical approach for the study of collective behavior of multicomponent mixture of charged particles has been developed. In particular, this approach has been applied to the comparison of binary mixtures of neutral and charged particles.

- It is shown that due to the long-range character and "antiferromagnetic" type of Coulombic interactions the hydrodynamic behavior of binary mixture of charged particles has several very specific features. Among them one can mention the dominant role of optic-like propagating kinetic modes in charge density fluctuations and specific k-dependence of generalized thermal conductivity in small k region.

- Optic-like collective excitations are not a unique feature of binary mixture of charged particles. Such modes can also be found in binary mixtures of neutral particles. However, the behavior of mode contributions to time correlation functions in small k range in these two cases is quite different. In particular, amplitude of optic-like modes to the "mass concentration" autocorrelation function tends to zero for the latter case, whereas for the former one these modes produce the finite contribution even in the hydrodynamic limit.

The overall conclusion from these studies is that, in spite of essential progress in understanding the collective behavior of ionic liquids, many problems are still open for further work. In particular, it is important to perform an additional study of a binary mixture of neutral particles with "antiferromagnetic" type of iterations and to compare the results obtained. Some new effects both in equilibrium and nonequlibrium properties can be expected for strongly asymmetric models of molten salts. Much more complicated for the investigation is the case of ternary mixtures opening new possibilities in modelling the realistic ionic liquids. Some of these issues are the subject of our current interest.

Acknowledgements

Part of this work has been supported by the Fonds zur Förderung der wissenschaftlichen Forschung (Austria) under Project No. P15247. Authors also wish to thank the Editors of this book for the possibility to present our results.

Appendix: Some Relations for Static Correlation Functions

The matrix $(\hat{\mathcal{N}}_k, \hat{\mathcal{N}}_k^+)_{\alpha\gamma}$ is simply related to the matrix of partial static structure factors $S_{\alpha\gamma}(k)$, namely,

$$\left(\hat{N}_{k,\alpha}, \hat{N}_{-k,\gamma}\right) \equiv \left(\hat{\mathcal{N}}_k, \hat{\mathcal{N}}_k^+\right)_{\alpha\gamma} = N(c_\alpha c_\gamma)^{1/2} S_{\alpha\gamma}(k) \tag{A.1}$$

with the definition,

$$
\begin{aligned}
S_{\alpha\gamma}(k) &= \delta_{\alpha\gamma} + 4\pi(n_\alpha n_\gamma)^{1/2} \int_0^\infty dr \, r^2 \left(g_{\alpha\gamma}(r) - 1\right) \frac{\sin kr}{kr} \\
&= \delta_{\alpha\gamma} + (n_\alpha)^{1/2} H_{\alpha\gamma}(k)(n_\gamma)^{1/2},
\end{aligned}
\tag{A.2}
$$

where $\mathbf{H} = \|H_{\alpha\gamma}(k)\|$ is the matrix of the pair correlation functions. Hence, for $(\hat{\mathcal{N}}_k, \hat{\mathcal{N}}_k^+)$ one has,

$$\frac{1}{V}\left(\hat{\mathcal{N}}_k, \hat{\mathcal{N}}_k^+\right) = \mathbf{n} + \mathbf{n}\,\mathbf{H}\,\mathbf{n}, \tag{A.3}$$

where $\mathbf{n}$ is the diagonal matrix with the elements $\|n_\alpha \delta_{\alpha\beta}\|$. In many practical applications it is needed to know the inverse matrix $(\hat{\mathcal{N}}, \hat{\mathcal{N}}^+)^{-1}$. As it follows from (A.3) one gets,

$$V\left(\hat{\mathcal{N}}, \hat{\mathcal{N}}^+\right)^{-1} = \left[\mathbf{I} + \mathbf{H}\,\mathbf{n}\right]^{-1}\mathbf{n}^{-1}. \tag{A.4}$$

Using the Ornstein-Zernike equation,

$$\mathbf{H}(k) = \mathbf{C}(k) + \mathbf{H}(k)\,\mathbf{n}\,\mathbf{C}(k), \tag{A.5}$$

which gives the connection between pair and direct correlation functions, we can easily derive that,

$$V\left(\hat{\mathcal{N}}, \hat{\mathcal{N}}^+\right)^{-1} = \mathbf{n}^{-1} - \mathbf{C}, \tag{A.6}$$

or

$$V\left(\hat{\mathcal{N}}_k, \hat{\mathcal{N}}_k^+\right)_{\alpha\gamma}^{-1} = n_\alpha^{-1}\delta_{\alpha\gamma} - C_{\alpha\gamma}(k). \tag{A.7}$$

Applying this last relation to the expressions (23) and (24), one gets,

$$\frac{1}{\kappa_T(k)} = k_B T \left[n - \sum_{\alpha,\gamma=1}^{\nu} n_\alpha \, C_{\alpha\gamma}(k) \, n_\gamma \right], \tag{A.8}$$

and

$$\frac{v_\gamma(k)}{\kappa_T(k)} = k_B T \left[1 - \sum_{\alpha=1}^{\nu} n_\alpha \, C_{\alpha\gamma}(k) \right]. \tag{A.9}$$

References

[1] Rovere, M., and Tosi, M.P. *Rep. Prog. Phys.*, 1986, **49**, p. 1001.

[2] March, N.H., and Tosi, M.P. (1984). *Coulomb Liquids*. London, New York: Academic Press.

[3] Parrinello, M., and Tosi, M.P. Structure and Dynamics of Simple Ionic Liquids. *Riv. Nuovo Cimento*, 1979, **2**, No. 6, p. 1–69.

[4] Hansen, J.-P., and McDonald, I.R. *Phys. Rev. A*, 1975, **11**, p. 2276.

[5] Bosse, J., and Munakata, T. Mode-coupling theory of a simple molten salt. *Phys. Rev. A*, 1982, **25**, No. 5, p. 2763–2777.

[6] Zubarev, D.N., and Tokarchuk, M.V. Nonequilibrium statistical hydrodynamics of ionic systems. *Theor. Math. Phys.*, 1987, **70**, No. 2, p. 164–178; *Teor. Mat. Fiz.*, 1987, **70**, p. 234–254.

[7] Barthel, J.M.V., Krienke, H., and Kunz, W. (1998). *Physical Chemistry of Electrolyte Solutions*. New York: Modern Aspects, Steinkopff.

[8] Giaquinta, P.V., Parrinello, M., and Tosi, M.P. Hydrodynamic correlation functions for molten salts. *Phys. Chem. Liq.*, 1976, **5**, p. 305–324.

[9] Irving, H., and Kirkwood, J.G. *J. Chem. Phys.*, 1950, **18**, p. 187.

[10] MacGowan, D., and Evans, D.J. Heat and matter transport in binary liquid mixtures. *Phys. Rev. A*, 1986, **34**, No. 3, p. 2133–2142.

[11] Koishi, T., Kawase, S., and Tamaki, S. A theory of electrical conductivity of molten salt. *J. Chem.Phys.*, 2002, **116**, No. 7, p. 3018–3026.

[12] Zubarev, D.N. Modern methods of the statistical theory of nonequilibrium processes. *Itogi Nauki i Tekhniki, Sovremennyje Problemy Matematiki. VINITI*, 1980, **15**, p. 131–226 (in Russian).

[13] Mryglod, I.M., and Hachkevych, A.M. On non-equilibrium statistical theory of a fluid: Linear relaxation theory. *Cond. Matt. Phys.*, 1995, **5**, p. 105–118.

[14] Mryglod, I.M. Generalized statistical hydrodynamics of fluids: Approach of generalized collective modes. *Cond. Matt. Phys.*, 1998, **1**, No. 4(16), p. 753–796.

[15] Boon, J.P., and Yip, S. (1980). *Molecular Hydrodynamics*. New York: McGraw-Hill.

[16] Hansen, J.-P., and McDonald, I.R. (1986). *Theory of Simple Liquids*. London: Academic Press.

[17] Mryglod, I.M. Generalized hydrodynamics of multicomponent fluids. *Condens. Matter Phys.*, 1997, No. 10, p. 115–135.

[18] Vogelsang, R., Hoheisel, C., Paolini, O.V., and Ciccitti, G. Soret coefficient of isotropic Lennard-Jones mixtures and Ar-Kr system as determined by equilibrium molecular dynamics simulations. *Phys. Rev. A*, 1987, **36**, No. 8, p. 3964–3974.

[19] Perronace, A., Ciccotti, G., Leroy, F., Fuchs A.H., and Rousseau, B. Soret coefficient for liquid argon-krypton mixtures via equilibrium and nonequilibrium molecular dynamics: A comparison with experiments. *Phys. Rev. E*, 2002, **66**, p. 031201:1–15.

[20] Mryglod, I.M., and Omelyan, I.P. Generalized mode approach: 3. Generalized transport coefficients of a Lennard-Jones fluid. *Mol. Phys.*, 1997, **92**, No. 5, p. 913–927.

[21] Bryk, T., and Mryglod, I. Collective excitations and generalized transport coefficients in a molten metallic alloy Li_4Pb. *Condens. Matt. Phys.*, 2004, **7**, No. 2(38), p. 285–300.

[22] Bryk, T., and Mryglod, I. Generalized hydrodynamics of binary liquids: Transverse collective modes. *Phys. Rev. E.*, 2000, **62**, No. 2, p. 2188–2199.

[23] Bryk, T., and Mryglod, I. Optic-like excitations in binary liquids: Transverse dynamics. *J. Phys.: Cond. Matt.*, 2000, **12**, p. 6063–6076.

[24] Bryk, T., and Mryglod, I. Longitudinal optic-like excitations in binary liquid mixtures. *J. Phys.: Cond. Matt.*, 2002, **14**, No. 25, p. L445–451.

[25] Bhatia, A.B., Thornton, D.E., and March, N.H. Dynamic structure factors for a fluid binary mixture in the hydrodynamic limit. *Phys. Chem. Liq.*, 1974, **4**, p. 97–111.

[26] March, N.M., and Tosi, M.P. (1976). *Atomic Dynamics in Liquids*. London: Macmillan.

[27] Bosse, J., Götze, W., and Lücke, M. Mode-coupling theory of simple classical liquids. *Phys. Rev. A.*, 1978, **17**, No. 1, p. 434–446.

[28] Bruin, C., Michels, J.P.J., van Rijs, J.C., de Graaf, L.A., and de Schepper, I.M. Extended hydrodynamic modes in a dense hard sphere fluid. *Phys. Lett. A.*, 1985, **110**, No. 1, p. 40–43.

[29] Kamgar-Parsi, B., Cohen, E.G.D., and de Schepper, I.M. Dynamic processes in hard-sphere fluids. *Phys. Rev. A.*, 1987, **35**, No. 11, p. 4781–4795.

[30] de Schepper, I.M., Cohen, E.G.D., Bruin, C., van Rijs, J.C., Montrooij, W., and Graaf, L.A. Hydrodynamic time correlation functions for Lennard-Jones fluids. *Phys. Rev. A.*, 1988, **38**, No. 1, p. 271–287.

[31] Cohen, E.G.D., and de Schepper, I.M. Effective eigenmode description of dynamic processes in dense classical fluid mixtures. *Nouvo Cimento D.*, 1990, **12**, No. 4–5, p. 521–542.

[32] Bertolini, D., and Tani, A. Generalized hydrodynamics and the acoustic modes of water: Theory and simulation results. *Phys. Rev. E.*, 1995, **51**, No. 2, p. 1091–1118.

[33] Bosse, J., Jacucci, G., Ronchetti, M., and Schirmacher, W. Fast sound in two-component liquids. *Phys. Rev. Lett.*, 1986, **57**, No. 26, p. 3277–3279.

[34] Westerhuijs, P., Montfrooij, W., de Graaf, L.A., and de Schepper, I.M. Fast and slow sound in a dense gas mixture of helium and neon. *Phys. Rev. A*, 1992, **45**, No. 6, p. 3749–3762.

[35] Smorenburg, H.E., Crevecoeur, R.M., and de Schepper, I.M. Fast sound in a dense helium argon gas mixture. *Phys. Lett. A*, 1996, **211**, p. 118–124.

[36] Mryglod, I.M., Omelyan, I.P., and Tokarchuk, M.V. Generalized collective modes for the Lennard-Jones fluid. *Mol. Phys.*, 1995, **84**, No. 2, p. 235–259.

[37] Mryglod, I.M., and Omelyan, I.P. Generalized collective modes for a Lennard-Jones fluid in higher mode approximations. *Phys. Lett. A*, 1995, **205**, p. 401–406.

[38] Bryk, T.M., Mryglod, I.M., and Kahl, G. Generalized collective modes in a binary $He_{0.65}-Ne_{0.35}$ mixture. *Phys. Rev. E.*, 1997, **56**, No. 3, p. 2903–2915.

[39] Mori, H.A Continued-fraction representation of the time correlation functions. *Prog. Theor. Phys.*, 1965, **34**, No. 3, p. 399–416.

[40] Mryglod, I.M., and Hachkevych, A.M. Simple iterating scheme for evaluation of memory functions: Shoulder problem for generalized shear viscosity. *Ukr. Fiz. Zhur.*, 1999, **44**, No. 7, p. 901–907 (in Ukrainian).

[41] Ciccotti, G., Jacucci, G., and McDonald, I.R. Transport properties of molten alkali halides. *Phys. Rev. A*, 1976, **13**, p. 426–436.

[42] Bryk, T., and Mryglod, I. Charge density autocorrelation functions of molten salts: Analytical treatment in long-wavelength limit. *J. Cond. Matt. Phys.*, 2004, **16**, p. L463–L469.

CRITICALITY OF IONIC LIQUIDS IN SOLUTION

Review of recent progress in the theory, simulations and experiment

W. Schröer
Institut für Anorganische und Physikalische Chemie,
Universität Bremen, D–28359 Bremen, Germany

Abstract   The long-range nature of the Coulomb interactions and the high energy at small ion separations determine the properties of ionic systems. Theory and experiments yield a chemical equilibrium involving ions, ion pairs and higher clusters. Theory predicts a fluid-phase transition at low reduced temperatures (ratio of thermal energy to Coulomb energy at contact), which corresponds to the liquid-liquid phase transitions in ionic solutions. The reduced critical temperatures agree with the prediction in non-polar solvents, but increase with the dielectric constant of the solvent, as solvophobic interactions become important. The long-range nature of the Coulomb potential suggests questioning the applicability of the Ising model, but experiments state Ising criticality in all systems.

Keywords: Ionic liquids, phase diagrams, critical properties, corresponding state, light scattering, ionic solutions, ion pairs

1.   Introduction

Ionic liquids (ILs) are salts that unlike the typical inorganic salts as NaCl are liquid at ambient conditions [1, 2]. Some of them stay fluid at temperature as low as –70°C [3]. They offer new properties for applications in preparative chemistry and chemical engineering as reaction media and for separation processes. Reactions have been proposed for ILs that, taking advantage of phase transitions due changes of temperature or composition, enable elegant separation of products, educts and catalyst. ILs can be designed for different solvent properties so that they are also called designer liquids. A great number of ILs is commercially available.

The ILs are stable chemicals and can be recycled effectively. Their vapor pressure is too small to be measurable, so that the hazards in handling such compounds in the laboratory are small. Because of those properties they are advocated as solvents in Green Chemistry. Clearly, knowledge of basic physi-

143

D. Henderson et al. (eds.), Ionic Soft Matter: Modern Trends in Theory and Applications, 143–180.
© 2005 *Springer. Printed in the Netherlands.*

cal properties, e.g. solubilities and coexistence curves is essential in the design of technical processes. The variety of ILs calls for systematic studies of specific and general properties of pure ILs and their solutions.

Ionic fluids such as molten salts and electrolyte solutions have always been of central interest in Chemical Physics, Physical Chemistry and many applied fields such as electrochemistry, chemical engineering or the geosciences. It is the aim of this review to connect the knowledge about structure and thermodynamic properties of ionic fluids and electrolyte solutions with that of the ILs. Liquid-liquid phase transition of ionic solutions are the main topic of this paper.

The fundamental aspect of the research on ionic systems is founded in the long-range nature of the Coulomb interactions. It is now well established that liquid-gas as well as liquid-liquid phase transitions both belong to the Ising universality class [4, 5] independent of the molecular details of the system. However, this universality applies only if the phase transition is driven by short range interactions. The Ising model assumes only next neighbor interactions, and has been proven to apply in 3-dimensinal space for transitions driven by r^{-p} interactions with $p > 4.97$, termed short range interactions. Therefore, phase transitions driven by long range Coulomb forces may be expected to behave differently [6–8]. Thus phase transitions in ionic systems became a prominent subject in statistical mechanics [7, 8] and related experiments [6, 9–11].

It has been speculated that r^{-1} Coulomb interactions give rise to van der Waals (vdW) mean-field critical behavior [6], as obtained from the analytical solution of a one dimensional system for a infinitesimal small long-range potential. Pioneering experiments on liquid/vapor coexistence suggested a vdW mean-field-like liquid/vapor coexistence curve for NH_4Cl [12] that is a rare example for a molten salt, where the critical point can be reached. Friedman [13] suggested that liquid-liquid phase transitions in electrolyte solutions near room temperature should be more convenient to investigate. In the late 1980s, Pitzer and others followed this hint and started investigating the liquid-liquid phase-transition of ionic solutions in organic solvents, indeed finding evidence for an apparent mean-field behavior [14–16]. Later experiments, however, yielded Ising criticality on Pitzerós system [17] and other ionic solutions [10].

Liquid-liquid phase transitions in ionic solutions, although rare, are known for quite a long time [18–20] but remained unnoticed till the recent interest in criticality. The major difficulty to find liquid-liquid phase separation in ionic solutions arises from the interference of crystallization, driven by the high melting points of salts. Certainly, liquid-liquid phase separation cannot be expected in conventional electrolyte solutions of inorganic salts. Low-melting salts, which usually comprise large organic ions, now enable the systematic design of suitable systems [9–11]. A major flaw in former investigations is the limited chemical stability of the salts considered. Chemical stability, however, is necessary to ensure the stability of the critical data. ILs may allow for reliable experiments as they are expected to be more stable than, e.g. picrates. In fact, an enormous number of solutions of ILs showing liquid-liquid separation has been discovered recently [21].

Concerning theory, two properties of ionic fluids provide major challenges, if compared to non-ionic fluids: the long-range nature of the Coulomb interactions and the high figures of the Coulomb energy at small ion separations [7, 8]. The long-range Coulomb interactions challenge the universality of the Ising model and cause severe problems in theory and simulations. For example, integrals as the 2nd virial coefficient that determine the thermodynamics of normal fluids diverge for Coulomb interactions [4, 22].

The second striking difference between ionic and nonionic fluids concerns the strength of the intermolecular forces. In normal fluids at ambient conditions, the size of the intermolecular interactions is of the order of the thermal energy. For simple neutral fluids the reduced temperature $T^* = kT/\phi_{min}$, which is the ratio between the thermal energy kT and the depth of the interaction potential $\phi(r)$ at its minimum, is of the order of unity. In contrast, in typical molten salts the inter ionic interactions are more than one order of magnitude larger than kT and can even reach the magnitude of chemical bonds, i.e. $T^* \ll 1$. In electrolyte solutions, the strength of the Coulomb interactions depends on the dielectric constant ϵ of the solvent. In aqueous solutions T^* may come close to unity but in the solvents of low dielectric constant, one has $T^* \ll 1$ like in molten salts [22–24]. Such strong interactions have major consequences for the structure of ionic liquids. While ordinary liquids can in good approximation be described in the vdW picture as hard core bodies in a sea of an average potential, the Coulomb interactions may give rise to the formation of ion pairs and higher ion clusters. These long-living molecule-like entities separate into free ions by rather mild changes of the conditions, e.g. by dilution or change of the solvent. One way to describe this ion distribution therefore resorts to a "chemical picture", in which the fluid is regarded as a system of ions, ion pairs and higher clusters in chemical equilibrium. This is true even for the simplest model fluid, which is the model of equal-sized, charged hardspheres in a dielectric continuum, termed restricted primitive model (RPM). Therefore, ionic liquids and even the RPM are complex liquids.

In the "physical picture" ion-pairs are just consequences of large values of the Mayer f-functions that describe the ion distribution [22]. The technical consequence, however, is a major complication of the theory: the high-temperature approximations of the f-functions applied, e.g. in the mean spherical approximation (MSA) or the Percus-Yevick approximation (PY) [25], suffice in simple fluids but not in ionic systems.

A solution to these difficulties is a blend of the chemical picture in which clustered ion configurations are described by the mass action law, while the interactions between the various entities are treated by methods applying the high-temperature approximations of the f-functions, e.g. by the MSA. The Debye-Hückel (DH) theory [26], although derived from classical electrostatics, is also a high-temperature approximation, whose range of applicability can be extended by supplementing a mass action law for ion pair formation [27].

Because phase transitions driven by Coulomb interactions fall into the regime, where ion pairs and higher clusters play a major role, an adequate electrolyte theory for this regime is of crucial importance. Clearly, simulations do not rely on such artificial entities as ion pairs. Quite on the contrary, the analysis of simulation results enables to identify ion pairs and higher clusters, which merely appear as a consequence of the ion-ion potential. However, this does not devalue analytical theories, which provide insight and may form the basis for a renormalization (RG) group analysis [7]. It should be noted that both, the theoretical methods and the common simulations concern the mean field level, therefore do not allow conclusions on the nature of the critical point. Predictions on the nature of the critical point require either a RG analysis or the application of finite size simulation techniques that evaluate the dependence of the predicted properties from the size of the simulation boxes and allow extrapolation to infinite box size. The critical data and the critical exponents result from the extrapolation procedure [28–30]. The RG analysis would start from a mean-field free energy density that adequately describes the inter ionic forces and correlations in non-uniform systems [4]. This calls for *analytical* expressions. At present, such analytical expressions are only provided by DH-type approaches, which have been developed further by including the interactions of dipolar ion pairs [31, 32]. In fact, the need for a transparent analytical theory has triggered a revival of DH theory.

In the present article, we review recent progress in this subject area. In Sec. 2, we give a short overview on the chemical composition of the low melting salts and ILs. In Sec. 3 we address the problem of the electrolyte solution structure at conditions of low reduced temperature, where phase separations are known to occur. In Sec. 4, we consider experimental and theoretical results concerning the location of the two-phase regime in solutions of ionic fluids. In Sec. 5 we finally review theoretical and experimental results on near-critical behavior of ionic fluids.

2.   Ionic compounds

Commonly, inorganic salts like NaCl are regarded as typical ionic compounds. The lattice energy is determined by Coulomb forces. Some salts are soluble in water, others like $CaCO_3$ are not. Solubility in organic solvents is generally extremely low. The melting point is of the order 1000 K the liquid-gas critical point is even higher say 3000 K as in the case of NaCl [33].

Organic salts have a much lower melting point. The bulky ions enlarge the separation of the ions, thus reducing the coulomb energy of the lattice. Furthermore, the asymmetric shape of the ions is also favoring the disordered state. Both, the lower Coulomb energy and the hydrocarbon body of the ions enhance the solubility in organic solvents. Complete mixing at higher temperatures and phase separation at lower temperatures are common features. Nevertheless the Coulomb forces are strong enough that the vapor pressure becomes so small that it is virtually not measurable at ambient conditions [1]. At higher temperatures, say 600 K, chemical decomposition occurs, so that the liquid-gas critical point is not accessible for those compounds. In the following we will name some compounds and ions which have been investigated.

The first ionic liquid that was discovered as early as 1914 by Walden [34], is ethyl ammonium nitrate (EAN), $[NH_3C_2H_5]^+[NO_3]^-$. It is an explosive liquid, which is soluble in water and alcohols. The liquid structure is determined by the Coulomb interactions but to a large extend also by a network of hydrogen bonds [9].

The salt, used in the investigations that yielded classical exponents [14–16] was Triethylhexyl ammonium triethlhexyl borate ($N^+_{2226} B^-_{2226}$), $[N(C_2H_5)_3(C_6H_{13})]^+[B(C_2H_5)_3(C_6H_{13})]^-$,

The compound is rather difficult to handle, it changes color with light even in flame-sealed samples and is self-igniting on air.

Many investigations [9–11] have been carried out using Tetraalkyl ammonium picrates ($NR_4^+Pic^-$). Picrates are well known explosives,

The new ILs [1, 2] consist in substituted pyridinium, imidazolium, ammonium, or phosphonium cations, with anions chloride, bromide, tetrafluoro borate, hexafluoro phosphate, dicyanamide, bis(trifluoromethansulfonyl)amide (BTA), alkylsulfate, trifluormethylsulfate: Cl^-, Br^-, BF_4^-, PF_6^-, $(CN)_2N^+$, $(CF_3SO_2)_2N^-$, RSO_4^-, $CF_3SO_4^-$. This list is by no means complete. In principle some acids like acetic acid could also be regarded as an ionic liquid, where, however, chemical bonding gives a major contribution to the inter-ionic interaction, so that even in solvents with high dielectric permittivity the dissociation is low.

3.   Particle distribution in ionic liquids

3.1   *Experimental observations*

The basic idea, picturing the electrolyte as a system in chemical equilibrium of free ions, ion pairs and higher associates goes back to Arrhenius [35], who introduced a mass-action law for a chemical equilibrium between free ions and salt molecules. The mass action law implies complete dissociation in the limit of infinite dilution and a decrease of the degree of dissociation with increasing salt concentration.

Arrhenius theory applies well to solutions of weak acids and bases in water, but fails in the case of strong electrolytes such as ordinary salts. Debye and Hückel [26] solved this problem assuming complete dissociation, but considering the Coulomb interactions between the ions by a patchwork theory based on both macroscopic electrostatics and statistical mechanics.

Bjerrum (Bj) combined the Arrhenius and DH approaches by assuming a chemical equilibrium between ion-pairs and free ions [27]. This concept takes into account interactions of ions at short range, which are not adequately described in DH theory. It also includes a theory for the mass action constant as a function of the dielectric constant ϵ of the solvent. Many experimental investigations of the electrical conductance Λ, e.g. reviewed by Kraus [36], have confirmed Bjerrum's concept, which is the basic concept of many modern approaches.

In systems showing phase separation, a minimum in the concentration dependence of the equivalent conductance Λ is a major feature [36–39]. Arrhenius suggested that the degree of dissociation α at any concentration is equal to the ratio of the observed equivalent conductance Λ to its limiting value Λ^∞ at infinite dilution, i.e. $\alpha = \Lambda/\Lambda^\infty$. The minimum in Λ is mainly determined by the concentration dependence of α, which follows a mass action law. Bjerrum's expression for the association constant is [27],

$$K(T) = \int_{\sigma}^{R} \exp\left[\frac{R}{r}\right] 4\pi r^2 \mathrm{d}r, \tag{1}$$

where $R = \beta q^2/\epsilon$ is the so-called Bjerrum length and σ is the collision diameter. Quite generally the molar conductivity Λ is proportional to the inverse of the shear viscosity η. In order to analyze the concentration dependence of α, it is appropriate to discuss the Walden product (WP), which is the product of the molar conductivity Λ and the shear viscosity η.

We consider first the low-concentration branch, where the WP decreases rapidly with increasing molar concentration c of the salt. In contrast to the situation encountered for aqueous systems, in low-ϵ solvents one does not reach the limiting regime of the Kohlrausch-Debye-Hückel-Onsager $c^{1/2}$ law [40], which results from long-range inter-ionic interactions between free ions. Even at the lowest concentrations α controls the concentration dependence of the WP, thus enabling the determination of the association constant $K(T)$ for ion pair formation.

The interpretation of the branch, where the WP increases, more sophisticated. A minimum of WP implies a redissociation of the pairs and/or the formation of charged ion clusters. Fuoss and Kraus [41] assumed the formation of charged ion triplets. Recent theory attributes the increase of the conductivity to redissociation, resulting from the interactions of the free ions with the ion pairs and the increase of the dielectric permittivity due to the formation of ion pairs that causes a decrease of the association constant [38].

If the data are normalized to the values at the critical points, the data sets of different systems fall onto a common curve [24]. This indicates that corresponding states behavior of α. The critical concentration is located near $\alpha = 0.9$ on the branch characterized by the redissociation. Therefore, the coexisting phases may be described as a phase of high concentration, where due to redissociation the conductivity is high and vice versa.

Another important source of information is dielectric spectroscopy, because dipolar ion pairs contribute to the static dielectric constant of the solution [42, 43]. In polar solvents the dielectric spectra reflect two modes caused by the reorientation of solvent and of ion-pairs. In non-polar solvents one solely observes ion pair reorientation. For Bu_4N-iodide (Bu_4NI) in dichloromethane (CH_2Cl_2) an increase of the total dielectric constant ϵ with the concentration of the salt is found as result of the ion pair formation. A decrease in the particle density of the solvent causes a minute decrease of the solvent contribution. The dielectric constant does, however, not increase linearly with the salt concentration. A decreasing slope at high salt concentrations may result from the redissociation of the ion pairs but at a quantitative level, redissociation alone is not sufficient to account for this effect. Rather, one has to account for correlations between the ion pairs with a tendency to antiparallel orientations. Then, the well-known Kirkwood g factor for orientation correlations [44, 45] may be used to interpret the data, implying Kirkwood factors $g < 1$. In the cluster picture, this result can be rationalized as the formation of quadruple ions formed by antiparallel pairs.

3.2   *Computer simulations*

The ions of the salts, which mix with organic solvents have a rather complicated structure. Simulations of solutions of such salts appropriate to criticality are not feasible at present. Therefore simulations of generic models may suffice, which, however, can only provide a cartoon picture. Quantitative agreement with measured data, certainly, can only be expected from simulations considering the molecular details.

The generic model for a ionic fluid is the "primitive model" (or RPM) of charged hard spheres in a dielectric continuum with the dielectric constant ϵ. Thus, the potential is just the Coulomb potential between the ions labelled i and k, which is cut-off at the collision diameter σ_{ik},

$$\phi_{ik}(r) = \begin{cases} q_i q_k / \epsilon r, & r > \sigma_{ik}, \\ \infty, & r \leqslant \sigma_{ik}. \end{cases} \tag{2}$$

Primitive models satisfy the corresponding states principle [23, 24]. The thermodynamic state is completely specified by a reduced temperature T^* and a reduced ion density ρ^*,

$$T^* = \frac{kT\epsilon\sigma_{+-}}{|q_+q_-|} \quad \text{and} \quad \rho^* = \rho\sigma_{+-}^3. \tag{3}$$

The energy scale used for defining T^* is set by the Coulomb energy at contact, ρ is the total ion density. Most simulations concern the RPM fluid, which consists of equal-sized charged hard spheres with charges $q_+ = |q_-|$.

The ion distribution is obtained from *simulation* data by counting ions separated by $(1-2)\sigma$ as being paired [46, 47]. Data are available for the low-density regime at $\rho = 0.004$, which is one order of magnitude below the critical density of the RPM. Here the density of free ions decreases when lowering the reduced temperature T^*, and near $T^* = 0.1$ the concentrations of ion pairs and free ions become equal. At $T^* = 0.05$ (which is roughly the critical temperature of the RPM) the free ions practically vanish. Moreover, below $T^* = 0.1$ higher clusters begin to appear. Thus, when lowering T^*, the ion pair concentration passes a maximum. This maximum corresponds to a maximum in the specific heat and a decrease of the dielectric constant.

In real systems, cations and ions are usually of different size and often also of different charge. Few simulations now concern primitive models with ions differing in the size or/and charges [48–50]. In general, these simulations appear to show a more rich variation of clusters than the RPM. While for the RPM at $T^* = 0.05$ and $\rho^* = 0.002$ clusters of 2, 4 or 6 particles are found, much larger clusters occur, if for similar conditions the sizes of the ions differ. Moreover, in addition to the predominant chain-like clusters of the RPM, other configurations, e.g. rings are found. If the ions differ in the charge, even more cluster configurations are observed in the simulation.

3.3 Analytical theories

Analytical liquid-state statistical mechanics provides a number of means for determining pair correlation functions [25]. Most successful are the methods based on the Ornstein-Zernike (OZ) integral equation. Once the pair correlation function is available, the Helmholtz free energy A can be calculated by thermodynamic integration from the internal energy, the pressure, or the compressibility. Thus the reduced free energy density Φ is obtained from the energy density u by thermodynamic integration with respect to $\beta = 1/kT$, carried out for a fixed distribution of particles,

$$\Phi = \beta A/V = \int_0^\beta u \mathrm{d}\beta. \qquad (4)$$

The three routes for calculating the Helmholtz free energy are equivalent in principle. However, due to approximations involved in the different closures of the OZ equation, this is not the case and thermodynamic consistence can hardly be achieved. The energy route is usually superior. From the various closures, the mean spherical approximation (MSA), supplemented by a mass action law is quite apt for treating electrolyte solutions in media of low dielectric constant. Thereby, Bjerrum's expression (1) for the association constant has occasionally been replaced by a slightly different expression going back to Ebeling [51], which results from the virial expansion and remains correct in the high temperature limit. Based on these concepts, various versions of the MSA have been applied to systems at low T^* [52, 53]. All versions predict the minimum in the degree of dissociation observed.

The need for an analytical expressions for the equation of state have led to a revival of the *macroscopic electrostatic theory* due to Debye, Hückel and Bjerrum. DH theory becomes exact for large particles. In pilot work by Fisher and Levin (FL) [31], DH-Bj theory is extended by considering the interactions of the pairs with the free ions. Weiss and Schröer (WS) [32] have supplemented this theory accounting for dipole-dipole interactions between pairs and the ϵ-dependence of the association constant.

In the following we give a short account on this theory for demonstrating that interactions between dipolar ion pairs and free ions and/or other pairs can be incorporated in a natural and transparent way [31, 32]. The theories rest on the Poisson-Boltzmann equation. With the presumption of electro neutrality, the expansion in first order of β yields the Helmholtz equation or linearized Poisson-Boltzmann equation,

$$\nabla^2 \Psi = \kappa^2 \Psi, \tag{5}$$

for the electrostatic potential Ψ in a medium with the dielectric permittivity ϵ [26]. Considering charge-symmetrical systems, the reciprocal Debye length, κ, is given by

$$\kappa^2 = 4\pi\beta q^2 \frac{(\rho_+ + \rho_-)}{\epsilon}, \tag{6}$$

or in terms of reduced quantities by

$$x = \kappa\sigma = \sqrt{\frac{4\pi\rho^*}{T^*}}. \tag{7}$$

For $\kappa = 0$, Eq. (5) reduces to the Laplace equation, which is used to calculate the potential inside the particle that is modeled as dielectric sphere with central multipoles. For this case the solution in spherical coordinates is given by a series of Legendre polynomials $P_l(x)$ and related powers of r,

$$\Psi(r, \theta) = \sum_l \left(A_l r^{-l-1} + B_l r^l \right) P_l \cos\theta. \tag{8}$$

This series arises naturally, when expressing the Coulomb potential of a charge separated by a distance s from the origin in terms of spherical coordinates. The positive powers result when $r < s$, while for $r > s$ the potential is described by the negative powers. Similarly the solutions of the linearized Poisson-Boltzmann equation are generated by the analogous expansion of the shielded Coulomb potential $\exp[\kappa r]/r$ of a non-centered point charge. Now the expansion for $r > s$ involves the modified spherical Bessel-functions $k_l(x)$, while for $r < s$ the functions are the same as for the unshielded Coulomb potential,

$$\Psi(r, \theta) = \sum_l \left(C_l k_l(\kappa r) + D_l r^l \right) P_l \cos\theta. \tag{9}$$

The coefficients A_l, B_l, C_l, D_l are calculable from the boundary conditions and from the known behavior at $r \to 0$ and $r \to \infty$. The potential of an ion is given by the $l = 0$ terms, while the potential of a dipole is determined by the $l = 1$ terms.

B_0 is the potential inside the sphere caused by charge distribution induced the by the central charge. The potential energy of this charge is therefore qB_0. This potential energy separates into a contribution due to the ion-ion interactions that is determined by κ and a term due to the dipole-dipole interactions, which is governed by $\epsilon - \epsilon_0$.

B_1 is the field inside the sphere resulting from the distribution of the external charges induced by the permanent central dipole and the polarization of the sphere. In the language of dielectric theory, B_1 defines the reaction field factor [45],

$$R = \frac{2(\epsilon - \epsilon_0)(1 + \kappa\sigma_D) + \epsilon(\kappa\sigma_D)^2}{\sigma_D^3(2\epsilon + \epsilon_0)(1 + \kappa\sigma_D) + \epsilon(\kappa\sigma_D)^2}. \tag{10}$$

The potential energy of the dipole is $\mu B_1 = -\mu^2 R$. The part depending on $(\epsilon - \epsilon_0)$ represents the dipole-dipole interactions while the $(\kappa\sigma_D)^2$ term concerns the dipole-ion interactions. For $\kappa = 0$, Eq. (10) reduces to the reaction-field expression of dielectric theory [54] derived from Laplace equation. Interestingly, even for $\epsilon = \epsilon_0$ a reaction field results because of the polarization of the ionic cloud.

Clearly, the sum of the ion-dipole interactions must be equal to the sum of the dipole-ion interactions. By equating the dipole-ion and the ion-dipole parts of B_1 and B_0 we get for $\sigma = \sigma_D$,

$$y = \frac{(\epsilon - \epsilon_0)\left((2\epsilon + \epsilon_0)(1 + x) + \epsilon x^2\right)}{9(\epsilon(1 + x))} = \frac{4\pi\epsilon\rho_D^*}{9T^*}, \tag{11}$$

where $\rho_D^* = \rho_D\sigma^3$. Eq. (11) is a generalization of the well-known Fröhlich-Onsager-Kirkwood equation [57] for a mixture of ions and dipoles. It reduces to the standard expression known from dielectric theory for the ion-free case. Eq. (11) becomes slightly more complicated when $\sigma \neq \sigma_D$ [32]. The ionic correction to the Fröhlich-Onsager-Kirkwood equation may formally be written as a Kirkwood g-factor [44, 45],

$$g = \left(1 + \frac{\epsilon x^2}{(2\epsilon + \epsilon_0)(1 + x)}\right)^{-1}. \tag{12}$$

The shielding by the charges results in $g < 1$ as does anti-parallel correlation in the absence of ions. Both effects reduce the dielectric constant. Near criticality we have $x = \kappa\sigma \cong 1$, and the reduction is of the order of 25 .

If the ion pairs are the only dipoles in the fluid, the energy density is given by

$$u = \frac{1}{2}\left[(\rho_+ + \rho_-)qB_0 + \rho_D\mu B_1\right]. \tag{13}$$

The densities of the ions and dipoles are connected by the mass action law.

Again, for calculating the Helmholtz energy thermodynamic integration of the energy according to Eq. (4) is required. The charging process applied by Debye and Hückel [26] is an approximation to this thermodynamic integration for fixed dielectric permittivity. Actually, the thermodynamic integration is a rather involved task, because both ϵ and κ are functions of β, and κ is a function of ϵ. Therefore, Weiss and Schröer [32] performed an expansion in $(\epsilon - \epsilon_0)$ up to second order, which includes the ϵ-dependence of the Bjerrum constant.

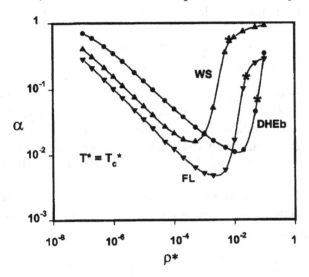

Figure 1.   Degree of dissociation α as a function of the reduced density ρ^* along the critical isotherms according DHEb, FL, and WS Theories The stars indicate the critical density of the respective theories.

A great simplification occurs, when the contribution of the ion pairs to the dielectric constant is neglected as in FL [31]. In this approximation, B_0 represents the ion-ion interactions while B_1 gives the interaction of an ion-pair with the surrounding charges only.

Clearly, Eq. (13) concerns the electrostatic interactions only, so that a suitably chosen hard-core contribution, e.g. of Carnahan-Starling type [25] must be added to the free energy densities. Differentiation with respect to the densities of the species finally yields the chemical potential and the activity coefficients required for evaluating the mass action law determining the concentrations of free ions and ion pairs.

Using these procedures, Weingärtner, Weiss and Schröer [38] have calculated the degree of dissociation, α, over a wide range of conditions from subcritical states at $T^* = 0.04$ up to distinctly supercritical states at $T^* = 0.15$ for different approximations of the outlined electrostatic model. They considered the WS, FL and the DHBj models, but choose the Ebeling expression for the association constant. As shown in Fig. 1, all theories yield a conductance minimum.

DI and DD interactions considered in FL and WS theory, are essential ingredients for rationalizing the observed conductance behavior near the two-phase regime. Thus one has not to resort to the assumption of charged triple ions which are not seen in simulations [46, 47] Pure pairing theories such as DHEb fail at $T^* < 0.08$, which is distinctly above criticality. However, it appears that both FL and WS theory overestimate dissociation.

In WS theory, this high ionicity is a consequence of the increase of the dielectric constant induced by dipolar pairs. One can expect that an account for neutral quadruple ions, as predicted by the MC studies, will improve the performance of DH-based theories, because, the coupled mass action equilibria reduce dissociation, and, furthermore, quadrupoles yield no contribution to the dielectric constant. Such an effect is suggested from dielectric measurements for electrolyte solutions at low T^* [43].

4. Phase diagram and location of the critical point

4.1 Monte Carlo simulations

Simulations of the RPM predict a phase transition for the RPM at low reduced temperature and low reduced density. It was difficult to localize because of the low figures of the critical data. By corresponding states arguments this critical point corresponds to the liquid/vapor transition of molten salts and to some liquid/liquid transitions in electrolyte solutions in solvents of low dielectric constant [23, 24].

Simulations of critical properties are extremely involved and require much more computer force than, e.g. the calculation of the pair-correlation functions. Methodological developments in Monte Carlo (MC) techniques were addressed in a recent review [55]. Most simulations use the Gibbs ensemble technique, which enables direct simulation of phase equilibrium. A further important step, which reduces computer time in large-scale simulations, was to recognize that MC simulations can be carried on a lattice [56]. If the separation of the lattice points is chosen about one order of magnitude smaller than the diameter of the particles, results of such simulations prove to be almost identical with those of MC simulations in continuous space. The virtue of this technique is not only the technical advantage, but also the possibility to investigate the change from a lattice, for which exact analytical theories are available [56, 57], to the fluid.

The first question to be answered by the simulations concerns the existence of the phase transition. The early calculations for the RPM of Stell et al. [58] and Ebeling [59] based on the MSA, other integral equations and the DH theory have indeed indicated the existence of a two-phase regime with an upper critical point. However, such conclusion may be misleading. For hard sphere dipole fluids it was taken for granted for a long time that this system has a gas-liquid phase transition as estimated from a generalized vdW theory. However, simulations showed the formation polymer like chain associates preempting the phase transition [60]. Now, this conclusion is challenged again as recent simulations yield a fluid phase transition also in this model system at rather low densities and temperatures [61].

Meanwhile, a liquid/vapor transition of the RPM is well established, but over the years, the figures for the critical parameters have changed appreciably [7, 9, 10]. One reason for these inconsistencies is associated with finite-size scaling effects, as the critical fluctuations exceed the dimension of the simulation box. Therefore, conventional simulations can be regarded as "mean-field simulations". To obtain the real critical data and critical exponents, the size L of the simulation box is varied and the results are extrapolated by scaling methods to an infinite sample sizes [62]. For the RPM the finite-size scaling technique leads to a reduction of the critical temperature and shifts the critical density to higher values. There is now a series of accurate MC studies [63, 66], which, applying finite-size scaling corrections, locate the critical point of the RPM near

$$T_c^* = 0.049, \qquad \rho_c^* = 0.08. \tag{14}$$

Simulations are now also conducted for investigating the influence of the relative size of the ions and a charge asymmetry on critical parameters [48–50]. It is found that the critical temperature is reduced by the asymmetry of the particle size, as is the critical density. The charge asymmetry causes a reduction of the critical temperature, but an enhancement of the critical density.

Although free ions are certainly present in ionic fluids, it is questionable if their concentration is really dominating the thermodynamics. Following an old idea of Stillinger [67], Camp and Patey [68, 69] and Panagiotopoulos with coworkers [70] considered a fluid of dumb-bells representing the ion pairs. It turned out that the coexistence curve comes out almost the same as for the RPM.

4.2   Analytical theories

The determination of the phase diagram and of the critical point by analytical theories usually requires the calculation of the excess part of the reduced free energy density Φ^{ex}, which is then supplemented by an appropriately chosen hard-core contribution. For any temperature the spinodal is defined by the stability condition $\left(\partial^2 \Phi / \partial \rho^2 \right) = 0$, and the maximum of the spinodal with respect to T defines the critical point. The coexistence curve can be obtained from the expansion of the free energy density about the critical point and solving for ρ at a given temperature. Alternatively, one can exploit the condition that at coexistence, pressure and chemical potentials are equal in the two phases.

Interestingly, in both DH and MSA theory, the excess free energy density of the ions supplemented by the ideal gas free energy density predicts already a phase transition without accounting for pair formation. However, there are gross deviations from simulation results. In DH theory the analytical results are $x_c = 1$, $T_c^* = 1/16$ and $\rho_c^* = 1/64\pi$. While the critical temperature comes out almost correctly, the density is too small by more than an order of magnitude. This flaw is corrected by including the pairing concept and employing the mass action law with the association constant proposed by Bjerrum. Actually, it turns out [31, 32] that at the critical point $x = \kappa\sigma \cong 1$ in all DH-based theories. Recalling that is evaluated at the free ion density, all effects that increase the free ion density will displace the critical density towards lower values and *vice versa*. Including the Bjerrum pairing the critical point is calculated in fair agreement with those simulations for the RPM, which do not use the finite size scaling technique. The coexistence curve, however, has an unnatural banana-shaped form [7, 31].

This flaw is remedied by FL theory. Taking into account the interactions of the ion pairs with the surrounding ion cloud, FL theory leads to good agreement of the phase-diagram with the mean-field simulations. However, cancellation of errors may contribute to the impressing good performance of the FL theory. For example, if the ideal gas free energy density is replaced by more appropriate hard core expressions, the good agreement is lost [32, 52].

It is instructive to consider some other possible versions [7, 31, 32, 52]. The combination of the DH limiting law for the ionic free energy with the ideal gas free energy does not yield a phase transition. This is also true, if the electrostatic energy is used instead of the free energy. In both cases, however, phase transitions are obtained with the vdW hard-core free energy or the Carnahan-Starling hard sphere term. However, then the critical density and the critical temperature come out nearer to that of vdW fluids [7].

Finally, one may complete the theory by taking into account the dipole-dipole interactions between ion pairs [32, 52] as well as the resulting change of the dielectric permittivity. This approximation, however, shifts the critical density to figures, which are too small. The physical origin of this defect seems related to the overestimate of the ionicity discussed in Sec. 3.3. An account for the presence of anti-parallel ion pairs may largely remove the discrepancy.

4.3   *Experimental results*

We turn now to the question, whether the critical points predicted for the RPM have some relation to critical points of real ionic systems, the liquid-vapor critical point of molten salts and liquid-liquid critical points of ionic solutions [23, 24].

The RPM critical parameters, $T_c^* \cong 0.049$ and $\rho_c^* \cong 0.08$, are much lower than observed for simple nonionic fluids, for example $T_c^* = 1.31$ and $\rho_c^* = 0.32$ for the Lennard-Jones fluid [7, 24]. Low values of T_c^* are typical signatures for phase transition driven by Coulomb interactions [23].

Let us first consider the liquid/vapor phase transition of NaCl. PVT data have been recorded up to about 2000 K, which is still far below the critical point. Extrapolations guided by simulations and theory predict $T_c = 3300$ K and the critical mass density $d_c = 0.18$ g cm^{-3} [33]. With $\sigma \cong 0.276$ nm and $\epsilon = 1$, this maps onto $T_c^* \cong 0.05$ and $\rho_c^* \cong 0.08$.

Considering liquid/liquid transitions the RPM critical parameters suggest that for 1:1-electrolytes such transition should occur in solvents of low dielectric constant. Meanwhile many solutions of low-melting salts in organic solvents are known that are in general agreement with the critical data predicted by the RPM [21, 37].

However, there are systems with gross deviations from these predictions. Liquid-liquid immiscibility was observed with some of these salts even in aqueous solutions [37, 71]. In such cases, the ionic forces are not expected to drive the phase separation. Rather, solvophobic effects of salts with large ions in solvents of high cohesive energy density may be responsible for these transitions [37].

Instructive examples for the interplay of Coulomb and solvophobic forces are tetraalkyl ammonium picrates dissolved in alcohols. Extending earlier work [37], Kleemeier et al. [72] observed critical points of tetra-n-butyl ammonium picrate (Bu$_4$NPic) in a series of 10 alcohols with dielectric constants ranging from 3.6 (1-tetradecanol) to 16.8 (2-propanol). For mapping the critical data onto reduced variables, we estimate the separation of the ionic charges to be $\sigma = 6.6$ E, based on the Stokes radius of the cation and the vdW radius of the oxygen, which is assumed to be the center of the charges in the anion. For the almost non-polar long chain alcohols one indeed approaches this "Coulomb limit" $T_c^* \cong 0.05$ of the RPM. However, for shorter chain length, i.e. higher dielectric constants of the solvents, T_c^* increases, and one moves continuously from a Coulomb mechanism to a non-Coulomb mechanism for phase separation. A survey investigation on the phase separation of ILs in alcohols including methanol, hexadecanol, and water [21] shows the same regularity. The ILs investigated were alkyl methyl imidazolium salts (Rmim) with butyl, hexyl, and octyl side chains and the anions BF$_4^-$ and PF$_6^-$.

Considering the ILs, at first glance, the separation temperatures depend in a rather specific way on the molecular details. As observed by other authors, who considered C$_2$mim$^+$CF$_3$SO$_2$)$_2$N$^-$ [73] and C$_n$mim$^+$ PF$_6^-$ [74, 75], an increase of the critical temperature with decreasing polarity of the alcohols is found, which is in accordance with the RPM that predicts a proportionality of the critical temperature with $1/\epsilon$. Only the separation temperature of the mixture with water does not follow this trend. Here, because of the hydrophobic interactions, the separation temperature is higher than in methanol and in the same region as for the hexanol solutions. Comparing the separation temperatures in the solutions of a given alcohol, we see that the separation temperatures are reduced, if the length of the side chains of the cation is increased. Obviously, with increasing chain length the solvation of the ions by the alcohols is improved. In water and in dialcohols (in the case of the BF_4^- compounds) this trend is reversed as one can expect for hydrophobic unmixing. Comparing the BF_4^- and the PF_6^- salts, we note that for all salts and all alcohols the separation temperatures of the PF_6^- salts are higher than that of the BF_4^- salts.

This picture simplifies considerably in terms of reduced RPM variables. For this purpose the relevant ion separation has to be estimated. Using vdW radii and bond lengths and assuming that the charge centre of the cation is the centre of the imidazolium ring the shortest ion-ion separation is estimated to 4.6 Ě for the BF_4^- salts and 4.93 Ě for the PF_6^- salts, respectively. In this estimate it is assumed that the anion is located on top of the ring. Infact, reanalyzing the X-ray data of the $C_1mim^+Cl^-$ crystal [76] shows (see Fig. 2) that the chlorine ion is located above the ring plane.

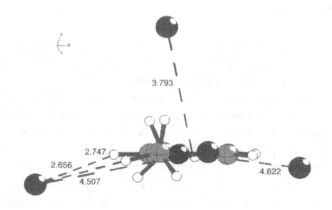

Figure 2. Coordination of the dimethyl imidazolium cation by chloride ions in the lattice.

Our crude estimate turns out to be in fair agreement to estimates based on structural data from simulations [77] or scattering experiments [78]. The RPM reduced temperatures increase almost linearly with the dielectric constant of the solvent. There seems to be a continuous transition from the low-ϵ alcohols to water. For different salts the n-alcohols, the branched alcohols, the secondary alcohols, and even water the reduced temperatures are represented as function of ϵ in a master plot, shown in Fig. 3.

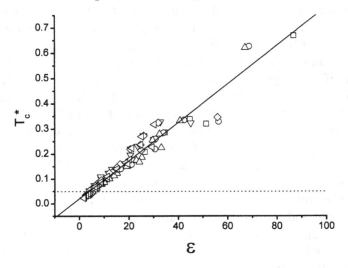

Figure 3.   RPM reduced temperatures of the phase separation as function of the dielectric constant of the corresponding solvent. The different salts are distinguished as follows $C_4mim^+BF_4^-$ ($\square$), $C_6mim^+BF_4^-$ ($\bigcirc$), $C_8mim^+BF_4^-$ ($\triangle$), $C_4mim^+PF_6^-$ ($\triangledown$), $C_6mim^+PF_6^-$ ($\lozenge$), and $C_8mim^+PF_6^-$ ($\triangleleft$). The dotted line is the critical temperature of the RPM, the solid line the best linear fit to the data.

In respect to the critical density, the behavior is more irregular [21]. The figures are below that of the RPM. The linear alcohols and the secondary alcohols form different sets. The results show that, the RPM is a reasonable guide to start a systematic study. However, it becomes clear that RPM is not the best possible generic model for solutions of ionic liquids. Models taking explicitly the solvent into account are called for. First simulations on solutions of charged hard spheres in hard spheres [79, 80] and dipolar hard spheres [80] are consistent with the trends found in the experiments. Using RPM reduced variables the critical temperature of mixtures of charged hard spheres with hard spheres is somewhat higher than for the RPM, and is increased further in mixtures with hard sphere dipoles.

References [9] and [10] summarize several other systems, where specific non-Coulomb interactions come into play, so that the critical point is largely displaced from what is expected for an ionic system. One system is molten NH_4Cl [12] which, although highly conducting, exhibits a too low critical temperature $T_c^* \cong 0.02$ and a too high critical density $\rho_c^* \cong 0.2$. This salt decomposes, however, in the gaseous phase into HCl and NH_3, which largely affects the equation of state. Another system is ethyl ammonium nitrate ($EtNH_3NO_3$)

dissolved in 1-octanol, which, as summarized in Ref. [9], has been used in many investigations of ionic criticality. Here the critical temperature corresponds to the expected value, but the critical density is located in the salt-rich regime. A rationale, supported by conductance data, is that hydrogen bonds between cations and anions stabilize the pairs in excess to what is expected from electrostatic interactions.

5. Criticality of ionic fluids

5.1 *Survey on the theory*

Considering a liquid-gas or liquid-liquid coexistence curve, any analytical equation of state such as the vdW equation predicts a parabolic top, which is the consequence of the presumption that the free energy is an analytic function of temperature and density. Already van der Waals was aware that the top of such coexistence curves is approximately cubic. Today we know that this anomaly is but one example for the universal Ising behavior of fluid properties near critical points [4] to be distinguished from "classical" or "mean-field" behavior. Based on the available experimental findings [14–16] and simple theoretical arguments [13], in the late 1980s Pitzer [14, 15] conjectured, however, that ionic fluids might behave classical.

Critical exponents. Theoretically the criticality in systems where Coulomb interactions drive the phase transition poses serious problems. According to renormalization group (RG) theory [4], Ising universality rests on the short-range nature of the interaction potential $\phi(r)$. Short-range means that $\phi(r)$ decays as r^{-p} with $p = d+s$ and $s > 2-\eta$ [81], where η is the so-called Fisher exponent, which for short range interactions in 3D space assumes $\eta \cong 0.03$. For fluids with a leading term of $\phi(r) \propto r^{-6}$, the experimental verification of Ising-like criticality is unquestionable. The bare Coulomb interaction ($s = -2$) and interactions of charges with rotating dipoles ($s = 1$) do not fall into this category.

For $s < 0$ the thermodynamic limit does not exist [4]. Potentials with $0 < s < 2 - \eta$ are termed long-range potentials. Renormalization group calculations yield for potentials with $0 < s < d/2$ the following set of critical exponents [81] $\nu = 1/s$, $\eta = 2 - s$, and $\gamma = 1$ that are termed mean field exponents. The exponent ν and γ determine the temperature dependence of the correlation length ξ and of the susceptibility χ. Other exponents may be calculated using relations between exponents [4, 5]. Assuming the validity of the hyper scaling relation $d\nu = 2 - \alpha$ one gets $\alpha = 2 - d/s$, $\beta = (d - s)/2s$, and $\delta = (d + s)/(d - s)$, where α is the critical exponent of the specific heat, β describes the temperature dependence of the coexistence curve, and δ determines the divergence of the osmotic susceptibility $\chi \propto |x - x_c|^{1-\delta}$, if, at the critical temperature, the composition variable x approaches the critical value. For $s = d/2$ the exponents α, β, and δ agree with vdW mean field coefficients. Only at the dimension $d = 4$, termed critical dimension d_c, all exponents agree with the values of the vdW mean field theory. The vdW mean-field exponents, which conventionally are called mean field exponents, result from a mean field theory for fluids with particles interacting by a short-range potential.

The applicability of the hyper scaling relation is, however, questionable if long range interactions drive the phase transition. RG analysis [4] requires that the critical dimension d_c depends on the power of the potential according $d_c = 2s$ and that the dimension d in the hyper scaling relation has to be replaced by d_c. Therefore, if $0 < s < 2$ the coefficients α, β, δ and γ take the vdW mean field values, while ν, μ assume the vdW mean field values only for $s = 2$. Simulations [82, 83] of fluids with long range potentials $0 < s < 2$ yield $\beta/\nu = 0.8$ for $s = 1$ and $d = 3$. This result is between the estimates $\beta/\nu = 1$ assuming the validity of the hyper scaling relation and $\beta/\nu = 1/2$ obtained with $d_c = 2$. The Ising value is $\beta/\nu = 0.515$.

Thus, non-Ising behavior may be expected in systems determined by Coulomb and charge dipole interactions. However, due to the screening by counter ions the potential of the average force becomes short range. Therefore, Ising-like criticality may be restored as in liquid metals, where the electrons screen the interactions of the Coulomb interactions of the cores [84].

Simulations of critical properties are a formidable task. In spite of the periodic boundary conditions the size of the boxes is too small to treat long-range fluctuations adequately in standard simulations. Therefore, special techniques are required to elucidate the nature of the critical point even in normal fluids. The finite scaling technique provides the adequate means [62] to yield reliable values for the critical point and to allow conclusions about the universality class of model fluids such as of the square well fluid or the RPM [63]. Recent simulations [65, 66, 85] agree in suggesting an Ising-like critical point for the RPM. Lujiten et al. [85] even claim an accuracy of the exponents obtained from the simulations, which even allows distinguishing between Ising criticality and other models with similar exponents as the XY model and the self-avoiding walk model.

Crossover. Generally, crossover from an Ising-like asymptotic behavior to mean-field behavior further away from the critical point [86, 87] may be expected. Such a behavior is also expected for nonionic fluids, but occurs so far away, that conditions close to mean-field behavior are never reached. Reports about crossover [88] and the finding of mean field criticality [14–16] suggest that in ionic systems the temperature distance of the crossover regime from the critical point, as characterized by the so-called Ginzburg temperature ΔT_{Gi}^*, is much smaller than observed for nonionic fluids. Actually, such a behavior was concluded from coexistence curve data for the system $Na + NH_3$ [89]. There is a transition to metallic states in concentrated solutions, but in dilute solutions and near criticality ionic states prevail.

Theoretically the Ginzburg temperature can be calculated from the free energy density. For the global density ρ_c^* at criticality, the local free energy density can be written as an expansion in powers of the density deviations $\tilde{\rho}(r) = (\rho^*/\rho_c^* - 1)$ from this composition plus the corresponding gradients. Considering the critical composition and keeping only the leading terms, one gets,

$$\Phi\left(T^*, \tilde{\rho}\left(r^*\right)\right) = \frac{1}{2}c_2\left[\tilde{\rho}\left(r^*\right)\right]^2 + \frac{1}{4!}c_4\left[\tilde{\rho}\left(r^*\right)\right]^4 + \frac{1}{2}c_{2\text{g}}\left(\nabla^*\tilde{\rho}\left(r^*\right)\right)^2 + \dots \quad (15)$$

The star on the gradient indicates that the coordinates are scaled by the particle diameter σ, and index g denotes quantities related to the gradient terms. The system-dependent coefficients c_2 and c_4 are the second and fourth derivatives of the free energy density, respectively. Quite generally, $c_2 = c_2^\circ \cdot \tau$, where $\tau = |T - T_c|/T_c$ characterizes the temperature distance from the critical point. The gradient theory, given by Eq. (15) is the classical approach to fluctuations in mean-field theory. It is often termed after Landau and Ginzburg or Cahn and Hilliard, but actually it goes back to the vdW theory of surface tension [90]. Gradient theory is the starting point of the renormalization group (RG) theory [4] of critical phenomena and of crossover theories [86]. For the critical isochore, variation of Φ yields an exponential decay of density fluctuations determined by the mean field correlation length $\xi^* = \xi_0^* \cdot \tau^{-1/2}$, where $\xi_0^* = \sqrt{c_{2\text{g}}/c_2^\circ}$. Similarly, for $\tau \leqslant 0$ the coexistence curve is given by $\tilde{\rho} = B^* \cdot \tau^{1/2} = (3!c_2^\circ/c_4)^{1/2} \cdot \tau^{1/2}$. Finally, the temperature independent amplitude of the susceptibility is $\chi_0^* = 1/c_2^\circ$.

Whenever an appropriate mean field theory is available, it is straightforward to work out the coefficients c_2 and c_4. By functional differentiation, it can be shown that $c_{2\text{g}}$ is the second moment of the direct correlation function [91],

$$c_{2\text{g}} = \frac{1}{6}\int r^{*2}c(r^*)4\pi r^{*2}\mathrm{d}r^* \,. \quad (16)$$

For $d = 3$ the ΔT_{Gi}^* becomes,

$$\Delta T_{\text{Gi}}^* = \frac{c_4^2}{64 \cdot c_2^{\circ 2} \cdot c_{2\text{g}}^3} = \frac{9}{16} \cdot \frac{\chi_0^{*2}}{B^{*4}\xi_0^{*6}} \,. \quad (17)$$

The numerical factor is to some degree arbitrary [86].

However, the direct correlation function $c(r^*)$ is subject to approximations. In van der Waals theory $c(r^*)$ is equal to $-\beta\phi(r^*)$ outside the excluded volume and equal to -1 inside. A straightforward generalization of the van der Waals approach is then [92],

$$c(r^*) = -\int g(r^*)\,\phi(r^*)\,\mathrm{d}\beta,\qquad (18)$$

which corresponds to the method by which the free energy is calculated on the energy route. The flaw in this approach is that $g(r^*)$ is largely density dependent and the particular variation of the density should be considered in evaluation of the integral (18). A second approach employs a reformulation of the direct correlation function in terms of the total correlation function $h(r^*)$, which is derived from graph theory [93]. The leading terms are,

$$c(r^*) = -\beta\cdot\phi(r^*) + \frac{1}{2}h(r^*)^2 + \dots .\qquad (19)$$

For the RPM, the first term cancels, and the second is dominant. In principle, $c(r^*)$ defined in this way may also vary locally. Third, one may also calculate c_{2g} by working backward from the Laplace transformation of the pair correlation function, as done by Lee and Fisher [94] for the pair correlation function of their generalized Debye-Hückel theory (GDH).

Considering now results from DH theory, Eq. (18) leads to a very small Ginzburg number [92], apparently confirming rapid crossover. However, the GDH theory leads to a Ginzburg number, which is large, even if compared to values for ordinary non-ionic fluids [94]. The same result is obtained when applying Eq. (19) or when the density dependence of the pair correlation function in Eq. (18) [95] is taken into account. It is quite tricky that the predictions for ΔT_{Gi}^* depend sensitively on the approach by which c_{2g} is calculated. Note, the differences of the estimates of c_{2g} become negligible in non-ionic and non-polar fluids because, $\beta\phi(r)$ dominates, which vanishes in ionic and polar fluids.

Tricritical point.   The experiments yield indications of a crossover in the region of, $\Delta T_{\mathrm{Gi}}^* \approx 0.01$ [88]. This figure is neither very small, of order 10^{-4} say, as resulting from the straightforward use of Eq. (18), nor large, of the order of unity say, as estimated by using Eq. (19).

Thus Ginzburg analysis just described does not provide an explanation for the observed crossover and apparent mean field behavior in ionic systems. At present, the search for a solution to this problem is focused on the possibility of a tricritical point [96]. Crossover may be controlled by the approach toward a real or virtual tricritical point which in $d = 3$ is mean-field-like [4, 5].

Generally, a tricritical point is defined as the point, where three phases become identical. Alternatively a tricritical point can be defined as the point where a second order λ-line of an order transition becomes first order or else as the point, where a λ-line cuts a coexistence curve at its critical point [4, 5]. The last definition is relevant for the ionic fluids.

Notably, there are models closely related to the RPM that have indeed a tricritical point. In particular, the lattice analogue of the RPM, which is the Coulomb gas with cations and anions on lattice sites interacting by Coulomb interactions, has no liquid-gas critical point but a tricritical point [97, 98]. This is quite unexpected, because for uncharged fluids the nature of the critical point is unchanged when replacing the continuum by a lattice. In fact, much of our knowledge about phase transitions of fluids comes from investigations of the Ising lattice model. The lattice RPM yields a first-order transition between a disordered and an antiferroelectrically ordered phase, terminating at a tricritical point. Above the tricritical point, a line of second-order transitions between a disordered and antiferroelectric phase is predicted. Thus, there is a tricritical point at $T^* = 0.38$ and $r^* = 0.36$ rather than a critical point, and there is no liquid/vapor transition. However, in an anisotropic lattice obtained by stretching a cubic lattice in one direction, the low-density gas-liquid phase separation reappears and the phase diagram exhibits critical, tricritical and triple points [99]. MC simulations of the RPM on a lattice show by varying the grid that the order-disorder transition vanishes, if the grid of the lattice becomes smaller than the diameter of the particles by one order of magnitude [56, 57].

Concerning the fluid some possibilities for tricriticality are discussed. The GDH theory predicts an order-transition between the isotropic fluid and a phase with periodic charge density waves [94]. The corresponding transition line cuts the liquid-liquid line near their critical point, so that a tricritical point is a possibility. The possible presence of both, gas-liquid and tricritical points in ionic fluids has been predicted by Ciach and Stell [96] for a model, in which additional short-range forces supplement the lattice Coulomb forces.

Furthermore, the $2d$ RPM also yields a tricritical point, which, however, has a different physical basis [100]. Here, tricriticality is founded on the insulator-conductor transition, which changes from second to first order. Notably, in real ionic solutions the conductivity shows two points of inflection one at low densities, which corresponds to the conductor insulator transition in $2d$, and one near the criticality [38]. Although accompanied by a maximum of the specific heat [68, 69], those changes of the conductivity are soft transitions determined by the mass action law and not cooperative λ-transitions, required to allow for a tricritical point.

Concluding we state, different theoretical arguments suggest a possibility of a tricritical point in ionic systems.

5.2   *Experimental results for binary systems*

Coexistence of binary systems.   Coexisting phases are characterized by different figures of the order parameter M. In pure fluids, one identifies M with the density difference of the coexisting phases. In solutions, M is related to some concentration variable, where theory now advocates the number density or the closely related volume fraction [101]. At a quantitative level, these divergences are described by crossover theory [86, 87] or by asymptotic scaling laws and corrections to scaling, which are expressed in the form of a so-called Wegner series [104]. The two branches of the coexistence curve are described by

$$M_{+,-} = \pm B_0/2\tau^\beta \left(1 + B_1\tau^\Delta + \ldots\right) - A\tau - D\tau^{1-\alpha}, \qquad (20)$$

where β, α and Δ are universal critical exponents, B_i, A and D are a substance-specific amplitudes. Experimentally, one determines the difference $\Delta M = M_+ - M_-$ and the diameter $(M_+ + M_-)/2$. For Ising-like systems holds $\beta \cong 0.326$ and $\alpha \cong 0.11$, while in the mean-field case $\beta = 0.5$ and $\alpha = 0$ exactly. The correction terms depend on the universal crossover exponent $\Delta \cong 0.5$. In a mean-field system the diameter is "rectilinear". In the asymptotic regime ΔM reduces to $\Delta M = B_0\tau^\beta$. For practical data evaluation, one often characterizes the deviations from asymptotic behavior by an effective exponent.

Measurements of the liquid gas transition of salts, e.g. of NH_4Cl at 1150 K [12] that yielded the mean field value of β can hardly achieve the mK accuracy required in work on critical phenomena. Therefore, measurements of the liquid-liquid coexistence occurring near ambient conditions are more apt for measuring critical exponents.

Coexistence curves can accurately be determined from measurements of the refractive index in the two phases of a flame sealed critical sample. Criticality of the sample is ensured by the equal volume criteria, requiring equal volume of the phases at the cloud point. The Lorenz-Lorentz relation is employed to determine the composition in the coexisting phases. Note, measurements of cloud points for samples of different compositions usually do not suffice to reach the necessary accuracy as minute uncontrollable impurities influence the cloud point temperature.

Pitzer and coworkers observed parabolic coexistence curves for two liquid-liquid phase transitions with critical points near 414 K [14] and 318 K [15]. In particular, the critical point of the latter system, consisting of the IL $N_{2226}^+ B_{2226}^-$ ("Pitzer's salt") dissolved in biphenyl ether, should be close enough to ambient conditions to perform accurate measurements. However, the critical temperature $T_c \cong 318$ K could not be reproduced in later work [17], which, depending on the sample, yielded values between 288 and 309 K (see also Tab. 2 of [10]). These differences seem to indicate a considerable chemical instability of Pitzer's salt, so that decomposition products displace T_c. Because of these observations, coexistence curve measurements were repeated with a sample, which was tempered until stability of T_c was reached. These experiments, as well as determinations of other critical exponents discussed later, yielded plain Ising behavior [17] for this system.

Measurements on solutions of Tetrabutyl ammonium picrate (Bu_4NPic) in long-chain alcohols are better reproducible, as the salt is chemically more stable [37, 72, 102]. The systems show a gradual shift from Coulomb to solvophobic behavior of the phase transitions, thus allowing investigating the influence of the change of the forces determining the phase transition on the criticality and the nature of the critical point. Asymptotic Ising behavior is found, but positive deviations occur away from T_c. Although the deviations are quite small, they show a systematic increase with increasing chain length of the alcohol, thus suggesting an increased tendency for crossover to the mean-field case, when the Coulomb contribution becomes essential.

The first accurate measurements with ILs concern the salt C_6mimBF_4 and two isomers of the alcohols butanol, and pentanol [103]. The coexistence curves shown in Fig. 4, which include the measurements of the solutions in water, show corresponding states behavior.

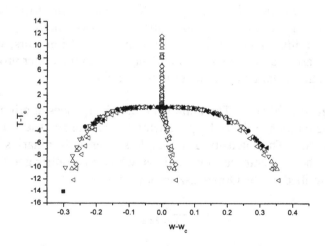

Figure 4. Comparison of the phase diagrams determined from the refractive index measurements with that obtained by the visual method. The weight fraction w is the concentration variable. Empty symbols are used, when the compositions are estimated from refractive index data ($C_6mim^+BF_4^-$ in 1-butanol ($\triangle$), 1-pentanol ($\triangledown$), 2-butanol ($\lozenge$), and 2-pentanol ($\triangleleft$)), full symbols, when the data are obtained by the visual method observing the separation temperature in samples of different composition ($C_6mim^+BF_4^-$ in water ($\square$), 1-propanol ($\bigcirc$), 1-butanol ($\triangle$), 2-butanol ($\triangledown$), 1-pentanol ($\lozenge$), 2-pentanol ($\triangleleft$), and 1-hexanol ($\triangleright$)).

Agreement is found between the optical measurements in sealed samples and the direct observation of the cloud points in different samples. The composition of the critical sample agrees with that of the sample yielding the maximum of the phase diagram, which indicates sufficient purity of the IL. In the systems investigated Coulomb and hydrophobic interactions both drive the phase separation. Unfortunately it was not possible to measure higher alcohols, where the Coulomb forces are expected to determine the behavior. The difference between the refractive index of the salt and the alcohol became too small to allow accurate measurements. Measurements of the critical exponents in water solutions were also not feasible because of hydrolysis of the saltduring the measurements. Critical exponents near to the Ising value are found. Here the deviations are negative. This indicates a non-monotonous change from Ising to mean-field criticality [87]. It should be checked if the observation of monotonous crossover in Coulomb systems and of non-monotonous crossover in cases of unmixing due to solvophobic interactions is a general rule.

Now we discuss the diameter $(M_1 + M_2)/2$ of the coexistence curve. For a long time, the diameter anomaly in nonionic systems was a matter of controversy, because the deviations from rectilinear behavior are small, and there is an additional spurious 2β contribution, when an improper order parameter is chosen in data evaluation [104]. The investigations of the picrate systems [72] and of the IL solutions [103] both yielded a substantial anomaly, consistent with an $(1 - \alpha)$ anomaly. Large diameter anomalies are expected, when the intermolecular interactions depend on the density [84]. In the systems considered here, the dilute phase is essentially composed of ion pairs, while the concentrated phase is an ionic melt, which may explain the rather pronounced deviation from the rectangular diameter in the ionic systems.

Scattering and turbidity. The non-analytical divergences at critical points result from fluctuations of the order parameter, which can be observed by scattering experiments. The intensity I of single scattering in binary systems is determined by the concentration fluctuations, which in a rather good approximation are described by the Ornstein-Zernike equation,

$$I = \frac{\chi}{1 + q^2 \xi^2} .\tag{21}$$

The OZ equation depends on the scattering vector q and involves the osmotic compressibility χ and the correlation length ξ of the fluctuations, which temperature dependencies are given by scaling laws of the form analogous to Eq. (20),

$$\chi = \chi_0 \tau^{-\gamma} \left(1 + \chi_1 \tau^\Delta \ldots \right), \quad \text{and} \quad \xi = \xi_0 \tau^{-\nu} \left(1 + \xi_1 \tau^\Delta \ldots \right) .\tag{22}$$

The exponents γ and ν satisfy the equality $\gamma = \nu(2 - \eta) \cong 2\nu$, where $\eta \cong 0.03$ is the Fisher exponent. In the mean-field case $\gamma = 1$ and $\nu = 0.5$ and $\eta = 0$, so that $\gamma = 2\nu$ exactly. In the Ising case, $\gamma = 1.24$ and $\nu = 0.63$. Experimental intensity data may be evaluated in two ways. Most experiments are carried out for a fixed scattering vector q analyzing the temperature dependence of I by fitting the data by Eq. (21). Alternatively the OZ-plot in which $1/I$ is plotted as function of q^2 enables independent determination of χ and ξ for a given temperature. The turbidity is the integral of I over all scattering vectors. Thus, static light or neutron scattering and turbidity measurements enable to determine the exponents γ and ν. Moreover, dynamic light scattering yields the time correlation function of the concentration fluctuations, which decays as $\exp\left[-2D(q)q^2t\right]$. The diffusion coefficient D depends on the temperature and the scattering vector and can be converted to the correlation length, thus providing the exponent ν.

The intensity, as given in Eq. (21) is obtained from the experimental intensity after correcting for the angular dependence of the scattering volume, the loss of light due to scattering on the light path. In strong scattering media the additional intensity due to multiple scattering should also be considered. In order to account for the multiple scattering, simulation techniques can be applied [105]. Cross-correlation techniques were also developed allowing separating the singly scattered light from the other contributions [106, 107].

Following Singh and Pitzer's report on the parabolic coexistence curve [15], there was large interest in the exponent γ, because a theory assumed the so-called spherical model [108] for describing criticality in ionic systems. The spherical model implies $\beta = 1/2$, but the other exponents differ largely from the mean-field exponents, e.g. with $\gamma = 2$ instead of $\gamma = 1$. Dynamic light scattering measurements in solutions of Bu_4NPic in 1-tridecanol [102] ruled out the spherical model by showing that the results are consistent with $\gamma = 1$ or the Ising exponent $\gamma = 1.24$ rather than $\gamma = 2$.

The discrimination between Ising and mean-field behavior by scattering experiments proved to be more subtle. Turbidity measurements on Pitzer's system [16] confirmed plain mean-field criticality without noticeable crossover. However, turbidity measurements performed later with another sample in the same laboratory yielded Ising behavior [17]. With this second sample, Ising-like criticality was consistently found by light-scattering experiments [17], measurements of the coexistence curve [17] and of the viscosity [115]. Two technical aspects of the latter measurements should be emphasized. To ensure homogeneity of the sample and a stability of the critical temperature the sample was tempered for more than a week before the measurement started. Furthermore, careful corrections for losses due to turbidity and additional intensity due multiple scattering were carried out by means of simulation [105].

Again, on grounds of chemical stability, experiments with picrate solutions may be more convenient. Narayanan and Pitzer therefore performed turbidity experiments with solutions of Bu_4NPic in 1-undecanol, 1-dodecanol, and 1-tridecanol [88]. The data show quite sharp crossovers from mean-field criticality away from T_c to asymptotic Ising criticality, which occurs almost within one decade of the reduced temperature. Crossover is closest to T_c for the highest homologue, where Coulombic interaction is expected to be strongest. Adding 1,4-butanediole to 1-dodecanol shifts the crossover region further away from T_c, and for pure 1,4-butanediol clear Ising behavior is observed. Later, Kleemeier [109] performed turbidity measurements for the picrates in conjunction with light scattering experiments. His results confirm the essential features of Pitzer's data, but crossover seems to be smoother than observed earlier. Clearly, these results confirm the picture developed above from coexistence curve data [72]. Ising-like criticality was also deduced from both, measurements of the coexistence curve and light-scattering experiments on systems in nonpolar solvents as cyclohexane and toluene [109, 110].

Light scatttering measurements were carried out also on the C_6mimBF_4 solutions in alcohols [111], for which the coexistence was measured. Again, measurements on solutions with higher alcohols were not feasible, because, the difference of the refractive index of salt and solvent are so small that the scattering intensity was not measurable with sufficient accuracy. In the measured systems the turbidity was also too small to be measurable and therefore, it was not necessary to correct the intensity of the scattered light for the losses due to scattering and multiple scattering. However, other effects were perturbing the measurements. Because of the small scattering intensity it was necessary to use a laser power of 10 mW instead of 2 mW commonly used in such investigations. The first analysis of the temperature dependence of the scattering intensity yielded classical exponents! However as shown in Fig. 5 the temperature dependence of the parameters in extracted from the OZ plots showed deviations from the exponential laws.

Approaching the critical temperature a bending was observed indicating that, due to laser heating, the temperature in the beam was above that of the bath. The temperature reached in the beam at long times was calibrated by measuring the time dependence of the scattering intensity. This was still not sufficient to get consistence between the results of static and dynamic scattering. An increase of the amplitude of the correlation function when approaching the critical temperature was observed that indicated a contribution of a noncritical scattering process. Estimating this contribution from the measurement of the amplitude enabled to calculate the critical scattering. Taking all those obstacles into account the scattering intensity is finally found consistent with Ising criticality but it is not possible to allow further conclusions.

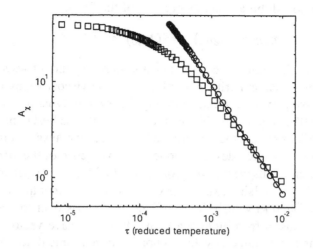

Figure 5. Scattering intensity A_χ for $q = 0$ as obtained from the OZ-plots as function of the reduced temperature. The original data ($\square$) are corrected for local laser heating and background scattering to give the scattering of the critical fluctuations ($\bigcirc$).

In concluding this section, we like to draw attention to the amplitudes ξ_0 derived in the scattering experiments. As outlined in Sec. 5.1, ξ_0 enters into theoretical expressions for the crossover temperature. Thereby, large ξ_0-values favor a small Ising regime. In simple non-electrolyte mixtures ξ_0 is generally found to be of the order of the molecular diameters, $\xi_0 = 0.1 - 0.4$ nm say. It is therefore notable that in the studies mentioned above considerably larger amplitudes were observed, in parts yielding $\xi_0 > 1$ nm. Large ξ_0-values seem to be a signature of ionic systems, not related to the mean-field versus Ising problem in an obvious way.

Small exponents. Evidence for Ising criticality can be provided by some properties showing weak divergences, which are absent in the mean-field case. One such case is the specific heat, which diverges with the exponent α. Kaatze and coworkers [112] have indeed shown the presence of such an a anomaly in $EtNH_3NO_3 + n$-octanol, but as already mentioned, this system shows an anomalous location of the critical point, indicating that non-Coulomb interactions play a considerable role in driving the phase separation.

Another property related to α is the electrical conductance, which diverges as $\tau^{1-\alpha}$. Bonetti and Oleinikova [113] proved the existence of an Ising-type $(1 - \alpha)$ anomaly for some picrate systems, finding no essential difference between Coulomb and solvophobic systems. Finally, the viscosity of Ising-like systems is known to exhibit a weak divergence of the form,

$$\eta/\eta_b = (Q_0\xi)^z = (Q_0\xi_0)^z \tau^{-y}, \tag{23}$$

where η_b is the background viscosity, Q_0 a system-dependent wave vector, $(Q_0\xi_0)^z$ a system-dependent critical amplitude, y the viscosity exponent and $z = \nu y$. For an Ising system, mode coupling theory and dynamic RG theory both yield $z = 0.065$, i.e. $y = 0.041$ in agreement with the best experimental data. For mean-field systems there seems to be no decisive answer, but probably mean-field behavior excludes an exponential divergence of the viscosity.

Viscosity measurements were first reported for Bu_4NPic + 1-tridecanol [114]. The scaling law (23) with Ising exponents was satisfied. Further away from T_c, the anomaly was found to vanish, in accordance with crossover to mean-field behavior. With ξ_0 taken from light scattering data, the wave vector, Q_0 was found to fall in the broad range of values reported for nonionic fluids. Wiegand et al. [115] observed a viscosity anomaly of Ising-type for Pitzer's system as well. Similarly, measurements on solutions of the IL C_6mim^+ BF_4^- in pentanol yielded Ising criticality with crossover to regular behavior, well described by the mode coupling theory [116]. Presuming the absence of a critical viscosity anomaly for mean-field systems, these experiments clearly prove the Ising-like character of the critical points in ionic solutions.

5.3   Tricriticality in ternary systems

One option to explain classical critical behavior and unusual crossover, is the existence of a tricritical point, which in $d = 3$ is mean-field like [4, 5]. In ternary systems, tricritical behavior is generally obtained, if three phases have a common critical point.

For mixtures of 3-methyl pyridine + H_2O + NaBr Jacob et al. [117] reported light scattering measurements showing a change of the criticality with the salt content. When increasing the salt concentration, the tendency to approach mean-field behavior at T_c becomes more pronounced, and eventually, at 17 mass% NaBr pure mean-field behavior was obtained. As viewed from the Coulombic vs. solvophobic dichotomy, the mean-field-like behavior of this ternary system is unexpected. Thermodynamic properties of the salt-free systems point towards a hydrophobic phase separation mechanism. One would expect the ions to enhance the forces already present in the binary mixture, which in turn, would imply that the critical point is Ising-like. In fact, further thorough examinations of the NaBr + water + 3-methylpyridine system did not confirm the reported results. Checking carefully the homogeneity of the samples in the concentration range of 10–18 mass% NaBr, viscosity [118], coexistence curves [119], turbidity [120], and light scattering data [121, 122] provided neither evidence for the pronounced crossover nor the observed mean-

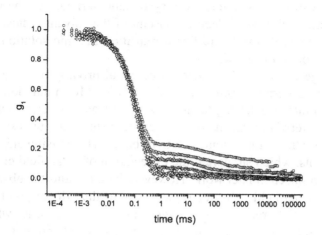

Figure 6. The autocorrelation function depends on the waiting time (2.5 h (□), 8 h (○), 20(△) , 27 h (▽), 63 h (◇) and 90 h (◁)). The amplitude of the slow non-equilibrium process decays with the waiting time.

field behavior reported for 17 mass% NaBr. It appears by now, that long living non-equilibrium inhomogeneities are responsible for the apparent mean field behavior [121, 122]. In dynamic light scattering the correlation functions, shown in Fig. 6, display two processes, where the slow process corresponds to the non-equilibrium inhomogeneities that vanish within a waiting time of one week.

6. Summary

Two properties render electrolyte theories difficult, namely the long-range nature of the Coulomb interactions and the high figures of the Coulomb energy at small ion separations.

In solvents of low dielectric constant, where the Coulomb interactions are particularly strong, electrical conductance and dielectric spectra suggest that the ion distribution involves dipolar ion pairs, which then interact with the free ions and with other dipolar pairs. The ion pairs cause an increase of the dielectric constant, which in turn stabilizes the free ions, thus leading to redissociation at high salt concentrations. Extending the approach of Debye-Hückel and Bjerrum, theory accounts for ion pairing, ion-ion pair and ion pair-ion pair interactions and rationalizes the basic features of the ion distribution in accordance with experiments and MC-simulations.

Analytical theory also predicts a fluid-phase transition at low reduced temperatures, which is in close agreement with simulation results and agrees, using corresponding states considerations, with the experimentally observed liquid-

liquid phase transitions. Nevertheless, a systematic variation of the reduced critical temperature with the dielectric constant of the solvent indicates limitations of the analogy of the liquid-liquid transition in ionic solutions to the liquid-gas transition of the RPM.

The long-range nature of the Coulomb potential driving these transitions raises questions concerning their universality class. MC-simulations of the RPM involving finite size scaling techniques are at least consistent with Ising criticality. The overwhelming majority of experiments also suggest that the liquid-liquid phase transition in ionic solutions belongs to the Ising universality class. Experiments, which supported the expectation of mean field criticality, could not be reproduced in later work. However, there remains the observation of crossover to mean-field behavior rather close to the critical point, which is not understood. Different theories suggest a crossover scenario, which involves a tricritical point. However, the decisive experiments proving, e.g. the hypothesis of charge density waves are not available.

Acknowledgements

My thanks go to Prof. Weingärtner for many discussions and the successful long standing collaboration and to my coworkers Drs. S.Wiegand, M.Kleemeier, V.Weiss, M.Wagner, O.Stanga, who, over the years, have contributed to this research. The discussions and the pre- and reprints received from Profs. M.E.Fisher, J.Sengers, M.Anisimov are much appreciated.

References

[1] Rogers, R.D., Seddon, K.R. (2003). *Ionic Liquids as Green Solvents-Progress and Prospects*. Oxford: UP.

[2] Wasserscheid, P., and Keim, W. Ionic liquids: new solvents for transition metal catalysis. *Angew. Chem. Int. Ed.*, 2000, **39**, p. 3772–91.

[3] Holbrey, J.D., and Seddon, K.R. The phase behavior of 1-alky-3-methylimidazolium tetrafluoroborates; ionic liquids and ionic liquid crystals. *J. Chem. Soc, DaltonTrans.*, 1999, p. 2133–9.

[4] Pfeuty, P., and Tolouse, G. (1977). *Introduction to Renormalization Group and Critical Phenomena*. New York: Wiley.

[5] Anisimov, M.A. (1991). *Critical Phenomena in Liquids and Liquid Crystals*. Philadelphia: Gordon and Breach.

[6] Pitzer, K.S. Ionic fluids: Near-critical and related properties. *J. Phys. Chem.*, 1995, **99**, p. 13070–7.

[7] Fisher, M.E. The story of coulombic criticality. *J. Stat. Phys.*, 1995, **75**, p. 1–36.

[8] Stell, G. Criticality and phase transitions in ionic fluids. *J. Stat. Phys.*, 1995, **78**, p. 197–238.

[9] Weingärtner, H., Kleemeier, M., Wiegand, S., and Schröer, W. Coulombic and non-coulombic contributions to the criticality of ionic fluids. *J. Stat. Phys.*, 1995, **78**, p. 169–96.

[10] Weingärtner, H., and Schröer, W. Criticality of ionic fluids. *Adv. Chem. Phys.*, 2001, **116**, p. 1–66.

[11] Schröer, W., and Weingärtner, H. Structure and criticality of ionic fluids. *Pure and Applied Chemistry*, 2004, **76**, p. 19–28.

[12] Buback, M., and Franck, E.U. Measurements of vapor pressure and critical data of ammonium halides. *Ber. Bunsenges. Phys Chem.*, 1972, **76**, p. 350–4.

[13] Friedman, H.L. Discussion remark. *J. Solution Chem.*, 1972, **2**, p. 354.

[14] de Lima, M.C.P., Schreiber, D.R., and Pitzer, K.S. Critical point and phase separation for an ionic system. *J. Phys. Chem.*, 1985, **89**, p. 1854–5.

[15] Singh, R.R., and Pitzer, K.S. Near-critical coexistence curve and critical exponent of an ionic fluid. *J Chem. Phys.*, 1990, **92**, p. 6775–8.

[16] Zhang, K.C., Briggs, M.E., Gammon, R.W., and Levelt Sengers, J.M.H. The susceptibility critical exponent for a nonaqueous ionic binary mixture near a consolute point. *J. Chem. Phys.*, 1992, **97**, p. 8692–7.

[17] Wiegand S., Briggs M.E., Levelt Sengers, J.M.H., Kleemeier, M., and Schröer, W. Turbidity, light scattering, and coexistence curve data for the ionic binary mixture triethyl n-hexyl ammonium triethyl n-hexyl borate in diphenyl ether. *J. Chem. Phys.*, 1998, **109**, p. 9038–51.

[18] Walden, P., and Centnerszwer, M. IJber verbindungen des schwefeldioxyds mit salzen. *Z. Phys. Chem.*, 1903, **42**, p. 432–68.

[19] Friedman, H.L. Electrolyte solutions that unmix to form two liquid phases in benzene and in Diethyl ether. *J. Phys. Chem.*, 1962, **66**, p. 1595–9.

[20] Gordon, J.E. Characterization of quarternary ammonium salts, phase equilibria for salt-salt and salt-nonelectrolyte systems. *J. Am. Chem. Soc.*, 1965, **87**, p. 4347–58.

[21] Wagner, M., Stanga, O., and Schröer, W. Corresponding states analysis of the critical points in binary solutions of room temperature ionic liquids. *Phys. Chem. Chem. Phys*, 2003, **5**, p. 3943–50.

[22] Friedman, H.L., and Dale, W.D.T. Electrolyte solutions at equilibrium. *Modern Theoretical Chemistry*, 1977, **5A**, p. 85–135.

[23] Friedman, H.L., and Larsen, B. Corresponding states for ionic fluids. *J. Chem. Phys.*, 1979, **70**, p. 92–100.

[24] Weingärtner, H. Corresponding states for electrolyte solutions. *Pure Appl. Chem.*, 2001, **73**, p. 1733–48.

[25] Hansen, J.-P., and McDonald, I.R. (1986). *Theory of Simple Liquids*. New York: Academic Press.

[26] Debye, P., and Hückel, E. The theory of electrolytes. *Physik. Z.*, 1923, **24**, p. 185–206.

[27] Bjerrum, N. Ionic association. I. Influence of ionic association on the activity of ions at moderate degrees of association. *Kgl. Danske Videnskab. Selskab. Math.-fys. Medd.*, 1927, **7**, p. 1–48.

[28] Caillol, J.-M., Levesque, D., and Weis, J.-J. Critical behavior of the restricted primitive model revisited. *J. Chem. Phys.*, 2002, **116**, p. 10794–800.

[29] Orkoulas, G., Panagiotopoulos, A.Z., and Fisher, M.E. Criticality and crossover in accessible regimes. *Phys. Rev. E*, 2000, **61**, p. 5930–9.

[30] Orkoulas, G., and Panagiotopoulos, A.Z. Phase behavior of the restricted primitive model and square-well fluids from Monte Carlo simulations in the grand canonical ensemble. *J. Chem. Phys.*, 1999, **110**, p. 1581–90.

[31] Levin, Y., and Fisher, M.E. Criticality in the hard-sphere ionic fluid. *Physica A*, 1996, **225**, p. 164–220, (references cited therein).

[32] Weiss, V.C., and Schröer, W. Macroscopic theory for equilibrium properties of ionic-dipolar mixtures and application to an ionic model fluid. *J. Chem. Phys.*, 1998, **108**, p. 7747–57.

[33] Guissani, Y., and Guillot, B. Coexisting phases and criticality in NaCl by computer simulation. *J. Chem. Phys.*, 1994, **101**, p. 490–509.

[34] Walden, P. *Bull. Acad. Imper.Sci.*, 1914, p. 1800, (quoted from Bradaric, C.J., Downard, A., Kennedy, C., Robertson, A.L, and Zhou, Y. Phosphonium Ionic Liquids. *The Strem Chemiker*, 2003, **XX**, p. 1).

[35] Arrhenius, S. IJber die Dissoziation der in Wasser gelösten Stoffe. *Z. Phys. Chem.*, 1887, **1**, p. 631–48.

[36] Kraus, C.A. The ion-pair concept: its evolution and some applications. *J. Phys. Chem.*, 1956, **60**, p. 129–41.

[37] Weingärtner, H., Merkel, T., Maurer, U., Conzen, J.P., Glasbrenner, H., and Käshammer, S. Coulombic and solvophobic liquid-liquid phase-separation in electrolyte solutions. *Ber. Bunsenges. Phys. Chem.*, 1991, **95**, p. 1579–86.

[38] Weingärtner, H., Weiss, V.C., and Schröer, W. Ion association and electrical conductance minimum in Debye-Hückel-based theories of the hard sphere ionic fluid. *J. Chem. Phys.*, 2000, **113**, p. 762–70.

[39] Schreiber, D.R., de Lima, M.C.P., and Pitzer, K.S. Electrical conductivity, viscosity, and density of a two-component ionic system at its critical point. *J. Phys. Chem.*, 1987, **91**, p. 4087–91.

[40] Onsager, L. The theory of electrolytes. *Physik. Z.*, 1927, **28**, p. 277–98.

[41] Fuoss, R.M., and Kraus, C.A. Properties of electrolytic solutions. IV. The conductance minimum and the formation of triple ions due to the action of Coulomb forces. *J. Am. Chem. Soc.*, 1933, **55**, p. 2387–99.

[42] Gestblom, B., and Songstad, J. Solvent properties of dichloromethane. VI. Dielectric properties of electrolytes in dichloromethane. *Acta Chem. Scand. Ser. B*, 1987, **41**, p. 396–400.

[43] Weingärtner, H., Nadolny, H.G., and Käshammer, S. Dielectric properties of an electrolyte solution at low reduced temperature. *J. Phys. Chem. B*, 1999, **103**, p. 4738–43.

[44] Kirkwood, J.G. The theory of dielectric polarization. *J. Chem. Phys.*, 1936, **4**, p. 592–601.

[45] Schröer, W. Generalization of the Kirkwood-Fröhlich theory of dielectric polarization for ionic fluids. *J. Mol. Liquids*, 2001, **92**, p. 67–76.

[46] Caillol, J.M. A Monte Carlo study of the dielectric constant of the restricted primitive model of electrolytes on the vapor branch of the coexistence line. *J. Chem. Phys.*, 1995, **102**, p. 5471–5479.

[47] Caillol, J.M., and Weis, J.J. Free energy and cluster structure in the coexistence region of the restricted primitive model. *J. Chem. Phys.*, 1995, **102**, p. 7610–21.

[48] Yan, Q.L., and de Pablo, J.J. Phase equilibria and clustering in size-asymmetric primitive model electrolytes. *J. Chem. Phys.*, 2001, **114**, p. 1727–31.

[49] Yan, Q.L., and de Pablo, J.J. Phase equilibria of charge-, size-, and shape-asymmetrical models of electrolytes. *Phys. Rev. Lett.*, 2002, **88**, p. 095504/1–4.

[50] Romero-Enrique, J.M., Orkoulas, G., Panagiotopoulos, A.Z., and Fisher, M.E. Coexistence and criticality in size-asymmetric hard-core electrolytes. *Phys. Rev. Lett.*, 2000, **85**, p. 4558–61.

[51] Ebeling, W. Theory of ion-pair formation in electrolytes. *Z. Phys. Chem. (Leipzig)*, 1972, **249**, p. 140–2.

[52] Guillot, B., and Guissani, Y. Towards a theory of coexistence and criticality in real molten salts. *Mol. Phys.*, 1996, **87**, p. 37–86.

[53] Zhou, Y., Yeh, S., and Stell, G. Criticality of charged systems. I. The restricted primitive model. *J. Chem. Phys.*, 1995, **102**, p. 5785–95.

[54] Fröhlich, H. (1958). *Theory of Dielectrics*. Oxford: Oxford University Press.

[55] Panagiotopoulos, A.Z. Monte Carlo methods for phase equilibria of fluids. *J. Phys. Cond. Matter*, 2000, **12**, p. R25–R52.

[56] Panagiotopoulos, A.Z., and Kumar, S.K. Large lattice discretization effects on the phase coexistence of ionic fluids. *Phys. Rev. Lett.*, 1999, **83**, p. 2981–84.

[57] Kim, J.C., and Fisher, M.E. Discretization dependence of criticality in model fluids: A had core electrolyte, *Phys. Rev. Lett.*, **92**, p. 185703/1–4.

[58] Stell, G., Wu, K.C., and Larsen, B. Critical point in a fluid of charged hard spheres. *Phys. Rev. Lett.*, 1976, **37**, p. 1369–72.

[59] Ebeling, W. On the possibility of diffusion instability in non-aqueous weak electrolytes. *Z. Phys. Chem. (Leipzig)*, 1971, **247**, p. 340–2.

[60] Weis, J.J., and Levesque, D. Chain formation in low density dipolar hard spheres: a Monte Carlo study. *Phys. Rev. Lett.*1993, **71**, p. 2729–32.

[61] Camp, P.J., Shelley, J.C., and Patey, G.N. Isotropic fluid phases of dipolar hard spheres. *Phys. Rev. Lett.*, 2000, **84**, p. 115–8.

[62] Bruce, A.D., and Wilding, N.B. Scaling field and universality of the liquid gas critical point. *Phys. Rev. Lett.*, 1992, **68**, p. 193–6.

[63] Orkoulas, G., and Panagiotopoulo, A.Z. Phase behavior of the restricted primitive model and square-well fluids from Monte Carlo simulations in the grand canonical ensemble. *J. Chem. Phys.*, 1999, **110**, p. 1581–90.

[64] Panagiotopoulos, A.Z. Critical parameters of the restricted primitive model. *J. Chem. Phys.*, 2002, **116**, p. 3007–11.

[65] Orkoulas, G., Panagiotopoulos, A.Z., and Fisher, M.E. Criticality and crossover in accessible regimes. *Phys. Rev. E*, 2000, **61**, p. 5930–39.

[66] Caillol, J.-M., Levesque, D., and Weis, J.-J. Critical behavior of the restricted primitive model revisited. *J. Chem. Phys.*, 2002, **116**, p. 10794–800.

[67] Stillinger, F.H., and Lovett, R. Ion pair theory of concentrated electrolytes. I. Basic concepts. *J. Chem. Phys.*, 1968, **48**, p. 3858–68.

[68] Camp, P.J., and Patey, G.N. Ion association and condensation in primitive models of electrolyte solutions. *J. Chem. Phys.*, 1999, **111**, p. 9000–8.

[69] Camp, P.J., and Patey, G.N. Ion association in model ionic fluids. *Phys. Rev. E*, 1999, **60**, p. 1063–66.

[70] Romero-Enrique, J.M., Rull, L.F., and Panagiotopoulos, A.Z. Dipolar origin of the gas-liquid coexistence of the hard-core 1:1 electrolyte model. *Phys. Rev. E*, 2002, **66**, p. 041204/1–10.

[71] Japas, M.L., and Levelt Sengers, J.M.H. Critical Behavior of a Conducting Ionic Solution near its Critical Point. *J. Phys. Chem.*, 1990, **94**, p. 5361–68.

[72] Kleemeier, M., Wiegand, S., Schröer, W., and Weingärtner, H. The liquid-liquid phase transition in ionic solutions: Coexistence curves of tetra-n-butylammonium pricrate in alkyl alcohols. *J. Chem. Phys.*, 1999, **110**, p. 3085–99.

[73] Heintz, A., Lehmann, K.J., and Wertz, C. Thermodynamic properties of mixtures containing Ionic Liquids 3. Liquid-Liquid Equilibria of Binary Mixtures of 1ethyl-3-methylimidazolium Bistrifluoromethylsulfonyl)imide with propanol, butanol, and pentanol. *J. Chem. Eng. Data*, 2003, **48**, p. 472–4.

[74] Marsh, K.N., Deev, A., Wu, C.-T., Tran, E., and Klamt, A. Room temperature Ionic Liquids as relacement for conventional Solvents. *Kor. J. Chem. Eng.*, 2002, **19**, p. 357–62.

[75] Wu, C.-T., Marsh, K.N., Deev, A.V., and Boxall, J.A. Liquid Liquid equilibria of Room temperature ionic Liquids and butanol. *J. Chem. Eng. Data.*, 2003, **48**, p. 486–92.

[76] Arduengo III, A.J., Dias, H.V.R., Harlow, R.L., and Kline, M. Electronic stabilization of nucleophilic carbenes. *J. Am. Chem. Soc.*, 1992, **114**, p. 5530–4.

[77] Hanke, C.G., Price, S.L., and Lynden-Bell, R.M. Intermolecular potentials for simulations of liquid imidazolium salts. *Mol. Phys.*, 2001, **99**, p. 801–9.

[78] Hardacre, C., Holbrey, J.D., JaneMcMath, S.E., Bowro, D.T., and Soper, A.K. Structure of molten 1,3-dimethylimidazolum chloride using neutron diffraction. *J. Chem Phys.*, 2003, **118**, p. 273–8.

[79] Kristof, T., Boda, D., Szalai, I., and Henderson, D. A Gibbs ensemble Monte Carlo study of phase coexistence in the solvent primitive model. *J. Chem. Phys.*, 2000, **113**, p. 7488–91.

[80] Shelley, J.C., and Patey, G.N. Phase behavior of ionic solutions: Comparison of the primitive and explicit solvent models. *J. Chem. Phys.*, 1999, **110**, p. 1633–7.

[81] Fisher, M.E., Ma, S.K., and Nickel, B.G. Critical exponents for long-range interactions. *Phys. Rev. Lett.*, 1972, **29**, p. 917–20.

[82] Luiten, E., and Blöte, H.W.J. Classical critical behavior of spin models with long range interactions. *Phys Rev. B*, 1997, **56**, p. 8945–58.

[83] Camp, P.J., and Patey, G.N. Coexistence and Criticality of fluids with long range potentials. *J. Chem. Phys.*, 2001, **114**, p. 399–408.

[84] Hensel, F. Critical behavior of metallic liquids. *J. Phys. Condens. Matter*, 1990, 2 (**Suppl. A**), p. SA33–SA45.

[85] Luijten, E., Fisher, M.E., and Panagiotopoulos, A.Z. Universality class of criticality in the restricted primitive model electrolyte. *Phys. Rev. Lett.*, 2002, **88**, p. 185701/1–4.

[86] Anisimov, M.A., and Sengers, J.V. (2000). The critical region. *Equations of State for Fluids and Fluid mixtures*, (eds. Sengers, J.V., Kayser, R.F., Peters, C.J., White, H.J.). Amsterdam: Elsevier.

[87] Gutkowski, K., Anisimov, M.A., and Sengers, J.V. Crossover criticality in ionic solutions. *J. Chem. Phys.*, 2001, **114**, p. 3133–48.

[88] Narayanan, T., and Pitzer, K.S. Critical phenomena in ionic fluids: A systematic investigation of the crossover behavior. *J. Chem. Phys.*, 1995, **102**, p. 8118–30.

[89] Chieux, P., and Sienko, M.J. Phase separation and the critical index for liquid-liquid coexistence in the sodium-ammonia system. *J. Chem. Phys.*, 1970, **53**, p. 566–70.

[90] van der Waals, J.D. Thermodynamische Theorie der Kapillarität unter Voraussetzung stetiger Dichteänderung. *Z. Phys. Chemie*, 1894, **13**, p. 657–725

[91] Rowlinson, J.S., and Widom, B. (1982). *Molecular Theory of Capillarity*. Oxford: Clarendon.

[92] Weiss, V.C., and Schröer, W. On the Ginzburg temperature of ionic and dipolar fluids. *J. Chem. Phys.*, 1997, **106**, p. 1930–40.

[93] Stell, G. (1964). Cluster expansion for classical systems. *The equilibrium theory of classical fluids*, (eds. Frisch, H.L, Lebowitz, J.L.), p. 171–267. New York: Benjamin.

[94] Lee, B.P., and Fisher, M.E. Density fluctuations in an electrolyte from generalized Debye-Hueckel theory. *Phys. Rev. Lett.*, 1996, **76**, p. 2906–9.

[95] Schröer, W., and Weiss, V.C. Ginzburg criterium for the crossover behavior of model fluids. *J. Chem. Phys.*, 1998, **109**, p. 8504–13.

[96] Ciach, A., and Stell, G. Criticality and tricriticality in ionic systems. *Physica A*, 2002, **306**, p. 220–9.

[97] Dickman, R.(1999). Unpublished work, cited by Stell, G. New results on some ionic-fluid problems. *New Approaches to Problems in Liquid State Theory*, (eds. Caccamo, C., Hansen, J.-P., Stell, G.), p. 71–89. Dordrecht:NATO ASI Series C, Kluwer.

[98] Kobelev, V., Kolomeisky, A.B., and Fisher, M.E. Lattice models of ionic systems. *J. Chem. Phys.*, 2002, **116**, p. 7589–98.

[99] Kobelev, V., and Kolomeisky, A.B. Anisotropic lattice models of electrolytes. *J. Chem. Phys.*, **117**, p. 8879–85.

[100] Levine, Y., and Fisher, M.E. Coulombic criticality in general dimensions. *Phys. Rev. Lett.*, 1994, **73**, p. 2716–19.

[101] Anisimov, M.A., Gorodetskii, E.E, Kulikov, V.D., and Sengers, J.V. Crossover between vapor-liquid and consolute critical phenomena. *Phys.Rev. E*, 1995, **51**, p. 1199–1215.

[102] Weingärtner, H., Wiegand, S., and Schröer, W. Near-critical light scattering of an ionic fluid with liquid-liquid phase transition. *J. Chem. Phys.*, 1992, **96**, p. 848–51.

[103] Wagner, M., Stanga, O., and Schröer, W. The liquid-liquid coexistence of binary mixtures of the roomtemperature ionic liquid 1-methyl-3-hexylimidazolium tetrafluorid with alcohols. *Phys. Chem. Chem. Phys.*, 2004, **6**, p. 4421–31.

[104] Kumar, A., Krishnamurthy, H.R., and Gopal, E.S.R. Equilibrium critical phenomena in binary liquid mixtures. *Phys. Reports*, 1983, **98**, p. 57–143.

[105] Bailey, A.E., and Cannell, D.S Practical method for calculating of multiple light scattering. *Phys. Rev. E*, 1994, **50**, p. 4853–64.

[106] Aberle, L.B., Hülstede, P., Wiegand, S., Schröer, W., and Staude, W. Effective suppression of multiply scattered Light in static and dynamic lightscattering. *Applied Optics*, 1998, **37** , p. 6511–25.

[107] Schröder, J.M., Wiegand, S., Aberle, L.B., Kleemeier, M., and Schröer, W. Experimental determination of singly scattered light close to the critical point in a polystyrene/Cyclohexane mixture. *Phys. Chem. Chem. Phys.*, 1999, **1**, p. 3287–92.

[108] Kholodenko, A.L., and Beyerlein, A.L. Comment on "Near-critical coexistence curve and critical exponent of an ionic fluid". *J. Chem. Phys.*, 1990, **93**, p. 8405.

[109] Kleemeier, M. (1999) *Untersuchungen zum kritischen Verhalten des Flüssig-Flüssig Phasenübergangs in ionischen Lösungen*, Ph. D. Thesis. University of Bremen.

[110] Schröer, W., Kleemeier, M., Plikat, M., Weiss, V., and Wiegand, S. Critical behavior of ionic solutions in non-polar solvents with a liquid liquid phase transition. *J. Phys.: Condensed Matter*, 1996, **8**, p. 9321–7.

[111] Wagner, M. (2004) *Untersuchung der Flüssig-flüssig Phasenübergänge von Lösungen flüssiger Salze mit Imidazolium Kationen*, Ph. D. Thesis. University of Bremen.

[112] Heimburg, T., Mirzaev, S.Z., and Kaatze, U. Heat capacity behavior in the critical region of the ionic binary mixture ethylammonium nitrate – n-octanol. *Phys. Rev. E*, 2000, **62**, p. 4963–76.

[113] Oleinikova, A., and Bonetti, M. Electrical conductivity of highly concentrated electrolytes near the critical solute point: A study of tetra-n-butylammonium picrate in alcohols of moderate dielectric constant. *J. Chem. Phys.*, 2001, **115**, p. 9871–82.

[114] Kleemeier, M., Wiegand, S., Derr, T., Weiss, V., Schröer, W., and Weingärtner, H. Critical viscosity and Ising-to-mean-field crossover near the upper consolute point of an ionic solution. *Ber. Bunsenges. Phys. Chem.*, 1996, **100**, p. 27–32.

[115] Wiegand, S., Berg, R.F., and Levelt Sengers, J.M.H. Critical viscosity of the ionic mixture triethyl n-hexyl ammonium triethyl n-hexyl borate in diphenyl ether. *J. Chem. Phys.*, 1998, **109**, p. 4533–45.

[116] Wagner, M., Stanga, O., and Schröer, W. Critical viscosity near the liquid-liquid phase-transition in the solution of the ionic liquid 1-methyl-3-hexylimidazolium tetrafluoroborate in 1-pentanol. *Phys. Chem. Chem. Phys.*, 2004, **6**, p. 1750–7.

[117] Anisimov, M.A., Jacob, J., Kumar, A., Agayan, V.A., and Sengers, J.V. Novel phase-transition behavior in an aqueous electrolyte solution. *Phys.Rev.Lett.*, 2000, **85**, p. 2336–9.

[118] Wagner, M., Stanga, O., and Schröer, W. Tricriticality in the ternary system 3-methylpyridine + water + NaBr? Measurements of the viscosity. *Phys. Chem. Chem. Phys.*, 2002, **4**, p. 5300–06.

[119] Wagner, M., Stanga, O., and Schröer, W. Tricriticality in the ternary system 3-methylpyridine + water + NaBr? The coexistence curves. *Phys. Chem. Chem. Phys.*, 2003, **5**, p. 1225–34.

[120] Gutkowski, K.I., Bianchi, H.L., and Japas, M.L. Critical Behavior of a ternary ionic system: A controversy. *J. Chem. Phys.*, 2003, **118**, p. 2808–14.

[121] Wagner, M., Stanga, O., and Schröer, W. Tricriticality in the ternary system 3-methylpyridine + water + NaBr? The light scattering intensity. *Phys. Chem. Chem. Phys.*, 2004, **6**, p. 580–9.

[122] Kostko, A.F., Anisimov, M.A., and Sengers, J.V. Criticality in aqueous solutions of 3-methylpyridine and sodium bromide. *Phys. Rev. E*, 2004, **70**, p. 026118/1–11.

LIQUID-VAPOR CRITICALITY IN COULOMBIC AND RELATED FLUIDS

What can be learned from computer simulations?

P.J. Camp [1], C.D. Daub [2], G.N. Patey [2]

[1] *School of Chemistry, University of Edinburgh,*
West Mains Road, Edinburgh EH9 3JJ, United Kingdom.

[2] *Department of Chemistry, University of British Columbia,*
Vancouver, BC, Canada, V6T 1Z1.

Abstract   The liquid/vapor criticality of Coulombic and other fluids characterized by long-range interactions is investigated using canonical and grand-canonical Monte Carlo computer simulations. It is shown that while mixed-field finite-size scaling methods appear consistent with the theoretically expected universality class (Ising or classical depending on the potential), the constant-volume heat capacity shows a very strong dependence on the particular ensemble employed. Furthermore, for Coulombic and related systems, the heat capacities do not show the expected Ising-like behavior for the largest systems that have been simulated. The ensemble dependence of the constant-volume heat capacity and its still puzzling behavior will be discussed in some detail.

Keywords: Criticality, ionic fluids, long-range interactions, computer simulations

1.   Introduction

Over the last decade considerable effort has been directed towards characterizing and understanding the behavior of ionic fluids close to the critical point of the liquid/vapor transition. The motivation for this activity was the experimental observation that some ionic fluids seemed to exhibit classical critical exponents, rather than the more usual short-range or Ising exponents that characterize almost all three-dimensional liquid/vapor critical points [1, 2]. Renormalization group techniques [3] and effective Hamiltonian approaches [4] put the Ising-like criticality of simple fluids almost beyond doubt. Unfortunately, the direct application of such methods to ionic fluids is precluded by the lack of a reliable effective Hamiltonian, although recent developments using collective variables look promising [5, 6]. Although the bare Coulomb interaction is long

D. Henderson et al. (eds.), Ionic Soft Matter: Modern Trends in Theory and Applications, 181–197.
© 2005 *Springer. Printed in the Netherlands.*

ranged (in the sense that it decays slower than $r^{-4.5}$ in three dimensions and hence should give rise to classical criticality [7–10]) screening can render the effective interactions between ions short-ranged. Strong ion pairing – which is particularly significant in the vapor phase at and below the critical temperature – clearly has an impact on the screening length in the near-critical fluid. It is clear that a reliable effective Hamiltonian should provide a good account of the complex correlations and screening effects in ionic fluids at low temperature.

In the absence of a reliable theory, computer simulations have become the most important means of tackling the question of ionic criticality. In the last ten years there have been numerous attempts to identify the universality class of what is perhaps the most basic model of ionic fluids, the restricted primitive model (RPM). The RPM consists of an equimolar mixture of positively and negatively charged hard spheres with diameter σ, immersed in a dielectric continuum with dielectric constant D. The pair potential is,

$$u_{\mathrm{RPM}}(r) = \begin{cases} \infty, & r < \sigma, \\ q_1 q_2 / Dr, & r \geqslant \sigma, \end{cases} \tag{1}$$

where $q_i = \pm q$ is the charge on ion i, and r is the interionic separation. For this system we define the reduced temperature by $T^* = k_{\mathrm{B}} T D \sigma / q^2$ and the reduced ion number density by $\rho^* = \rho \sigma^3$. Over the last ten years numerous simulation studies have been devoted to locating and characterizing the liquid/vapor transition in the RPM [11–27]. The liquid/vapor coexistence envelope of the RPM in the density-temperature ($\rho^* - T^*$) plane is sketched in Fig. 1; for the purposes of illustration, the coexistence data are taken from simulations by Romero-Enrique *et al.* of the "discretized" RPM with discretization parameter $\zeta = 10$ and conducting boundary conditions [28] for which $T_{\mathrm{c}}^* = 0.0495(2)$ and $\rho_{\mathrm{c}}^* = 0.079(5)$.

The main features of the phase diagram are that the critical temperature and critical number density are both very low when compared to those for "normal" critical points, and that the vapor phase contains a high proportion of strongly associated ion pairs. These properties give rise to problems for the simulator because of pronounced finite-size effects in the very dilute vapor phase, and the difficulty in sampling the phase space efficiently when ions are strongly associated. The critical point has been located by various authors, mostly by mixed-field finite-size scaling (MFFSS) analyses [30] of data from grand-canonical ($\mu V T$) Monte Carlo simulations. Perhaps the most reliable current estimates have been provided by Caillol *et al.* who carried out simulations with system sizes up to $L = 34\sigma$ [15], the largest to date:

$$T_{\mathrm{c}}^* = 0.04917 \pm 0.00002, \tag{2}$$
$$\rho_{\mathrm{c}}^* = 0.080 \pm 0.005. \tag{3}$$

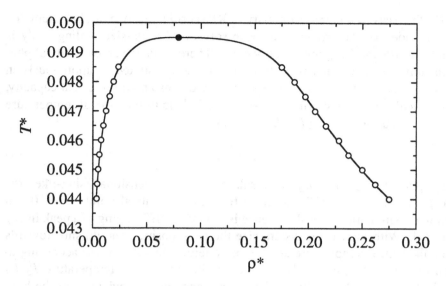

Figure 1. Liquid/vapor coexistence envelope for the RPM. The coexistence data (open circles) and critical point (filled circle) were obtained for the "discretized" RPM by Romero-Enrique et al. [28]. The solid line is the result of separate fits to the vapor (−) and liquid (+) branches with functions of the form $\rho^*_\pm = \rho^*_c + a_\pm (T^*_c - T^*)^\beta + b_\pm (T^*_c - T^*) + c_\pm (T^*_c - T^*)^2$, where $\beta = 0.326$ is the Ising order-parameter exponent [29], $T^*_c = 0.0495$, and $\rho^*_c = 0.079$ [28]. The fit parameters are $a_- = -0.5846(95)$, $b_- = 6.76(79)$, $c_- = -189(96)$, $a_+ = 0.982(22)$, $b_+ = -11.2(1.8)$, $c_+ = -2570(220)$.

The MFFSS analysis as formulated by Bruce and Wilding [30] requires that the universality class is known *a priori*, but one can at least check that the measured finite-size scaling is consistent with the initial premise. The Ising universality class is well characterized, and since almost all three-dimensional liquid/vapor critical points belong to this class, it is reasonable to work on the assumption that the RPM will be no different. All MFFSS simulation studies of the RPM have assumed Ising criticality, and such analyses have shown that RPM criticality is consistent with this assumption inasmuch as the finite-size scaling function of the order parameter can be superimposed on that for the Ising model. In addition, exponent ratios such as β/ν can be extracted from the MFFSS analysis, and these are in agreement with those for the Ising model.

Recently, simulation methods that do not require any prior information regarding the universality class have been devised and applied to the RPM [19, 20, 22]. These approaches furnish estimates of the critical parameters and critical exponents free of any assumptions; the most recent application to the RPM yields $T^*_c = 0.04933 \pm 0.00005$ and $\rho^*_c = 0.075 \pm 0.001$ [22]. In all cases the apparent critical exponents – and other quantities such as the critical Binder ratio – exclude universality classes that are "nearby" the Ising class.

On the basis of this brief summary of RPM criticality, one might be tempted to conclude that the problem has been solved; all finite-size scaling analysis point towards the Ising universality class. There is, however, one critical phenomenon which does not seem to have been demonstrated unambiguously in the RPM. This is the critical divergence of the constant-volume heat capacity, C_V. Recall that on the critical isochore and close to the critical temperature where the parameter $t \equiv (T - T_c)/T_c$ is small,

$$C_V \sim |t|^{-\alpha}, \tag{4}$$

the exponent $\alpha \geqslant 0$ being universal. For three-dimensional Ising-like critical points $\alpha \simeq 0.11$ [29], whereas for classical critical points $\alpha = 0$. In finite-size simulations this divergence is "rounded off" leading to a peak in C_V [31, 32]. With increasing system size the peak position should shift towards the bulk critical temperature and the peak height should increase according to well-known scaling laws. The apparent finite-size critical temperature $T_c(L)$ is identified with the position of the peak, and should deviate from the bulk critical temperature to leading order by $T_c(\infty) - T_c(L) \sim L^{-1/\nu}$ where ν is the appropriate correlation-length exponent (for Ising critical points in three dimensions $\nu = 0.630$ [29]); the form of the corrections to scaling depend on the particular ensemble (see, e.g., [33]). In addition, the height of the peak in C_V at $T = T_c(L)$ scales like $C_V \sim L^{\alpha/\nu}$. Therefore, in principle, an Ising-like critical point might be distinguished from a classical one by the presence of a system-size dependent "divergent" peak in C_V.

Although by most measures the RPM belongs to the Ising class, studies of its heat capacity have yielded ambiguous results. On the one hand, Valleau and Torrie's simulations in the canonical (NVT) ensemble suggest that there is no peak at all [25, 26]. On the other hand, the μVT simulations by Luijten et al. give evidence for a very pronounced finite-size dependent peak [21]. Caillol et al. did find C_V peaks in their μVT simulations of systems with linear dimensions $L \leqslant 34\sigma$: with increasing system size the peak positions shift towards the bulk critical temperature and the peak widths decrease as expected, but the peak heights do not scale in any known way [15]. In Sec. 2 we shall show that the discrepancy between C_V obtained by Valleau and Torrie [25, 26] and by Luijten et al. [21] is due to an extreme ensemble dependence near criticality [34]. In Sec. 3 we investigate the effect of ion pairing on the apparent criticality of ionic fluids by studying a charged hard dumbbell (CHD) model in which cation-anion pairs are fused together, thus removing the opportunity for conventional ion screening [35]. We have examined the ensemble dependence of C_V in CHDs, and have carried out a MFFSS analysis of the critical point which is consistent with Ising criticality [35].

To assess properly the significance of these results, and to test simulation methods, a benchmark fluid system is required which can exhibit classical or non-classical criticality, depending on the parameters. To this end, we have examined the liquid/vapor criticality in a fluid of hard spheres with algebraically decaying attractive interactions; we will refer to this system as "attractive hard spheres" (AHSs). The pair potential is,

$$u_{\text{AHS}}(r) = \begin{cases} \infty, & r < \sigma, \\ -\epsilon(\sigma/r)^a, & r \geqslant \sigma, \end{cases} \tag{5}$$

where ϵ is the attractive well depth, and a controls the range of the interactions and must be greater than the spatial dimension, d, to ensure that the thermodynamic limit can be taken. For this system we define the reduced temperature by $T^* = k_{\text{B}}T/\epsilon$ and the reduced number density by $\rho^* = \rho\sigma^3$. Renormalization group calculations and simulations indicate that attractive interactions of this type give rise to classical criticality when $d < a < 3d/2$, short-range Ising-like criticality when $a > d + 2 - \eta_{\text{SR}}$ where η_{SR} is the short-range correlation function exponent ($\eta_{\text{SR}} = 0.0335 \pm 0.0025$ in $d = 3$ [29]), and exponents that interpolate between classical and short-range values when $3d/2 < a < d + 2 - \eta_{\text{SR}}$ [7, 36, 8]. In three dimensions, the "long-range" (classical) regime corresponds to $3 < a < 4.5$, and the "short-range" regime to $a > 4.97$. In Sec. 4 we report the results of a MFFSS analysis, and an investigation of C_{V}, for systems with $a = 3.1$ (long-range), $a = 4$ (long-range), and $a = 6$ (short-range) [37, 34, 38].

The main aim of this contribution is to demonstrate the severe ensemble dependence of near-critical thermodynamic properties of the RPM and AHS fluid, and to examine the effects of ion pairing – or rather the lack thereof – on the criticality in ionic fluids. We will deal with these topics in turn, and then present a summary and discussion in Sec. 5.

2. Restricted primitive model

We wish here to summarize results for the constant-volume heat capacity, C_{V}, of the RPM. We have performed NVT and μVT MC simulations on systems confined to cubic cells of side L. The long-range Coulombic interactions were handled using Ewald sums with conducting boundary conditions [39]. C_{V} was computed using the appropriate fluctuation formula [40],

$$C_{\text{V}} = \frac{1}{k_{\text{B}}T^2} \left[X(U, U) - \frac{X(N, U)^2}{X(N, N)} \right], \tag{6}$$

where $X(A, B) = \langle (A - \langle A \rangle)(B - \langle B \rangle) \rangle$, and in the NVT simulations the second term in square brackets is omitted. In Fig. 2 we show C_{V} as a function of $t = (T - T_c)/T_c$ along a near-critical isochore $\rho^* = 0.068$ as obtained from NVT and μVT MC simulations of systems with $L = 10\sigma$. In the μVT simulations, the chemical potential was tuned to give an average density $\langle\rho^*\rangle = 0.068$. For this density and system size – which contains, on average, only $N = 68$ ions – Luijten *et al.* observed a very strong peak in C_V as obtained from μVT MC simulations [21]. Valleau and Torrie, on the other hand, find no peak in their NVT simulations of $N = 192$ ions at a density $\rho^* = 0.086$ [25, 26]. Returning to Fig. 2, at reduced temperatures $t > 0.2$ the NVT and μVT are essentially identical, whereas at $t < 0.2$ the two ensembles give very different results. The μVT results are suggestive of a critical "divergence" characteristic of non-classical criticality, whereas the NVT results show no sign of a peak. We have checked that our μVT results agree with those of Luijten *et al.* [21] and that NVT simulations at $\rho^* = 0.086$ give identical results to those of Valleau and Torrie [25, 26]. Since the top of the liquid/vapor coexistence curve is so "flat" (see Fig. 1) the differences between the results of Valleau and Torrie, and those of Luijten *et al.* are unlikely to be due to the difference between the (average) densities considered. Hence, the pronounced discrepancy between the two sets of results is due to an ensemble-dependence of near-critical thermodynamic properties. We will return to a discussion of these effects in Sec. 5.

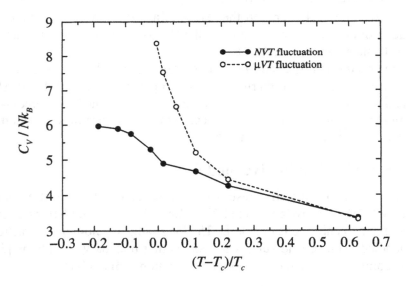

Figure 2. Constant-volume heat capacity, C_V, for the RPM as obtained from NVT (filled circles and solid line) and μVT (open circles and dashed line) MC simulations. The kinetic contribution (equal to $3Nk_B/2$) is included.

3.　Charged hard dumbbell fluid

We now consider the CHD fluid, which comprises an equal number of hard-sphere cations and anions – as in the RPM – but with cation-anion pairs fused together at contact to form dumbbells. This model was first studied in the light of observations that the vapor phase of the RPM consisted almost entirely of such ion pairs. The thermodynamics of the liquid phase is likely to be fairly insensitive to the presence of such ion pairing, and so the liquid/vapor coexistence envelope of the CHD fluid is expected to be similar to that of the RPM. Preliminary μVT MC simulations showed that the coexistence envelope of the CHD is indeed similar to that of the RPM [41]. We have examined the location of the critical point in more detail using the MFFSS technique [30] under the assumption of Ising criticality. In the case of the CHD fluid this assumption is justified because the asymptotic interactions between dumbbells are dipolar; the asymptotic angle-averaged attractions between the CHDs are therefore short-ranged ($\sim r^{-6}$) [42]. In any case the critical exponents of ferromagnets with long-range dipolar interactions are very similar to the short-range exponents [43, 44], and even in experiments it is difficult to distinguish between the two cases. Our simulations of the CHD fluid were carried out using cubic simulation cells of side L, and Ewald summations with conducting boundary conditions to treat the long-range Coulombic interactions. Using the Bruce and Wilding MFFSS analysis we have determined the critical parameters as,

$$T_c^* = 0.04911 \pm 0.00003, \tag{7}$$

$$\rho_c^* = 0.101 \pm 0.003, \tag{8}$$

where T^* is defined in exactly the same way as for the RPM, and ρ_c^* is the density of constituent ions (not the density of dumbbells) to facilitate comparisons with the RPM. The critical temperature is only about 0.1% lower than that for the RPM, while the critical density is about 25% higher than the corresponding RPM value. Clearly, fusing ions together does not affect the critical temperature very significantly; for a detailed discussion see [45]. With regard to the apparent universality class of the CHD fluid, the measured order parameter distribution at criticality does not deviate significantly from the universal Ising function. We therefore conclude that on the basis of the MFFSS analysis, deviations from Ising criticality in the CHD fluid are no more or less than those in the RPM.

On the basis that the asymptotic interactions between CHDs are dipolar, and that polar fluids are expected to exhibit short-range criticality [42], one might expect C_V along the critical isochore to exhibit a strong peak in the vicinity of T_c. In Fig. 3 we show C_V along the critical isochore from NVT

simulations of a system with $L = 17\sigma$ in which there are $N = 496$ ions. Two sets of results are shown, those being C_V computed from the fluctuation formula in Eq. (6), and the temperature derivative of a [5,5] Pade approximant in the variable $x = \sqrt{T^*}$ fitted to the reduced energy per ion, $UD\sigma/Nq^2$:

$$\frac{UD\sigma}{Nq^2} = \frac{\sum_{k=0}^{5} a_k x^k}{1 + \sum_{l=1}^{5} b_l x^l}. \tag{9}$$

There is good agreement between the results for C_V obtained from these two routes, but there is no sign of a strong near-critical peak. These results are consistent with the NVT results for the RPM summarized in Sec. 2.

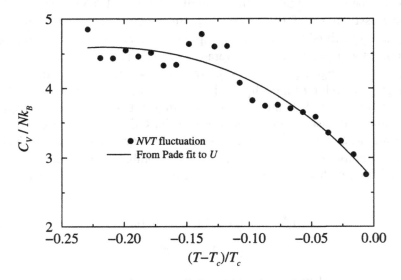

Figure 3.   Constant-volume heat capacity, C_V, for the CHD fluid on the critical isochore as obtained from NVT MC simulations: fluctuation formula (6) (points); from a [5,5] Pade approximant (9) fitted to the energy (solid line). The kinetic contribution is not included.

4.   Hard spheres with algebraic attractions

The results outlined above for the RPM and the CHD fluid clearly show a discrepancy between MFFSS analyses (which are consistent with Ising-like criticality) and calculations of C_V in NVT MC simulations. To investigate this discrepancy, we have studied a class of fluids defined by the potential in Eq. (5) with $a = 6$ (which should give rise to Ising-like criticality since $a > 4.7$) and $a = 4$ and $a = 3.1$ (which are classical since $a < 4.5$). All simulations reported in this section were carried out in cubic simulation cells of side L. The potentials were truncated at $L/2$, and a long-range correction was calculated by assuming that the pair correlation function $g(r) = 1$ for $r > L/2$ [39]. The liquid/vapor phase diagrams of the AHS systems are presented in Fig. 4 as deduced from Gibbs ensemble Monte Carlo (GEMC) calculations and MFFSS analyses assuming *Ising* criticality [37]. The MFFSS analysis of the $a = 6$ system appears entirely consistent with the assumption of Ising criticality; for instance, the results suggest that the exponent ratio $\beta/\nu = 0.54(1)$, to be compared with the Ising value $\beta/\nu = 0.517$. Clearly the assumption is not justified for the potentials with $a = 4$ and $a = 3.1$, and indeed the MFFSS results for these potentials show significant inconsistencies [37]. We interpret these deviations as a sign of classical criticality, although other explanations have been put forward [46].

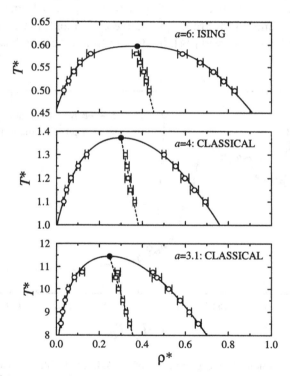

Figure 4. Phase diagrams for the AHS fluid with $a = 6$ (top), $a = 4$ (center), and $a = 3.1$ (bottom): coexistence densities from GEMC simulations (open circles); critical parameters from MFFSS analyses assuming Ising criticality (filled circles); fits of the form $\rho_{\pm}^* = \rho_c^* + A(T_c^* - T^*) \pm B(T_c^* - T^*)^{\beta_{\text{eff}}}$ (solid lines). For $a = 6$ the effective order parameter exponent is $\beta_{\text{eff}} = 0.36$, for $a = 4$ $\beta_{\text{eff}} = 0.46$, and for $a = 3.1$ $\beta_{\text{eff}} = 0.46$.

We now turn to measurements of the constant-volume heat capacity, C_V, along the critical isochore. In Fig. 5 we show C_V as measured in NVT and μVT simulations of a system with $a = 6$ and $L = 10\sigma$. The bulk critical parameters for the $a = 6$ system are $T^* = 0.5972(1)$ and $\rho^* = 0.3757(4)$ [37]. This system is expected to exhibit Ising criticality. Both sets of results were obtained using appropriate fluctuation Eq. (6), and exhibit features near T_c. While the μVT results show a strong "divergence" almost at T_c, the NVT peak is much more rounded and shifted to lower temperature. Hence, finite-size corrections appear to be more pronounced in the NVT results than in the μVT results. Very recently we have investigated the finite-size behavior of C_V obtained from NVT simulations of near-critical $a = 6$ systems with linear dimensions $L \leqslant 22\sigma$, and we find that the scalings of the peak position and peak height with system size are completely consistent with non-classical criticality [38]. The results of MFFSS analysis and the measurements of C_V in NVT MC simulations are consistent with Ising criticality [37, 38]; measurements of C_V in μVT MC simulations are in qualitative agreement, inasmuch as a near-critical peak has been observed [34], but the finite-size scaling has not yet been examined.

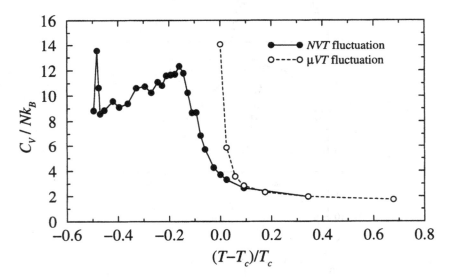

Figure 5. Constant-volume heat capacity, C_V, of the AHS fluid with $a = 6$ as obtained from NVT (filled circles and solid line) and μVT (open circles and dashed line) MC simulations of systems with $L = 10\sigma$ along the bulk critical isochore. The kinetic contribution (equal to $3Nk_B/2$) is included.

What of the classical critical points? The classical order parameter distribution function has not yet been determined, and so we have no choice but to employ the Ising distribution in the MFFSS analysis. We therefore have only very rough estimates of the critical parameters from MFFSS analyses of the systems with $a = 4$ [$T_c^* = 1.3724(1)$, $\rho_c^* = 0.2993(1)$] and $a = 3.1$

$[T_c^* = 11.452(8), \rho_c^* = 0.247(5)]$ [37]. Nonetheless, measurements of C_V along the approximate critical isochores should not show any strong critical "divergences", since the specific heat exponent $\alpha = 0$ for classical critical points. In Figs. 6 and 7 we show C_V along the critical isochores obtained from NVT and μVT simulations of systems with $a = 4$ and $a = 3.1$, respectively, and with linear dimension $L = 10\sigma$. For $a = 4$ (Fig. 6) the small peak in the NVT results at $(T - T_c)/T_c \simeq -0.4$ is very close to the triple point temperature $T_t^* = 0.830$ or $(T_t - T_c)/T_c = -0.40$ [47]. Hence, this feature is not connected with liquid/vapor criticality. The lack of a near-critical peak is not a reliable indicator of classical criticality, however, since the NVT simulations yield similar results for systems which otherwise appear firmly in the Ising universality class. Unfortunately, the μVT results also show the "wrong" behavior; as shown in Fig. 6 there is a very sharp "divergence" in C_V which is a characteristic of Ising critical points.

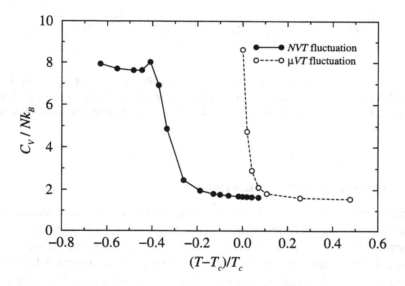

Figure 6. Constant-volume heat capacity, C_V, of the AHS fluid with $a = 4$ as obtained from NVT (filled circles and solid line) and μVT (open circles and dashed line) MC simulations of systems with $L = 10\sigma$ along the bulk critical isochore. The kinetic contribution (equal to $3Nk_B/2$) is included.

A very similar situation holds for the system with $a = 3.1$. The NVT results are completely featureless. This can be explained by the fact that the interactions are so long ranged that local density fluctuations within the simulation cell do not give rise to any appreciable energy fluctuations. Since the overall number density is held constant, energy fluctuations are almost completely suppressed. The μVT results, on the other hand, show a very sharp

Ising-like "divergence" at T_c; remember that the liquid/vapor criticality in this system is classical [7, 8].

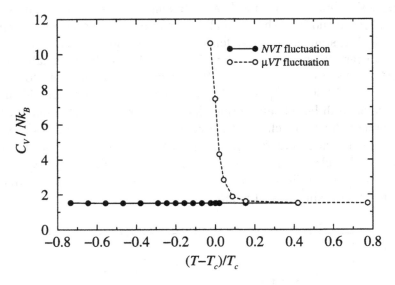

Figure 7. Constant-volume heat capacity, C_V, of the AHS fluid with $a = 3.1$ as obtained from NVT (filled circles and solid line) and μVT (open circles and dashed line) MC simulations of systems with $L = 10\sigma$ along the bulk critical isochore. The kinetic contribution (equal to $3Nk_B/2$) is included.

5.   Summary and discussion

In this contribution we have reviewed the results from recent simulations directed towards understanding the manifestation of critical phenomena and finite-size scaling in systems with long-range interactions, including the restricted primitive model of ionic fluids, the charged hard dumbbell fluid, and hard spheres with algebraic interactions. The main conclusions from our studies are as follows.

- Mixed-field finite-size scaling studies of the RPM point towards Ising criticality. C_V along the critical isochore should therefore exhibit a strong peak near the critical temperature: when calculated using μVT MC simulations a strong peak is present; when calculated using NVT MC simulations no peak is observed.

- Mixed-field finite-size scaling studies of the CHD fluid – in which cation-anion pairs are fused together – are consistent with Ising criticality. When calculated in NVT MC simulations, C_V along the critical isochore shows no sign of a near-critical peak.

- Mixed-field finite-size scaling studies of the AHS fluid with $a = 6$ are consistent with Ising criticality. Finite-size scaling analyses of the peak position and peak height in C_V obtained from NVT MC simulations are also consistent with Ising criticality.

- Mixed-field finite-size scaling studies of the AHS fluid with long-range parameters ($a = 4$ and $a = 3.1$) show significant deviations from Ising criticality. Although these long-range systems possess classical critical points and should therefore exhibit mere cusps in C_V, these functions as measured in μVT MC simulations show strong near-critical "divergences". NVT MC simulations show no such features, but this cannot be taken as a reliable indicator of classical criticality.

The overall conclusion from these studies is that the heat capacity, as measured in finite-size simulations, is strongly dependent on the particular statistical mechanical ensemble employed, regardless of the universality class. In the case of NVT simulations the near-critical peak in C_V is severely suppressed in systems with Coulombic and other long-range interactions. This is easy to understand since the constraints of constant N and V impose severe limitations on the density fluctuations which contribute to the energy fluctuations and other critical phenomena. In systems with long-range interactions this leads to pathological results since only local density fluctuations can be accommodated in small simulation cells, and these cannot have an impact on the measured energy fluctuations and heat capacity. With sufficiently short-range interactions, however, recent work [38] indicates that Ising-like finite-size scaling can be observed in NVT simulations.

So what of the μVT simulations? On the one hand they indicate Ising-like "divergences" in C_V for systems which are believed to belong to the Ising universality class, including the RPM, the CHD fluid, and the AHS fluid with $a = 6$. On the other hand, they also show strong "divergences" for *classical* critical points, such as those in the AHS fluid with $a = 4$ and $a = 3.1$. The μVT ensemble must be the best ensemble with which to study phase transitions since it can accommodate the density fluctuations (and compositional fluctuations in multicomponent fluids) that control critical phenomena. There is a catch when applied to finite-size simulations, however. One possible explanation of the anomalous results for classical critical points involves the fluctuation formula in Eq. (6). At criticality, $X(U, U)$, $X(N, U)$, and $X(N, N)$ must all diverge so that C_V diverges according to Eq. (4). For instance, $X(N, N)$ is related to the isothermal compressibility, χ_T, through the relation [48],

$$X(N, N) = \frac{\langle N^2 \rangle k_B T \chi_T}{V} . \tag{10}$$

χ_T itself is known to diverge along the critical isochore like $|T - T_c|^{-\gamma}$, where the exponent $\gamma = 1$ for classical critical points, and $\gamma = 1.24$ for Ising-like critical points in three dimensions [29]. Notice that χ_T diverges strongly regardless of whether the criticality is classical or not. The evaluation of C_V in μVT simulations therefore involves computing a ratio of two large numbers in the numerator and denominator of the second term in Eq. (6). Moreover, C_V itself is given by the difference between two large numbers. The extents to which $X(U, U)$, $X(N, U)$, and $X(N, N)$ are individually suppressed in finite-size simulations is unknown.

Some further work is required. (1) The ensemble dependence of near-critical fluctuations in finite-size systems needs to be carefully assessed. This might best be achieved by starting with an Ising model, of which simulations that allow the magnetization, M, to fluctuate yield "μVT" estimates, while simulations with $M = 0$ yield "NVT" estimates. (2) The suppression of density fluctuations, and hence energy fluctuations, in NVT simulations might be quantified in relation to the corresponding μVT simulations. As far as we are aware, all prior work directed towards this type of problem has excluded critical points and phase transitions [49]. (3) Unbiased simulation methods [19, 20, 22] should be applied to classical critical points, which amongst other things would allow the classical order parameter distribution function to be measured [50]. (4) The simulations of the AHS fluid reported here involved truncating the potential at $L/2$, and adding an approximate long-range correction. With this procedure, systems with different linear dimensions possess different potentials. A re-examination of the classical critical points should incorporate an appropriate treatment of the long-range interactions. Although Ewald sums may give rise to pathologies of their own, this might be the best available method. We are currently tackling some of these issues.

Acknowledgements

We are grateful to Professors M. E. Fisher, G. M. Torrie, and J. P. Valleau for numerous discussions and correspondence during the course of the investigations reported in this contribution. The financial support of the National Science and Engineering Research Council of Canada is gratefully acknowledged. Part of this research has been enabled by the use of WestGrid computing resources, which are funded in part by the Canada Foundation for Innovation, Alberta Innovation and Science, BC Advanced Education, and the participating research institutions. WestGrid equipment is provided by IBM, Hewlett Packard and SGI.

References

[1] Fisher, M.E. The nature of criticality in ionic fluids. *J. Phys.: Condens. Matter*, 1996, **8**, p. 9103–9109.

[2] Stell, G. Phase separation in ionic fluids. *J. Phys.: Condens. Matter*, 1996, **8**, p. 9329–9333.

[3] Fisher, M.E. The renormalization group in the theory of critical behavior. *Rev. Mod. Phys.*, 1974, **46**, p. 597–616.

[4] Hubbard, J. and Schofield, P. Wilson theory of a liquid-vapour critical point. *Phys. Lett.*, 1972, **40A**, p. 245–246.

[5] Patsahan, O.V. Ginzburg-landau-wilson hamiltonian for a multi-component continuous system: a microscopic description. *Condens. Matter Phys.*, 2002, **5**, p. 413–428.

[6] Patsahan, O.V. and Mryglod, I.M. Critical behaviour of the restricted primitive model. *J. Phys.: Condens. Matter*, 2004, **16**, p. L235–L241.

[7] Fisher, M.E., Ma, S., and Nickel, B.G. Critical exponents for long-range interactions. *Phys. Rev. Lett.*, 1972, **29**, p. 917–920.

[8] Sak, J. Recursion relations and fixed points for ferromagnets with long-range interactions. *Phys. Rev. B*, 1973, **8**, p. 281–285.

[9] Stell, G. Scaling theory of the critical region for systems with long-range forces. *Phys. Rev. B*, 1972, **5**, p. 981–985.

[10] Stell, G. Value of η for long-range interactions. *Phys. Rev. B*, 1973, **8**, p. 1271–1273.

[11] Caillol, J.M. A monte carlo study of the liquid-vapour coexistence of charged hard spheres. *J. Chem. Phys.*, 1994, **100**, p. 2161–2169.

[12] Caillol, J.M. A Monte-Carlo study of the dielectric constant of the restricted primitive model of electrolytes on the vapor branch of the coexistence line. *J. Chem. Phys.*, 1995, **102**, p. 5471–5479.

[13] Caillol, J.M., Levesque, D., and Weis, J.J. Critical behavior of the restricted primitive model. *Phys. Rev. Lett.*, 1996, **77**, p. 4039–4042.

[14] Caillol, J.M., Levesque, D., and Weis, J.J. A monte carlo finite size scaling study of charged hard-sphere criticality. *J. Chem. Phys.*, 1997, **107**, p. 1565–1575.

[15] Caillol, J.M., Levesque, D., and Weis, J.J. Critical behavior of the restricted primitive model revisited. *J. Chem. Phys.*, 2002, **116**, p. 10794–10800.

[16] Caillol, J.M. and Weis, J.J. Free energy and cluster structure in the coexistence region of the restricted primitive model. *J. Chem. Phys.*, 1995, **102**, p. 7610–7621.

[17] Camp, P.J. and Patey, G.N. Ion association and condensation in primitive models of electrolyte solutions. *J. Chem. Phys.*, 1999, **111**, p. 9000–9008.

[18] Camp, P.J. and Patey, G.N. Ion association in model ionic fluids. *Phys. Rev. E*, 1999, **60**, p. 1063–1066.

[19] Kim, Y.C. and Fisher, M.E. Discretization dependence of criticality in model fluids: A hard-core electrolyte. *Phys. Rev. Lett.*, 2004, **92**, p. 185703.

[20] Kim, Y.C., Fisher, M.E., and Luijten, E. Precise simulation of near-critical fluid coexistence. *Phys. Rev. Lett.*, 2003, **91**, p. 065701.

[21] Luijten, E., Fisher, M.E., and Panagiotopoulos, A.Z. The heat capacity of the restricted primitive model electrolyte. *J. Chem. Phys.*, 2001, **114**, p. 5468–5471.

[22] Luijten, E., Fisher, M.E., and Panagiotopoulos, A.Z. Universality class of criticality in the restricted primitive model electrolyte. *Phys. Rev. Lett.*, 2002, **88**, p. 185701.

[23] Orkoulas, G. and Panagiotopoulos, A.Z. Free energy and phase equilibria for the restricted primitive model of ionic fluids from monte carlo simulations. *J. Chem. Phys.*, 1994, **101**, p. 1452–1459.

[24] Orkoulas, G. and Panagiotopoulos, A.Z. Phase behavior of the restricted primitive model and square-well fluids from monte carlo simulations in the grand canonical ensemble. *J. Chem. Phys.*, 1999, **110**, p. 1581–1590.

[25] Valleau, J. and Torrie, G. Heat capacity of the restricted primitive model near criticality. *J. Chem. Phys.*, 1998, **108**, p. 5169–5172.

[26] Valleau, J. and Torrie, G. Further remarks on the heat capacity of the restricted primitive model. *J. Chem. Phys.*, 2002, **117**, p. 3305–3309.

[27] Yan, Q. and de Pablo, J.J. Hyper-parallel tempering monte carlo: application to the Lennard-Jones fluid and the restricted primitive model. *J. Chem. Phys.*, 1999, **111**, p. 9509–9516.

[28] Romero-Enrique, J.M., Orkoulas, G., Panagiotopoulos, A.Z., and Fisher, M.E. Coexistence and criticality in size-asymmetric hard-core electrolytes. *Phys. Rev. Lett.*, 2000, **85**, p. 4558–4561.

[29] Guida, R. and Zinn-Justin, J. Critical exponents of the n-vector model. *J. Phys. A: Math. Gen.*, 1998, **31**, p. 8103–8121.

[30] Bruce, A.D. and Wilding, N.B. Scaling fields and universality of the liquid-gas critical point. *Phys. Rev. Lett.*, 1993, **68**, p. 193–196.

[31] Barber, M.N. (1983). *Finite-size Scaling. Phase Transitions and Critical Phenomena*, vol. 8, p. 146–268, (Domb, C. and Lebowitz, J. L.). London: Academic Press.

[32] Privman, V.V. (1990). *Finite Size Scaling and Numerical Simulation of Statistical Systems*. London: World Scientific.

[33] Kim, Y.C. and Fisher, M.E. Asymmetric fluid criticality. ii. finite-size scaling for simulations. *Phys. Rev. E*, 2003, **68**, p. 041506.

[34] Daub, C.D., Camp, P.J., and Patey, G.N. The constant-volume heat capacity of near-critical fluids with long-range interactions: A discussion of different monte carlo estimates. *J. Chem. Phys.*, 2003, **118**, p. 4164–4168.

[35] Daub, C.D., Patey, G.N., and Camp, P.J. Liquid-vapor criticality in a fluid of charged hard dumbbells. *J. Chem. Phys.*, 2003, **119**, p. 7952–7956.

[36] Luijten, E. and Blöte, H.W.J. Boundary between long-range and short-range critical behavior in systems with algebraic interactions. *Phys. Rev. Lett.*, 2002, **89**, p. 025703.

[37] Camp, P.J. and Patey, G.N. Coexistence and criticality of fluids with long-range potentials. *J. Chem. Phys.*, 2001, **114**, p. 399–408.

[38] Daub, C.D., Camp, P.J., and Patey, G.N. Constant-volume heat capacity in a near-critical fluid from monte carlo simulations. *J. Chem. Phys.*, 2004, submitted.

[39] Allen, M.P. and Tildesley, D.J. (1987). *Computer Simulation of Liquids*. Oxford: Clarendon Press.

[40] Hill, T.L. (1987). *Statistical Mechanics: Principles and Selected Applications*. New York: Dover Publications, Inc.

[41] Shelley, J.C. and Patey, G.N. A comparison of liquid-vapor coexistence in charged hard sphere and charged hard dumbbell fluids. *J. Chem. Phys.*, 1995, **103**, p. 8299–8301.

[42] Stell, G. Critical behavior of polar fluids. *Phys. Rev. Lett.*, 1974, **32**, p. 286–288.

[43] Aharony, A. and Fisher, M.E. Critical behavior of magnets with dipolar interactions. Renormalization group near four dimensions. *Phys. Rev. B*, 1973, **8**, p. 3323–3341.

[44] Bruce, A.D. and Aharony, A. Critical exponents of ferromagnets with dipolar interactions: Second-order ϵ expansion. *Phys. Rev. B*, 1974, **10**, p. 2078–2087.

[45] Romero-Enrique, J.M., Rull, L.F., and Panagiotopoulos, A.Z. Dipolar origin of the gas-liquid coexistence of the hard-core 1:1 electrolyte model. *Phys. Rev. E*, 2002, **66**, p. 041204.

[46] Kim, Y.C. and Fisher, M.E. Fluid critical points from simulations: The bruce-wilding method and yang-yang anomalies. *J. Phys. Chem. B*, 2004, **108**, p. 6750–6759.

[47] Camp, P.J. Phase diagrams of hard spheres with algebraic attractive interactions. *Phys. Rev. E*, 2003, **67**, p. 011503.

[48] Hansen, J-P. and McDonald, I.R. (1986). *Theory of Simple Liquids*. London: Academic Press.

[49] Lebowitz, J.L., Percus, J.K., and Verlet, L. Ensemble dependence of fluctuations with application to machine computations. *Phys. Rev.*, 1967, **153**, p. 250–254.

[50] Hilfer, R. and Wilding, N.B. Are critical finite-size scaling functions calculable from knowledge of an appropriate critical exponent? *J. Phys. A: Math. Gen.*, 1995, **28**, p. L281–L286.

MACROIONS IN SOLUTION

Theory, experiment, and computer simulations

V. Vlachy[1], B. Hribar Lee[1], J. Reščič[1], Yu.V. Kalyuzhnyi[2]

[1]*Faculty of Chemistry and Chemical Technology,*
University of Ljubljana,
Aškerčeva 5, 1000 Ljubljana, Slovenia

[2]*Institute for Condensed Matter Physics*
of the National Academy of Sciences of Ukraine,
1 Svientsitskii Str., 79011 Lviv, Ukraine

Abstract   In this chapter we review models and theories suitable for studying polyelectrolytes in solution. The theories were applied to various problems: a) the first
is the catalytic effect on the reaction between small ions of the same charge
sign, caused by addition of a polyelectrolyte to an electrolyte solution. In the
next example b) we studied the stability of polyelectrolyte solutions. For highly
charged systems, and/or in presence of multivalent counterions, computer simulations and recent experimental data predict clustering of macroions. The new
computer simulation data support the assumption that a partial neutralization of
macroions is the first step in the process of macroion clustering. In the third example c) we considered low-charge polyelectrolytes where the effects of solvent
seemed to be important. We present results for a new model of a polyelectrolyte
solution with the chain-like polyions where the solvent molecules, approximated
as two fused charged hard spheres, are explicitly included in the calculation.
This model yields osmotic pressure results in good qualitative agreement with
experiment. In the last example d) we consider the Donnan membrane equilibrium in protein solutions. The theory is used to analyze experimental data for
the Donnan pressure in solutions of various proteins.

Keywords: Polyelectrolytes, globular proteins, osmotic pressure, Poissson-Boltzman theory,
Monte Carlo simulation, integral equation theory

1.   Introduction

Polymers with bound charges, micelles of ionic surfactants, globular proteins, and similar materials are referred to as polyelectrolytes, and have the
properties that separate them from normal low-molecular weight electrolytes

D. Henderson et al. (eds.), Ionic Soft Matter: Modern Trends in Theory and Applications, 199–231.

and also from uncharged macromolecules [1–7]. Polyelectrolytes are used in catalysis, separation processes, in waste-water treatment, medicine, cosmetics, they serve as nanoreactors, ion-exchangers, and in numerous other applications. On the other hand, many biologically important substances exhibit polyelectrolyte behavior; examples are nucleic acids, globular proteins, and polysaccharides. Better understanding of polyelectrolyte solutions is therefore important for technology and science in general.

Nature and synthetic chemistry have provided polyelectrolytes of different shapes: they can be rod-like as, for example, DNA, spherical or ellipsoidal (globular proteins) or flexible (chain-like) as are many of the synthetic polyelectrolytes. Poly (styrene) sulfonic acid and its salts is one of the most often studied representatives of the latter class. In aqueous solution these polyelectrolytes, due to repulsion between the charges on the polymer backbone, assume extended conformations. The properties of polyelectrolyte solutions seem to be dominated by interaction between large and highly charged macroions and the small ions of opposite charge, called counterions [7–10]. Attractive interaction between a macroion and counterions in a solution leads to the accumulation of counterions in the vicinity of the macroions [9, 11, 12], while the co-ions, i.e. small ions having charge of the same sign as the macroions, are pushed away from them. This situation is called the electrical double-layer. The osmotic pressure of these solutions often shows a large negative deviation from the ideal value. In dilute solutions, the activity of water is much higher than estimated from the solute concentration. Further, the mean activity coefficient of a simple electrolyte in a mixture with a polyelectrolyte is reduced considerably below its bulk value. When an external field is applied to such a solution, a fraction of the counterions moves as an integral part of a polyion [13]. Addition of a low molecular electrolyte containing divalent or multivalent counterions may destabilize the solution and result in precipitation [14, 15].

There are other features which distinguish low molecular weight electrolytes from polyelectrolytes; in electrolytes the nonideality decreases with decreasing concentration, while for solutions of strong linear polyelectrolytes the opposite trend is most often observed. For example, in solutions of poly(styrene)sulfonic acid the measured osmotic coefficient ϕ decreases with decreasing concentration and does not approach unity upon dilution in the experimentally accessible range [9]. For divalent salts of poly(styrene)sulfonic acid ϕ is very low and changes very little with concentration over a broad range [16]. The same holds true for cationic polyelectrolytes [17]; recent investigation of the osmotic properties of aqueous ionene solutions of different charge density confirms previous findings that the osmotic coefficient is rather insensitive to the concentration. Further, a compilation of osmotic coefficient data for various polyelectrolytes, presented as a function of the polyion charge density, indicates [17] that theoretically predicted values are systematically too high in comparison with experimental data. The experimental evidence suggests strong deviations from ideality even for weakly charged polyelectrolytes. Even more intriguing are experimental results for heats of dilution. Škerjanc and coworkers [18] measured the heats of dilution, ΔH, of various alkaline salts of poly(styrene)sulfonic acid in water. They found this quantity to be negative for the Li^+ salt at 40° C and positive (endothermic effect) for the Cs^+ salt solution under these conditions. Similarly, measurements of the enthalpy of dilution for cationic polyelectrolytes [19, 20] indicate positive values of ΔH. Note that electrostatic theories can only predict exothermic effect upon dilution. The above mentioned experimental results reflect specific effects related to the difference in solvation of the various counterion solutions. The results also indicate the important role of the solvent (water in this case) in these solutions; the latter in almost all theoretical studies has been approximated by a structureless continuum. Altogether, the properties of polyelectrolyte solutions are not adequately understood on the molecular level, which certainly has an impact on their efficient application in technology. In this paper, we present new computer simulations and theoretical results and review established theories and experiments.

2.   Modelling polyelectrolyte solutions

There are several approaches to treating polyelectrolyte solutions, each of them being of a different degree of complexity. The first class (a) are the so-called cell models, (b) the second are the multi-component models treating the spherical macroions and small ions on an equal level, and as a further idealization of this class (c) are the one-component models. Recently a new model treating a flexible polyelectrolyte as freely jointed tangent charged spheres has been developed and we call this (d) the chain-like model. All these theories ignore the discrete role of the solvent, treating it as continuous dielectric. The calculations are therefore performed at the so-called McMillan-Mayer level of theory. A more advanced, albeit still an approximate model, which includes the solvent as a separate species, is described farther in this section.

2.1   *Cell models*

The assumptions that highly charged macroions are mostly distributed at larger distances from each other, and that their mobility is considerably smaller than the mobility of small ions, lead naturally to the cell model approximation. In this theory each macroion is assigned its own cell and the solution as a whole is treated as a collective of noninteracting cells. Depending on the situation, the cell may be of spherical [21], cylindrical [22], or of ellipsoidal shape [23]. The cell model in spherical or cylindrical symmetries has been used by

numerous authors (see, for example, [9, 24–27]. The model can be examined
by computer simulation techniques [24, 26, 28, 29], or by potential approach
using the Poisson-Boltzmann (see, for example, [22, 24, 25]) or the modified
Poisson-Boltzmann theory [30, 31].

The key equation in potential approach is the Poisson-Boltzmann (PB) equation which reads,

$$\frac{1}{r^n}\frac{\mathrm{d}}{\mathrm{d}r}\left(r^n\frac{\mathrm{d}\psi}{\mathrm{d}r}\right) = -\frac{\rho_e}{\epsilon_0\epsilon_r}, \tag{1}$$

where $n = 2$ for spherical and $n = 1$ for cylindrical symmetry. As usual
$\epsilon_0\epsilon_r$ is the dielectric constant of solvent. In this equation $\psi(r)$ is the mean
electrostatic potential around the macroion of charge, $z_p e_0$ (e_0 is the charge on
a proton) and ρ_e is the charge density defined by

$$\rho_e(r) = \sum_a z_a e_0 \rho_a(R_{\text{cell}}) e^{-z_a e_0 \beta \psi(r)}, \tag{2}$$

where $\rho_a(R_{\text{cell}})$ is the number concentration of ionic species a at the cell radius
R_{cell}. As usual $\beta^{-1} = k_B T$, where k_B denotes the Boltzmann constant and
T is the absolute temperature. In the case of cylindrical symmetry the nonlinear PB can be solved analytically [22]. This is not the case for spherical
symmetry, where the PB equation (1) has to be solved numerically according
to the boundary conditions given by the Gauss Law [24, 25]. As a result we
obtain the mean electrostatic potential $\psi(r)$ around the macroion, relative to
the potential at the cell radius, where the zero of potential is chosen. From the
mean electrostatic potential the thermodynamic properties can be calculated.

2.2   *Multicomponent models*

A multicomponent model is often used to describe spherical micelles or
globular proteins in solution. In this case the ions are treated as charged hard
spheres immersed in a solvent of dielectric constant $\epsilon_0\epsilon_r$. In this way the micellar solution is depicted as an electrolyte where the ions grossly differ in size and
charge. The solvent averaged potential in this case is given by ($2\sigma_{ab} = \sigma_a + \sigma_b$)
the equation,

$$\beta u_{ab}(r) = \begin{cases} \infty, & r < \sigma_{ab}, \\ (z_a z_b L_B)/r, & r \geqslant \sigma_{ab}, \end{cases} \tag{3}$$

where a and b denote either macroion, counterion, or a co-ion. Further, z_a (z_b)
are the corresponding valences, and σ_a (σ_b) are the diameters. As usual L_B is
the Bjerrum length equal to $\beta e_0^2/(4\pi\epsilon_0\epsilon_r)$. Alternatively, the short-range part
of the potential may be represented, for example, by a $1/r^9$ function [32, 33].
The model, though treating the solvent as a continuum, represents a reasonable

starting point for studying solutions of highly asymmetric electrolytes [8, 10, 34].

2.3   *One-component fluid model*

In many colloidal and micellar systems the asymmetry in size is large enough for the experiment to measure only the macroion-macroion correlation [35]. For this reason various approximations, by which macroions are assumed to interact via an effective potential, are often applied. Macroions are assumed to be surrounded by a "cloud" of an opposite charge and it is assumed that the overlap of two clouds results in the repulsive interaction. In a popular theory, referred to as the one-component fluid (OCF) model, the macroions interact via the repulsive screened Coulomb potential in the form,

$$\beta u(r) = \begin{cases} \infty, & r < \sigma_{\mathrm{p}}, \\ z_{\mathrm{p}}^2 L_{\mathrm{B}} \frac{e^{-\kappa r}}{r}, & r \geqslant \sigma_{\mathrm{p}}, \end{cases} \tag{4}$$

where z_{p} is the number of charges on the macroion, and σ_{p} is the diameter of the macroion. As usual, κ is the Debye parameter, given by the ionic strength due to small ions only. Eq. (4) is obtained by solving the linear form of the PB equation. The approximations inherent to this approach have been discussed elsewhere [33, 36]. This model has been applied to interpret Donnan equilibrium results in solutions of globular proteins [37] (see Sec. 7).

In addition to the repulsive part of the potential given by Eq. (4), a short-range attraction between the macroions may also be present. This attraction is due to the van der Waals forces [17, 18], and can be modelled in different ways. The OCF model can be solved for the macroion-macroion pair-distribution function and thermodynamic properties using various statistical-mechanical theories. One of the most popular is the mean spherical approximation (MSA) [40]. The OCF model can be applied to the analysis of small-angle scattering data, where the results are obtained in terms of the macroion-macroion structure factor [35]. The same approach can also be applied to thermodynamic properties; Kalyuzhnyi and coworkers [41] analyzed Donnan pressure measurements for various globular proteins using a modification of this model which permits the protein molecules to form dimers (see Sec. 7).

2.4   *Chain-like model of a polyion*

Stretching of polyelectrolyte chains due to repulsion between the charges on the backbone is not complete for most polyelectrolytes and assuming a fully extended (rod-like) structure is clearly an approximation. For this reason a more realistic model in which the polyion can assume various conformations has been proposed. In this model the polyion is represented by a linear freely jointed tangent charged hard-sphere chain with diameter σ [42–44]. The model mimics flexible polyions and seems in many cases to be much closer to reality than the fully extended rod-like cell model. The solution contains short polyions represented as charged hard-sphere chains where each monomer unit carries a charge. The corresponding small ions, distributed in the solution, neutralize the charge on the polymer chains. The solvent is still represented as a dielectric continuum.

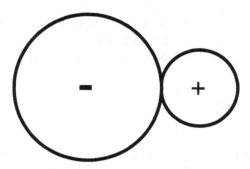

Figure 1.   Collins model of water [45].

The so-called product reactant Ornstein-Zernike approach (PROZA) for these systems was developed by Kalyuzhnyi, Stell, Blum, and others [46–54]. The theory is based on Wertheim's multidensity Ornstein-Zernike (WOZ) integral equation formalism [55] and yields the monomer-monomer pair correlation functions, from which the thermodynamic properties of the model fluid can be obtained. Based on the MSA closure an analytical theory has been developed which yields good agreement with computer simulations for short polyelectrolyte chains [44, 56]. The theory has been recently compared with experimental data for the osmotic pressure by Zhang and coworkers [57]. In the present paper we also show some preliminary results for an extension of this model in which the solvent is now treated explicitly as a separate species. In this first calculation the solvent molecules are modelled as two fused charged hard spheres of unequal radii as shown in Fig. 1 [45].

3.   Integral equation theories

In this section we introduce integral equation theories (IETs) and approximate closures applicable for various models of polyelectrolyte solutions. A theory for linear polyelectrolytes based on the polymer reference interaction site model has also been proposed [58, 59], but this approach will not be reviewed here.

3.1   *Ornstein-Zernike equation*

The basic relation in IET approach is the Ornstein-Zernike (OZ) equation [60],

$$h_{ab}(r_{ab}) = c_{ab}(r_{ab}) + \sum_k \rho_k \int c_{ak}(r_{ak}) h_{kb}(r_{kb}) \mathrm{d}\mathbf{r}_k, \qquad (5)$$

where the summation is performed over all the components. This equation is merely a definition of the direct correlation function c_{ab}, and another relation between the total correlation function, $h_{ab} = g_{ab} - 1$ and c_{ab}, is needed. This second relation reads,

$$\log_e [h_{ab}(r) + 1] = -\beta u_{ab}(r) + h_{ab}(r) + B_{ab}(r) - c_{ab}(r). \qquad (6)$$

The term $B_{ab}(r)$, known in the literature as the bridge graph term, cannot be written as a closed form function of the distribution functions $h_{ab}(r)$ and $c_{ab}(r)$. In the absence of a better approximation $B_{ab}(r)$ is often set to zero, which is called the hypernetted-chain (HNC) approximation. This approach was successfully applied in describing the properties of symmetric simple electrolytes (see, for example, [61, 62]). Belloni [63] successfully extended the HNC calculations to solutions of macroions and counterions. Monte Carlo (MC) simulations of model asymmetric electrolytes with asymmetry in size of 10:1, and in charge of −10:+1 and −15:+1 were performed to examine the accuracy of the HNC for these solutions [64]. The most important conclusion of this comparison was that the HNC closure predicts the counterions to be too close to each other, and also too close to the highly charged macroions, which eventually leads to an underestimation of the macroion-macroion repulsion. Unfortunately, for dilute solutions and for strong couplings the HNC closure diverges.

Other approximate closures often used in the theory of charged fluids may be obtained from Eq. (6). If we assume that $c_{ab}(r)$ is given by its asymptotic form $-\beta u_{ab}(r)$ for all r, then the Debye-Hückel limiting law result $g_{ab}(r) = 1 - (L_B e^{-\kappa r})/r$ [60] follows on substituting $c_{ab}(r)$ into Eq. (5). On the other hand, allowing the ions to have finite diameters, and assuming $g_{ab} = 0$ for $r_{ab} < \sigma_{ab}$, and $c_{ab} = -\beta u_{ab}(r)$ for $r_{ab} \geqslant \sigma_{ab}$, leads us to the mean spherical approximation (MSA) [65]. The latter theory can be solved analytically for an arbitrary mixture of ions yielding closed-form analytical expressions for thermodynamic quantities which are quite accurate in describing many electrolyte models [65–69].

3.2   *Two-density theory*

Spherical micelles or globular proteins in solution can be considered as an asymmetric electrolyte solution where the ionic species grossly differ in charge and size. Taking into account these asymmetries, an extension of WOZ equation has been proposed [70, 71]. In this theory, the potential energy between macroions and counterions, $U^{ab}(r)$ is divided into an associative (strongly attractive) part $U^{ab}_{ass}(r)$ and a nonassociative part $U^{ab}_{non}(r)$. The associative potential is of short range and usually located around the attractive minimum of the total potential; the Coulomb tail of the potential is included in the nonassociative part. Due to the asymmetry in size the counterions are treated as bondable to one macroion only, while each macroion can bond an arbitrary number of counterions. The two-density version of the WOZ equation is then derived [70, 71],

$$\mathbf{H}^{ab}(r_{12}) = \mathbf{C}^{ab}(r_{12}) + \sum_k \int \mathbf{C}^{ak}(r_{13}) \boldsymbol{\rho}^k \mathbf{H}^{kb}(r_{32}) \mathrm{d}r_3 , \qquad (7)$$

where $\mathbf{H}$, $\mathbf{C}$ and ρ are matrices defined as,

$$\mathbf{H}^{pp}(r) = h^{pp}(r), \qquad \mathbf{H}^{pc}(r) = (h^{pc}_0(r), h^{pc}_1(r)),$$

$$\mathbf{H}^{cp}(r) = \begin{pmatrix} h^{cp}_0(r) \\ h^{cp}_1(r) \end{pmatrix}, \qquad \mathbf{H}^{cc}(r) = \begin{pmatrix} h^{cc}_{00}(r) & h^{cc}_{01}(r) \\ h^{cc}_{10}(r) & h^{cc}_{11}(r) \end{pmatrix},$$

$$\boldsymbol{\rho}^p = \rho^p, \qquad \boldsymbol{\rho}^c = \begin{pmatrix} \rho^c & \rho^c_0 \\ \rho^c_0 & 0 \end{pmatrix}. \qquad (8)$$

The superscripts c and p denote counterions and macroions, respectively, and the subscripts 0 and 1 describe the state (0 for nonbonded and 1 for bonded) of the corresponding counterion. The partial correlation functions are related to the regular correlation functions, $h^{ab}(r) = g^{ab}(r) - 1$, by the relation $\rho^a h^{ab} \rho^b = [\rho^a \mathbf{H}^{ab} \rho^b]_{00}$, where 00 denotes the first element of the matrix. HNC-like [70, 71] and the MSA-like [72–74] closures have been proposed to solve the WOZ equation. For the former we have,

$$h^{pp}(r) = \exp[-\beta U^{pp}_{non}(r) + t^{pp}(r)] - 1, \qquad (9)$$

$$h^{cc}_{ij}(r) = [\delta_{i0}\delta_{j0} + \delta_{i1}\delta_{j0}t^{cc}_{10}(r) + \delta_{i0}\delta_{j1}t^{cc}_{01}(r)]$$
$$\times \exp[-\beta U^{cc}_{non}(r) + t^{cc}(r)] - \delta_{i0}\delta_{j0}, \qquad (10)$$

$$h^{pc}_i(r) = h^{cp}_i(r) = [\delta_{i0} + \delta_{i1}(t^{pc}_i(r) + f_{ass}(r))]$$
$$\times \exp[-\beta U^{pc}_{non}(r) + t^{pc}(r)] - \delta_{i0}. \qquad (11)$$

In the above equations, $t^{pp}(r)$, $t^{pc}_i(r)$, $t^{cc}_{ij}(r)$ are the elements of the matrix $\mathbf{T} = \mathbf{H} - \mathbf{C}$, $f_{\mathrm{ass}}(r) = \exp\left(-\beta U_{\mathrm{ass}}(r)\right) - 1$, and δ_{ij} is the Kronecker delta symbol. The two densities, ρ^c_0 and ρ^c, are related as $\rho^c_0 = \rho^c \left(1 + I\rho^p\right)^{-1}$, where I is given by

$$I = 4\pi \int_0^\infty g^{cp}_0(r) f_{\mathrm{ass}}(r) r^2 \mathrm{d}r. \tag{12}$$

The new IET approximations (associative HNC and associative MSA) provide good estimates for the correlation functions and the thermodynamic properties, even in the region of concentrations where the ordinary HNC does not give convergent results [70, 73].

3.3 *Product reactant Ornstein-Zernike approach*

The product reactant Ornstein-Zernike theory (PROZA) is designed for description of chain-like molecules. The theory is based on the integral equation, which in Fourier space assumes the form,

$$\hat{\mathbf{h}}^{ab}_{ij}(k) = \hat{\mathbf{c}}^{ab}_{ij}(k) + \sum_d \rho_d \sum_l \hat{\mathbf{c}}^{ad}_{il}(k) \boldsymbol{\alpha} \hat{\mathbf{h}}^{db}_{lj}(k), \tag{13}$$

and the closure relations connecting direct $c^{ab}_{ij}(r)$ and total $h^{ab}_{ij}(r)$ correlation functions. Here the indices i and j denote the species (position) of the monomers in the chain, subscripts a and b denote the species of the chain, $\hat{\mathbf{h}}^{ab}_{ij}(k)$ and $\hat{\mathbf{c}}^{ab}_{ij}(k)$ are matrices with elements that are Fourier transforms of the elements of the real-space matrices $\mathbf{h}^{ab}_{ij}(r)$ and $\mathbf{c}^{ab}_{ij}(r)$, defined by

$$\mathbf{f}^{ab}_{ij}(r) = \begin{pmatrix} f^{ab}_{i_0 j_0}(r) & f^{ab}_{i_0 j_A}(r) & f^{ab}_{i_0 j_B}(r) \\ f^{ab}_{i_A j_0}(r) & f^{ab}_{i_A j_A}(r) & f^{ab}_{i_A j_B}(r) \\ f^{ab}_{i_B j_0}(r) & f^{ab}_{i_B j_A}(r) & f^{ab}_{i_B j_B}(r) \end{pmatrix}, \quad \mathbf{f}^{ab}_{ij}(r) = \mathbf{h}^{ab}_{ij}(r), \mathbf{c}^{ab}_{ij}(r).$$

In this equation, $h^{ab}_{i_\alpha j_\beta}(r)$ and $c^{ab}_{i_\alpha j_\beta}(r)$, appearing in $\mathbf{h}^{ab}_{ij}(r)$ and $\mathbf{c}^{ab}_{ij}(r)$, respectively, represent the total and direct correlation functions for various interactions. The lower indices α and β each take the values 0, A and B, and denote the bonding states of the corresponding particles, i.e. the case of $\alpha = 0$ corresponds to an nonbonded particle, while $\alpha = A$ or $\alpha = B$ to an $A-$bonded or $B-$bonded particle, respectively.

Several closure conditions have been proposed (see [75] and references therein). In the present study the MSA-like closure [51, 52, 72],

$$\begin{cases} \mathbf{c}_{ij}^{ab}(r) & = -\mathbf{E}\beta U_{ij}^{(C)ab}(r) + \frac{\mathbf{t}_{ij}^{ab}}{2\pi\sigma_{ij}^{ab}}\delta(r - \sigma_{ij}^{ab}), \quad r \geqslant \sigma_{ij}^{ab} = \frac{1}{2}\left(\sigma_i^a + \sigma_j^b\right), \\ \mathbf{h}_{ij}^{ab}(r) & = -\mathbf{E}, \qquad\qquad\qquad\qquad\qquad r < \sigma_{ij}^{ab}, \end{cases}$$

(14)

is utilized. Here $\mathbf{t}_{ij}^{ab}$, $\boldsymbol{\alpha}$ and $\mathbf{E}$ are the matrices,

$$t_{i_\alpha j_\beta}^{ab} = \frac{\delta_{ab}}{2\rho_a}\left[\delta_{\alpha A}\delta_{\beta B}\frac{\delta_{ij+1}}{\sigma_{ii-1}} + \delta_{\alpha B}\delta_{\beta A}\frac{\delta_{ij-1}}{\sigma_{ii+1}}\right], \quad \alpha_{\alpha\beta} = 1 - \delta_{\alpha\beta} + \delta_{0\alpha}\delta_{0\beta}$$

$$E_{\alpha\beta} = \delta_{0\alpha}\delta_{0\beta},$$

(15)

while σ_i^a is the hard-sphere diameter and $\beta U_{ij}^{(C)ab}(r)$ is the Coulomb potential acting between the hard-sphere beads of the chain.

The site-site pair distribution function, $h_{ij}^{ab}(r)$, is related to the partial distribution functions $g_{i_\alpha j_\beta}^{ab}(r) = h_{i_\alpha j_\beta}^{ab}(r) + \delta_{\alpha 0}\delta_{\beta 0}$ by the relation,

$$g_{ij}^{ab}(r) = \sum_{\alpha,\beta=0} g_{i_\alpha j_\beta}^{ab}(r).$$

(16)

Solution of the MSA-like closure condition, given by Eq. (14), leads to a closed form analytical expression for the thermodynamic properties of the system [46, 48–50].

4.   Catalytic effect of polyelectrolyte addition

When a polyelectrolyte is added to a low-molecular weight electrolyte the probability of finding two counterions next to each other grossly increases. As mentioned before, the counterions are strongly attracted to macroions and their concentration near the macroion may be orders of magnitude larger than their bulk concentration [11]. In contrast to this the co-ions are repelled from the macroion's surface. In this way counterions and co-ions become spatially separated and the probability of finding these two species in contact decreases upon addition of the polyelectrolyte.

This effect has been known for quite some time [76–81] and used to influence the reaction rate between the charged particles. Examples include some hydrolysis reactions [80] where a small addition of polyelectrolyte causes a dramatic acceleration of the chemical reaction between equally charged divalent counterions in solution. The effect of a polyelectrolyte on ion-ion collision frequencies has also been used to probe the distribution of ions around the polyion. For example, Meares and coworkers [82] probed the electrostatic properties of biological polyelectrolytes using diffusion-enhanced energy transfer. Recent studies of this effect are due to Tapia and coworkers [83, 84].

The rate at which the rapid bimolecular chemical reaction occurs is influenced by several factors [85]. One of them [86, 87] is related to the rate at which potentially reactive encounters occur. In the approximate theoretical approach applied in [87] it was assumed that for the reaction to occur the two reactant molecules have to be in contact. In this way the enhancement of the reaction rate was related to the value of the pair correlation function of reacting particles a and b, $g_{ab}(\sigma_{ab})$, at contact. The effect of addition of a polyelectrolyte to an electrolyte solution can be "measured" by the ratio [86, 87],

$$\frac{k}{k^\circ} = \frac{\langle g_{ab}(\sigma_{ab})\rangle}{\langle g_{ab}^\circ(\sigma_{ab})\rangle},\tag{17}$$

where k and $g_{ab}(\sigma_{ab})$ are the rate and the value of the pair correlation function in contact, respectively, as obtained for the polyelectrolyte-electrolyte mixture, while superscript $^\circ$ denotes these values for the pure electrolyte solution. The triangular brackets denote the usual canonical average [60]. For two interacting particles having a charge of opposite sign to the macroion, a large enhancement in k, $(k \gg k^\circ)$ was found upon addition of polyelectrolyte [80]. On the other hand, for potentially reacting particles carrying a charge of the same sign as the macroion, only a weak increase in the reaction rate may be anticipated. Finally, for a reacting pair of ions having opposite charge to each other, a slight decrease in the ratio given by Eq. (17) was obtained [79]. The inhibition is due to the fact that upon addition of polyelectrolyte to the electrolyte solution the counterions and co-ions become spatially separated.

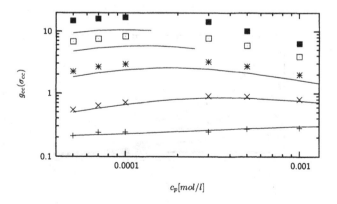

Figure 2. Comparison between MC data (symbols) and the two–density IET calculation (lines) for the ratio k/k°; $\sigma_c = \sigma_{co} = 0.4$ nm, $\sigma_p = 3.0$ nm. The concentration of added $+1 : -1$ electrolyte c_e was 0.005 mol dm^{-3}. The results apply, from bottom to top, to $z_p : z_c$ ratios $-10 : +1, -20 : +1, -30 : +1, -40 : +1$ and $-50 : +1$.

In a recent contribution [87] we studied the catalytic effect in polyelectrolyte-electrolyte mixtures by various theoretical techniques. For an isotropic model where the macroions, co-ions and counterions are pictured as charged hard spheres, we employed the HNC approximation, the modified PB and symmetric PB theories. The results for $k/k°$ were compared with the computer simulations for the same quantity. Note that this quantity is much more sensitive to the details of the model and theory than thermodynamic properties like osmotic pressure studied before. The conclusion was that these theories are not well-suited to treat the problem; they were capable of reproducing MC values only qualitatively and even this merely for low-charged macroions.

In this section we present new theoretical results for polyelectrolyte-electrolyte mixtures based on the two-density WOZ theory with the hypernetted-chain closure (Sec. 3). The focus is on the $k/k°$ ratio considered as a measure of the polyelectrolyte effect on the ion-ion reaction rate. Fig. 2 shows the comparison between the computer simulation results (symbols) and the IET calculation (lines). We take the same diameters of ions as in our previous study [87]: $\sigma_c = \sigma_{co} = 0.4$ nm, $\sigma_p = 3.0$ nm. The concentration of added $+1 : -1$ electrolyte c_e was 0.005 mol dm^{-3}. The results apply, from bottom to top, to $z_p : z_c$ ratios of $-10 : +1, -20 : +1, -30 : +1, -40 : +1$, and $-50 : +1$. For $z_p = -10$ and -20 the theory reproduces the machine calculations quite well; however, the agreement deteriorates for more negative z_p values. Similarly to the regular OZ/HNC theory, we have not been able to obtain convergent solutions of the two-density IET for high values of macroion charge and for all concentration ranges (cf Fig. 2). In any case, the two-density theory is more accurate and covers a considerably larger range of macroion charges than the regular OZ/HNC approach.

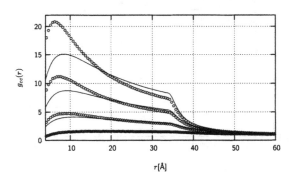

Figure 3.   The counterion-counterion pair-distribution function as obtained by theory (lines) and MC simulations (symbols). The calculations apply to $c_p = 0.0001$ and $c_e = 0.005$ mol dm^{-3}. The results apply, from bottom to top, to $z_p : z_c$ ratios $-20 : +1, -30 : +1, -40 : +1$ and $-50 : +1$.

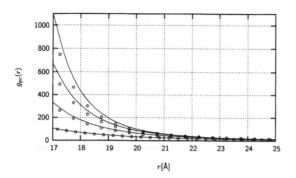

Figure 4. The macroion-counterion pair-distribution shown for the same set of parameters as before. Legend as for Fig. 3.

In the next two figures we discuss the pair-correlation functions as obtained from the two-density theory and computer simulations. First, in Fig. 3 we compare the counterion-counterion pair-distribution function as obtained theoretically (lines) and from simulations (symbols). The numerical calculations apply to $c_p = 0.0001$ and $c_e = 0.005$ mol dm^{-3}; the results show that the theory underestimates the counterion-counterion correlation. Next, in Fig. 4 the macroion-counterion pair-distribution is shown for the same set of parameters. Finally, in Fig. 5 the macroion-macroion pair-distribution functions are calculated by both theoretical approaches at $c_p = 0.0001$ mol dm^{-3} solution and for $z_p = -10$ and -30.

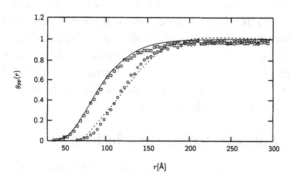

Figure 5. The macroion-macroion pair-distribution functions for $_p = -10$ (squares represent MC data and full line theory) and -30 (circles represent MC data and dashed line theory). Other parameters as for Fig. 3.

In the last figure to this section, Fig. 6, we present new MC data (symbols) for $z_p = -60$ and a concentration of added $+1 : -1$ electrolyte equal to 0.005 mol dm^{-3} as a function of the macroion concentration. In this case the increase of the probability of finding two counterions in contact is over 100 times! No convergent IET result could be obtained for multicomponent model.

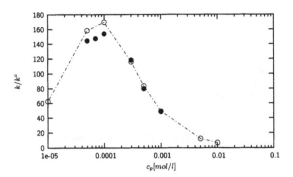

Figure 6.   MC results for multicomponent model (filled circles) and PB cell model results (open circles connected with line) for the ratio $k/k°$ at $z_\mathrm{p} = -60$ and for $c_\mathrm{e} = 0.005$ mol dm^{-3} as a function of the macroion concentration.

So the results were compared to the PB cell model calculation (cf Sec. 2). Considering that the PB theory treats the counterions and co-ions as a "cloud" of electrical charge around the macroion, the $k/k°$ was estimated by using the equation of Morawetz [77, 78],

$$\frac{k}{k°} = \frac{V^{-1} \int_v \rho_a(r)\rho_b(r)\mathrm{d}V}{V^{-1} \int_v \rho_a(r)\mathrm{d}V V^{-1} \int_v \rho_b(r)\mathrm{d}V} , \tag{18}$$

where $\rho_a(r)$ and $\rho_b(r)$ are local concentration of ions, obtained by solving multicomponent PB equations (1) and (2), V is the volume of the spherical cell, and a and b denote either counterion, or a co-ion. The electrostatic potential was obtained from the numerical solution of the non-linear PB equation in spherical symmetry, Eq. (1). For the distance of the closest approach of macroions to ions $\sigma_\mathrm{p-c} = \sigma_\mathrm{p-co}$ we took 1.7 nm. The PB cell model results presented here by lines are, as already found before [87], in surprisingly good agreement with the machine calculations.

5.   Stability of polyelectrolyte solutions

Though the theory of Derjaguin-Landau-Verwey-Overbeek (DLVO) [17, 18] was essentially designed for hydrophobic colloids, it is often applied to the analysis of the stability of polyelectrolyte solutions. According to this approach an overlap of the electrical double-layers of two charge-like colloidal spheres in an electrolyte solution always yields a repulsive screened Coulomb interaction, and the van der Waals forces are responsible for the attraction. A number of experiments in the recent decades, however, provide evidence that the effective interparticle potential shows a long-range attraction which cannot be ascribed to the van der Waals forces [15, 88–93]. In spite of numerous theoretical attempts to explain this phenomena (for a review see [7, 8, 10, 94,

28]), the existence of this long-range attraction is still one of the controversies in colloidal science. In a series of papers [96–102] we examined this phenomena in solutions of highly asymmetric electrolytes, using computer simulations and more analytical theories. We studied solutions in the primitive model approximation, see Eq. (3); the macroions were presented as charged spheres with 12, 20, or 24 negative elementary charges while the counterions were in the range from monovalent to four-valent ($z_c = +1, +2, +3$, or $+4$). In most cases we studied aqueous solutions and the solvent was treated as a dielectric continuum at 298 K. The computer simulations revealed that the properties of solutions with multivalent counterions differ qualitatively from those with monovalent ones. In particular, the presence of multivalent counterions in solution causes a nonuniform distribution of macroions. Multivalent counterions cause the macroions to come closer to each other; in some cases the distance with the highest probability of finding two macroions is the contact distance. This is clearly shown by the shape of the macroion-macroion pair distribution functions [96, 97, 102]. This phenomenon can be seen even more clearly in the snapshots from simulations. An example of a snapshot from an equilibrium configuration of the $-12 : +4$ electrolyte ($c_p = 0.005$ mol/dm^3) as obtained by molecular dynamics simulations is shown in Fig. 7 [103]. Here the correlations between macroions become so strong that most of the macroions aggregate into big clusters. The structure of the solution, as given by the macroion-macroion distribution function, is stable upon dilution [97], as well as upon an addition of a simple salt [99]. All these results seem to be consistent with the experimental observations mentioned above.

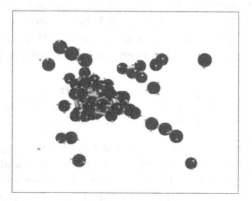

Figure 7. An example of clustering of macroions for a $-12 : +4$ electrolyte at $c_p = 0.005$ mol dm^{-3} as obtained by molecular dynamics computer simulation. Big spheres represent macroins and small spheres counterions, while "bonds" show the macroions connected into clusters.

Analysis of the cluster structure in solution [102] shows that the counterions may serve as bridges connecting the macroparticles. With increasing counterion charge the capability of counterions to neutralize macroion charge increases. The residence time (time spent by a counterion in the immediate vicinity of a macroion) indicates that the macroion-counterion "bonds" are more stable than the macroion-macroion "bonds". Such neutral or almost neutral formations of macroions and "condensed" counterions thus act as precursors in forming the aggregates in solution. The accumulation of the counterions around the macroion depends strongly on the macroion and counterion charge density, on the polyelectrolyte concentration, and the dielectric constant of the solvent, as investigated thoroughly by Reščič and Linse [104, 105]. The clusters of macroions in solution would dissolve upon addition of a solvent with a higher dielectric constant than water, or by substituting the multivalent counterions in solution with monovalent ones.

Due to the long-range of Coulomb forces, clustering of macroions in solutions with multivalent counterions is assumed to be a multiparticle effect. The range of interaction should therefore play an important role in these systems; such an analysis is presented in [101]. The results obtained using the screened Coulomb potential (instead of the bare Coulomb) suggest that macroions form clusters even in situations when, together with the related counterions, they interact via a potential which is of much shorter range than the bare Coulomb potential [101]. This finding supports the assumption that relatively short-range correlations are responsible for the observed clustering of macroions.

To investigate the role of possible ion-specific (non-Coulomb) forces in the processes of cluster formation, the following model was constructed. The ions of the asymmetric electrolyte were again modeled as charged hard spheres; the macroions had a negative charge of -12 and the counterions charge $+1$ or $+3$. The radius of the macroion was 1.0 nm and the radius of the counterion was 0.1 nm. An additional square-well attraction or repulsion, mimicking specific interaction, was supposed to act between macroions and counterions,

$$\beta u_{pc}(r) = \begin{cases} \infty, & r_{pc} < \sigma_{pc}, \\ \eta L_B, & \sigma_{pc} \leqslant r_{pc} \leqslant \lambda \sigma_{pc}, \\ 0, & r_{pc} > \lambda \sigma_{pc}, \end{cases} \qquad (19)$$

where $\sigma_{pc} = (\sigma_p + \sigma_c)/2$. As before, L_B is the Bjerrum length and the letters p and c denote macroions and counterions, respectively. Parameter η measures the depth of the square-well, and λ measures the range of interaction. The MC simulations were performed at constant volume and temperature with 64 macroions and an equivalent number of counterions. The standard Metropolis algorithm was applied, and to minimize the effects due to a small number of particles in the simulation cell, we used the Ewald summation method [106]. In calculating the statistics, the averages were collected over 70 mil-

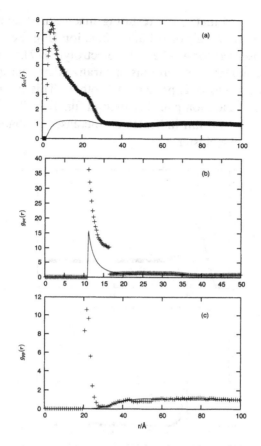

Figure 8. (a) Counterion-counterion, (b) macroion-counterion, and (c) macroion-macroion pair distribution function for a $-12 : +1$ electroltye at $c_p = 0.01$ mol dm^{-3} as obtained by MC computer simulation. Lines are for $\eta = 0$, and symbols for $\eta = -3.18$ nm, $\lambda = 1.5$.

lion MC moves, after an equilibration run of about 7 million configurations. The computer simulations were performed to evaluate the thermodynamic and structural properties, but here we only show the structure, as given by various distribution functions.

In Fig. 8 we show the results for a $-12 : +1$ electrolyte with the macroion concentration $c_p = 0.01$ mol/dm^3. The lines represent the results for the system where $\eta = 0$ (pure Coulomb potential), while the symbols represent the results for $\eta = -3.18$ nm and $\lambda = 1.5$ (cf Eq. (19)). The macroion-counterion pair distribution function (Fig. 8b) shows, as expected, that an attractive short-range potential causes higher accumulation of monovalent counterions around the macroions. As a consequence the counterions become strongly correlated as indicated by the shape of the counterion-counterion pair distribution func-

tion shown in Fig. 8a. Finally, the resulting macroion structure is reflected
in the macroion-macroion pair distribution function (Fig. 8c). This function
is similar to that observed for a $-12 : +3$ electrolyte [97]. The counterions,
being strongly attracted to the macroions, neutralize their charge to the degree
that the macroions can easily approach each other. This is indicated by the
peak in the macroion-macroion pair distribution function. The hump at a distance of approximately 4.0 nm indicates an increased probability of finding
three macroions in such a cluster.

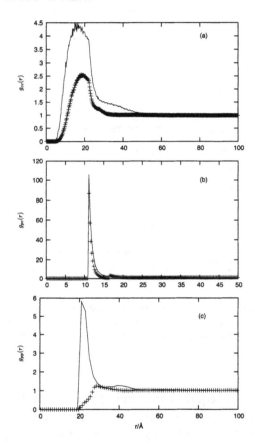

Figure 9.   (a) Counterion-counterion, (b) macroion-counterion, and (c) macroion-macroion
pair distribution function for a $-12 : +3$ electroltye at $c_m = 0.01$ mol dm^{-3} as obtained by
MC computer simulation. Lines are for $\eta = 0$, and symbols for $\eta = +3.18$ nm, $\lambda = 1.5$.

Just as an additional short-range attraction between macroions and counterions may lead to the clustering of macroions even for monovalent counterions,
an additional repulsion between macroions and multivalent counterions can
cause the macroion clusters to dissolve. The results for the macroion concen-

tration $c_p = 0.01$ mol/dm^3 are shown in Fig. 9. Again, the results for a -12 : $+3$ primitive model electrolyte with $\eta = 0$ are shown by lines and the results for an electrolyte with an additional square-well repulsion ($\eta = +3.18$ nm, and $\lambda = 1.5$ in both cases) are shown by symbols. Though the additional repulsion does not have a substantial effect on the macroion-counterion distribution, the changes in the macroion distribution are enormous. The peak in the macroion-macroion pair distribution function, indicating the clustering of macroions in the presence of trivalent counterions (continuous lines), completely disappeared when an additional repulsion between macroions and counterions (results shown by symbols) was operating. These results support the proposed mechanism of macroion clustering suggested in [102] in which the first step is a partial neutralization of the macroions. They also indicate the importance of the ion-specific macroion-counterion interaction in charge-like attraction phenomena.

6. Solvent primitive model calculations

In this section we present numerical results for a new model of polyelectrolyte solutions, which we call the "solvent primitive" model. According to this model the solvent molecule is approximated as two fused charged hard spheres of different radii similar to the Collins model of water [45]. The larger sphere (see Fig. 1) has a diameter 2σ and charge $0.2e$ and the smaller one a diameter 1.4σ and charge $-0.2e$. Here σ is the hard-sphere diameter of the chain molecule beads and e is the elementary charge. As explained in Sec. 2.4, the polyions are represented by a linear freely jointed tangent charged hard sphere chain. In these calculations the model solvent molecules are treated explicitly, but with a dielectric constant equal to that of bulk water ($\epsilon_0 \epsilon_r = 78.5$). The theory for this model, known under the acronym PROZA, was described in Sec. 3. This calculation considers the osmotic equilibrium which is established between an aqueous polyelectrolyte solution and pure water, when the two compartments are separated by a semipermeable membrane. The equilibrium requires the equality of chemical potential of water in both compartments. Osmotic pressure is then defined as a difference between pressures in the two compartments. The pressures and chemical potentials were calculated using the energy route.

Before presenting numerical results, it is worth summarizing the main characteristics of the experimental results for the osmotic pressure of polyelectrolyte solutions [9, 17, 18, 57, 107]. The measured osmotic coefficients most often exhibit strong negative deviations from ideality. The measured values are a) lower than it was predicted by the cylindrical cell model theory, b) rather (but not completely) insensitive to the nature of the counterions, and c) also insensitive to the polyelectrolyte concentration in a dilute regime and/or for

divalent counterions. An important recent analysis of osmotic coefficient behavior for rod-like and flexible polyions is based on molecular dynamics simulations [56]. Various approximations to the cylindrical cell model theory have
also been discussed by Ballauff in the next chapter. Their conclusions about
the role of the ion-ion correlations agree closely with our findings [11].

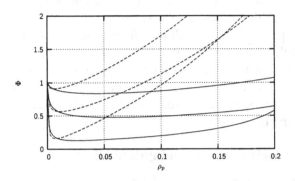

Figure 10.   Osmotic coefficient as a function of the reduced density of monomer units $\rho_p =$
$\rho_m \sigma^3$, where ρ_m is the number density of monomer units. Solvent primitive model (continuous
lines); McMillan-Mayer model results (broken lines). From top to bottom $\alpha = 0.125, 0.5$, and
1.0 (each bead is charged).

   In a first step toward more realistic modelling we present numerical results
for a mixture of chain-like polyelectrolytes with the degree of polymeriza-
tion equal to 32, spherical counterions, and the "water" molecules modelled as
fused charged hard spheres (Fig. 1). Numerical results for the osmotic coefficient as a function of the reduced density of monomer units and for the various
fractions α of the charge on the polion are shown in Fig. 10. Note that α
has a relative meaning only; it is defined to be between 0 (uncharged) and 1.0
(fully charged polion). These results are shown by continuous lines; curve a)
denotes the fully charged chain-like polion, in b) only every second bead on
the polion is charged ($\alpha = 0.5$), and c) belongs to $\alpha = 0.125$. As expected,
and also found experimentally (see, for example, [17]), the osmotic coefficient
decreases with increasing charge of the polion. For comparison, in the same
figure we present the results for the model with the solvent treated as dielectric
continuum [49, 50]. The latter data belong to the same state points and are
here given by broken lines. The difference between the two models, as clearly
seen in Fig. 10, is significant. The solvent primitive model approach (continuous lines) yields an osmotic coefficient which is considerably less sensitive to
the polion concentration than the McMillan-Mayer continuum solvent result
[49, 50] (broken lines).

7. Membrane equilibria

A broad variety of phenomena which are of particular interest in cell biology and colloidal science occur in systems consisting of two ionic solutions in equilibrium separated by a semipermeable membrane. The actual applications range from waste water purification to drug delivery. In addition, among experimental techniques used in order to identify the principal interactions in macromolecular solutions, measurements of the osmotic pressure play a very important role (see [109–112] and the references therein). The osmotic properties (in particular the second virial coefficient) may be related to the crystallization of proteins, i. e., an unsolved problem of determining the protein structure [109]. The literature data reveal that the agreement between theory and experiment is poor; in other words, we are not able to predict the osmotic pressure from the given molecular parameters with a sufficient degree of accuracy to be useful for crystallization and other purposes. We believe that there are several reasons for this failure, and among them most notably:

a) the effect of ion-ion correlations, not included in the conventional Poisson-Boltzmann theory (see Sec. 5); the effect may be important in relation to the non-uniform charge distribution on proteins [113];

b) clustering (most often dimerization) of protein molecules,

c) and most importantly, crude modelling of the "solvent", which is usually a concentrated mixture of ions in water.

Particularly, the osmotic pressure reflects the activity of water which is considered as a structure-less continuum in all simplified theories. The activity of water consequently depends on all interactions in the solution. Each of the three problems mentioned above deserves special study; however, in this work we shall mainly focus on the effects caused by the tendency of some protein molecules to form dimers.

It has been stressed by Haas and co-workers [114] that isotropic models are not adequate in describing protein solutions where the interactions may be strongly anisotropic. In one of our previous studies we studied a modification of a primitive model solution containing macroions (proteins) and counterions [115]. The protein molecules were assumed to carry 20 negative charges and the counterions were monovalent. Asymmetry in diameters between the two species was 3.0 nm : 0.4 nm. The main modification consisted in the inclusion of short-range directional attractive force acting between the macroions. The parameters of this attraction were chosen to result in the formation of pairs but not higher clusters. This model was examined using the two-density theory for associating fluids (Sec. 3). The results indicated that the formation of dimers considerably lowers the osmotic pressure of the model solution. Part of the effect is due to a stronger attraction between dimerized proteins and counterions, and part due to the smaller number of kinetic units in the solution. This model will be generalized in order to treat the macroion-electrolyte mixture in the next subsection.

7.1   *Donnan pressure: model and theory*

This section considers the Donnan equilibrium which is established by the equilibrium distribution of a simple electrolyte between an aqueous protein-electrolyte mixture and an aqueous solution of the same simple electrolyte, when the two phases are separated by a semipermeable membrane. A difference in osmotic pressure is established across the membrane permeable to all other species but proteins. This difference is measurable and provides important information about the protein-protein interaction in solution [37, 109–112, 116]. The principal goal of the theory is to explain how factors such as protein concentration, pH, protein aggregation, salt concentration and its composition, influence the osmotic pressure. At the moment this goal seems to be too ambitious; these systems are often complicated mixtures of highly concentrated electrolytes and protein molecules, and the principal forces are not easy to identify [117].

In one of our previous studies [41] we proposed a modification of the one-component model of protein solutions that accounts for the self-association of protein molecules in solution. In addition to the usual screened-Coulomb interaction the protein molecules were allowed to form dimers but no higher clusters. In this analytical theory the protein molecules are assumed to exist as monomers (charged hard spheres) or dimers (charged hard-sphere diatomics) distributed in the pseudo-solvent containing ions and water. The effect of the solvent is embedded in the Debye parameter κ, modifying the bare Coulomb potential between the charged sites. Thus the total pair potential between diatomics sites, a diatomic site and a charged monomer, and between charged monomers is essentially given by Eq. (4).

Here, we propose a more realistic model of protein-electrolyte mixture. In the present case all the ionic species (macroions, co-ions and counterions) are modelled as charged hard spheres interacting by Coulomb potential as for the primitive model (Sec. 2), but the macroions are allowed to form dimers as a result of the short-range attractive interaction. Numerical evaluation of this multicomponent version of the dimerizing-macroion model has been carried out using PROZA formalism, supplemented by the MSA closure conditions (Sec. 3).

7.2   *Analysis of experimental data*

To examine the potential of this new approach, we analyze the experimental data for the osmotic pressure of bovine serum albumin (BSA) in 0.15 mol dm^{-3} sodium chloride [112] and human serum albumin (HSA) solution in 0.1 mol× dm^{-3} phosphate buffer [111]. According to a previous experimental and theoretical study [111] the two solutions differ substantially in the degree of protein association. The theoretically determined osmotic coefficient can be fitted to the experimental results to obtain the fraction of dimers in the solution. The results of our analysis are presented in Figs. 11 and 12. The protein molecular weights used in these calculations were 69,000 g/mol for BSA and 66,700 g/mol for HSA. The hard-sphere diameter of spherical proteins was assumed to be 6.0 nm. For the case of the multicomponent model, the ions of the low-molecular weight $+1 : -1$ electrolyte were modelled as charged hard spheres with diameter 0.4 nm.

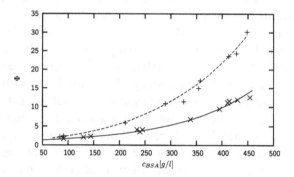

Figure 11.   Experimental osmotic coefficient data ($\times$) [41, 112] for bovine serum albumin (BSA) and the corresponding multicomponent model results (full line) at pH $= 5.4$. For pH $= 7.3$, experimental data are denoted by ($+$) and one-component model results by the dashed line.

Our analysis indicates no self-association of protein molecules for BSA solutions [112] at pH $= 5.4$ and 7.3 (Fig. 11). The fraction of dimers giving good agreement with experiment in this case is zero; this holds true for both the one-component [41] and the multicomponent model. The results obtained by the two theoretical models for pH $= 5.4$, where experimental results are denoted by ($\times$), practically coincide. Experimental data for pH$=7.3$, are denoted by ($+$) and one-component model results by the dashed line. In this case no IET results since the multicomponent model could be obtained for concentrations above 330 g/dm^{-3} and therefore only one-component calculations are shown.

In the case of HSA solutions strong negative deviations from ideal behavior at low concentration of the protein are apparent (Fig. 12). This trend in osmotic pressure can be partly explained by the formation of dimers. For a pH value of 8.0 a reasonably good agreement between theory and experiment ($+$) is obtained taking the fraction of dimers equal to be unity (full dimerization). This holds true for both approximate models. In the case of a lower pH value of 5.4 the results for the two models differ; the best fit is obtained for a fraction of dimers equal to 0.9 for the one-component model and 0.85 for the multicomponent model. Note that appreciable protein association in dilute HSA solutions has been confirmed experimentally [111]. Taking into account the differences in approximations inherent to the one-component and multicomponent level of description (cf Sec. 2), we found a very good agreement between the two theoretical approaches.

8.   Summary and prospects

The experimental and theoretical study of aqueous polyelectrolyte solutions has been an important topic in physical chemistry for the recent fifty years. Significant progress has been made since then, new theories developed, and some of them were reviewed in previous chapters. Yet, in spite of many efforts, polyelectrolyte solutions are considered to be poorly understood. Theoretical analysis became complicated by the fact that, in addition to the long-range Coulomb interaction between charges on the polyion (which may in addition assume various conformations) and small ions in solution, we need to consider the solvent effects. Due to the complexity of the problem, popular polyelectrolyte theories ignore water as a separate species; the solvent is treated merely as a dielectric continuum. The continuum approach had a considerable success, and purely electrostatic theories were capable of explaining many experimental results in dilute solutions (see, for example, [5, 9]). However, the electrostatic theories have their limitations; specific effects mediated by the solvent can be observed even for strong polyelectrolytes, and are not accounted for by these

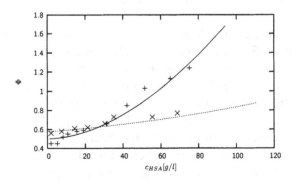

Figure 12.   Experimental osmotic coefficient data ($\times$) [111] for human serum albumin (HSA) and the corresponding results of multicomponent model of dimerizing macroions (dotted line), both at pH = 5.4. For pH = 8, experimental data are denoted by (+) and results of multicomponent model of dimerizing macroions by the full line.

approaches. We can illustrate the situation by two examples taken from the literature.

Osmotic pressure measurements of dilute linear polyelectrolytes indicate strong negative deviations from ideality [9, 17, 18, 57, 107, 108]. In other words, the activity of water is much higher than for an ideal solution. The deviations seem to be present for strong as well as for weakly charged polyelectrolytes in water [17]. In electrostatic theories the deviations are ascribed to the strong counterion-polyion association due to Coulomb forces. The PB theory, for example, shows correct trends, but the measured values are always lower than the theoretically predicted ones [9]. This purely electrostatic theory, however, has no explanation for the case of sparingly charged polyelectrolytes [17]. What makes the activity of water so high for these systems? While this effect has not been explained yet, it could be related to the hydrophobic interaction between water and the polymer backbone. There are interactions other than Coulombic ones that influence the activity of water and they are not included in these theories.

The osmotic pressure considered above is a relatively easy-to-interpret thermodynamic property reflecting the activity of water. A somewhat more informative quantity is the heat of dilution ΔH, which derives from the temperature derivative of the free energy. Here the specific effects of various counterions, brought about by differences in hydration, are clearly revealed; the measured heat of dilution in water is negative for the lithium salt of poly(styrene)sulfonic acid and positive for the cesium salt, if both are measured at 40°C [18]. Note that the ΔH values for both salts are negative at 25°C. The latter result is in good agreement with the PB theory which predicts an exothermic effect. Measurements of the enthalpy of dilution for cationic polyelectrolytes [19, 20] also indicate positive values of ΔH, again at variance with the purely electrostatic approach. Obviously, a more realistic description of water is needed to explain these phenomena. In order to make a small step ahead in polyelectrolyte modelling, this study proposes the so-called solvent primitive model of polyelectrolyte solutions. In this model the water molecules, approximated as two fused charged hard spheres, are explicitly included in the calculation of osmotic pressure. Numerical results for the osmotic coefficient, calculated as a function of the concentration of monomer units and for various degrees of ionization of a chain-like "polyion" with 32 beads, follow the experimental trends. Since aqueous polyelectrolyte solutions are still beyond the reach of all-atom computer simulations, such simplified models may represent a viable alternative [118].

One important class of biological polyelectrolytes is represented by solutions of globular proteins. These solutions differ from the synthetic chain-like (or more rod-like) polyelectrolytes mentioned above, not only in shape, but also in other aspects. One important distinction is that they may simultaneously possess both acidic and basic groups. In addition, the charged groups are not uniformly distributed on the surface of the protein, and therefore the molecules possess significant dipole moment. Further, hydrophobic effects are expected to play a significant role in these solutions. Simplified theories used to explain experimental data assume some net charge on the protein and a spherically symmetric Coulomb interaction with the small ions in solution. Protein interactions and association have been recently discussed in an important contribution of Piazza [117].

As a step toward an increasing understanding of the properties of protein solutions, our present work considers the model to account for association of globular proteins. In this simplified model all the ionic species including protein molecules are considered as charged hard spheres embedded in a continuous dielectric. In addition to the Coulomb force, the macroions interact via the directional short-range attractive interaction which causes the formation of macroion pairs. The appropriate theory needed in order to obtain the Donnan osmotic pressure is based on the two-density IET formalism. From comparison with experimental data for various proteins it is evident that the theory is capable of reproducing this behavior at least semi-quantitatively. The association of protein molecules may partly explain the low values of the Donnan pressure for some proteins.

In spite of the partial success in theoretical description, we believe that more realistic models are needed for the theory to have a predicting power. For example, measurements usually take place in the presence of a large excess of simple electrolyte. The electrolyte present is often a buffer, a rather complicated mixture (difficult to model *per se*) with several ionic species present in the system. Note that many effects in protein solutions are salt specific. Yet, most of the theories subsume all the effect of the electrolyte present into a single parameter, the Debye screening parameter κ. In the case of the Donnan equilibrium we measure the subtle difference between the osmotic pressures across a membrane permeable to small ions and water but not to proteins. We believe that an accurate theoretical description of protein solutions can only be built based on the models which take into account hydration effects.

Acknowledgements

We acknowledge the support of the Slovene Ministry of Education, Science and Sport through grants P1–0201, J1–6653, and Z1–3029–0103–03, the Ukraine-Slovenia Joint Grant 2003/5, and Slovenia-USA Joint Grant 2002/02.

References

[1] Dautzenberg, H., Jaeger, W., Kötz, J., Philipp, B., Seidel, Ch., and Steherbina, D. (1994). *Polyelectrolytes: Formation, Characterization and Application.* Germany, Munich: Hanser Publ.

[2] Frisch, K.C., Klempner, D., and Patsis, A.V. (1976). *Polyelectrolytes.* CT.

[3] Hara, M. (1993). *Polyelectrolytes: Science and Technology.* New York: Marcel Dekker Inc.

[4] Hunkeler, D., and Wandrey, C. Polyelectrolytes: Research, development, and applications. *Chimia*, 2001, **55**, No. 3, p. 223–227.

[5] Manning, G.S., and Ray, J. Counterion condensation revisited. *Journal of Biomolecular Structure & Dynamics*, 1998, **16**, No. 2, p. 461–476.

[6] Radeva, T. (2001). *Physical Chemistry of Polyelectrolytes.* New York: Marcel Dekker Inc.

[7] Schmitz, K.S. (1993). *Macroions in Solution and Colloidal Suspension.* New York: VCH.

[8] Bhuiyan, L.B., Vlachy, V., and Outhwaite, C.W. Understanding polyelectrolyte solutions: macroion condensation with emphasis on the presence of neutral co-solutes. *International Reviews in Physical Chemistry*, 2002, **21**, No. 1, p. 1–36.

[9] Dolar, D. (1966). *Polyelectrolytes*, Riedel D.. Dordrecht.

[10] Vlachy, V. Ionic effects beyond Poisson-Boltzmann theory. *Annual Review of Physical Chemistry*, 1999, **50**, p. 145–165.

[11] Bratko, D., and Vlachy, V. Distribution of counterions in the double-layer around a cylindrical polyion. *Chemical Physics Letters*, 1982, **90**, No. 6, p. 434–438.

[12] Wandrey, C., Hunkeler, D., Wendler, U., and Jaeger, W. Counterion activity of highly charged strong polyelectrolytes. *Macromolecules*, 2000, **33**, No. 19, p. 7136–7143.

[13] Špan, J., Bratko, D., Dolar, D., and Feguš, M. Electrical transport in polystyrenesulfonate solutions. *Polymer Bulletin*, 1983, **9**, No. 1–3, p. 33–39.

[14] Drifford, M., Dalbiez, J.P., Delsanti, M., and Belloni, L. Structure and dynamics of polyelectrolyte solutions with multivalent salts. *Berichte der Bunsen-Gesellschaft-Physical Chemistry Chemical Physics*, 1996, **100**, No. 6, p. 829–835.

[15] Gröhn, F., and Antonietti, M. Intermolecular structure of spherical polyelectrolyte microgels in salt-free solution. 1. quantification of the attraction between equally charged polyelectrolytes. *Macromolecules*, 2000, **33**, No. 16, p. 5938–5949.

[16] Vesnaver, G., Kranjc, Z., Pohar, C., and Škerjanc, J. Free enthalpies, enthalpies, and entropies of dilution of aqueous-solutions of alkaline-earth poly(styrenesulfonates) at different temperatures. *Journal of Physical Chemistry*, 1987, **91**, No. 14, p. 3845–3848.

[17] Arh, K., Pohar, C., and Vlachy, V. Osmotic properties of aqueous ionene solutions. *Journal of Physical Chemistry B*, 2002, **106**, No. 38, p. 9967–9973.

[18] Vesnaver, G., Rudež, M., Pohar, C., and Škerjanc, J. Effect of temperature on the enthalpy of dilution of strong poly-electrolyte solutions. *Journal of Physical Chemistry*, 1984, **88**, No. 11, p. 2411–2414.

[19] Arh, K., and Pohar, C. Enthalpies of dilution of aqueous ionene solutions. *Acta Chimica Slovenica*, 2001, **48**, No. 3, p. 385–394.

[20] Keller, M., Lichtenthaler, R.N., and Heintz, A. Enthalpies of dilution of some polycation solutions and exchange enthalpies of polycation counterions. *Berichte der Bunsen-Gesellschaft-Physical Chemistry Chemical Physics*, 1996, **100**, No. 6, p. 776–779.

[21] Wall, T.F., and Berkowitz, J. Numerical solution to the Poisson - Boltzmann equation for spherical polyelectrolyte molecules. *Journal of Chemical Physics*, 1957, **26**, p. 114–122.

[22] Fuoss, R.M., Katchalsky, A., and Lifson, S. The potential of an infinite rod-like molecule and the distribution of the counter ions. *Proceedings of the National Academy of Sciences of the United States of America*, 1951, **37**, p. 579–589.

[23] Bratko, D., and Dolar, D. Ellipsoidal model of poly-electrolyte solutions. *Journal of Chemical Physics*, 1984, **80**, No. 11, p. 5782–5789.

[24] Bratko, D., and Vlachy, V. An application of the modified Poisson-Boltzmann equation in studies of osmotic properties of micellar solutions. *Colloid and Polymer Science*, 1985, **263**, No. 5, p. 417–419.

[25] Linse, P., Gunnarson, G., and Jönsson, B. Electrostatic interactions in micellar solutions - a comparison between Monte-Carlo simulations and solutions of the Poisson-Boltzmann equation. *Journal of Physical Chemistry*, 1982, **86**, No. 3, p. 413–421.

[26] Rebolj, N., Kristl, J., Kalyuzhnyi, Yu.V., and Vlachy, V. Structure and thermodynamics of micellar solutions in isotropic and cell models. *Langmuir*, 1997, **13**, No. 14, p. 3646–3651.

[27] Wennerström, H., Jönsson, B., and Linse, P. The cell model for poly-electrolyte systems - exact statistical mechanical relations, Monte-Carlo simulations, and the Poisson-Boltzmann approximation. *Journal of Chemical Physics*, 1982, **76**, No. 9, p. 4665–4670.

[28] Deserno, M., and Holm, C. Theory and simulations of rigid polyelectrolytes. *Molecular Physics*, 2002, **100**, No. 18, p. 2941–2956.

[29] Nishio, T., and Minakata, A. Effects of ion size and valence on ion distribution in mixed counterion systems of a rodlike polyelectrolyte solution. 2. mixed-valence counterion systems. *Journal of Physical Chemistry B*, 2003, **107**, No. 32, p. 8140–8145.

[30] Bell, G.M., and Levine, S. (1966). *Chemical Physics of Ionic Solutions*, p. 409–461. New York: Wiley.

[31] Das, T., Bratko, D., Bhuiyan, L.B., and Outhwaite, C.W. Polyelectrolyte solutions containing mixed valency ions in the cell model: A simulation and modified Poisson-Boltzmann study. *Journal of Chemical Physics*, 1997, **107**, No. 21, p. 9197–9207.

[32] Hribar, B., Krienke, H., Kalyuzhnyi, Yu.V., and Vlachy, V. Dilute solutions of highly asymmetrical electrolytes in the primitive model approximation. *Journal of Molecular Liquids*, 1997, **73**, No. 4, p. 277–289.

[33] Linse, P. Highly asymmetric electrolyte - comparison between one-component and 2-component models at different levels of approximations. *Journal of Chemical Physics*, 1991, **94**, No. 5, p. 3817–3828.

[34] Reščič, J., Vlachy, V., Outhwaite, C.W., Bhuiyan, L.B., and Mukherjee, A.K. A Monte Carlo simulation and symmetric Poisson-Boltzmann study of a four-component electrolyte mixture. *Journal of Chemical Physics*, 1999, **111**, No. 12, p. 5514–5521.

[35] Liu, Y.C., Chen, S.H., and Itri, R. Ion correlations and counter-ion condensation in ionic micellar solutions. *Journal of Physics-Condensed Matter*, 1996, **8**, No. 25A, p. A169–A187.

[36] Belloni, L. Electrostatic interactions in colloidal solutions – comparison between primitive and one-component models. *Journal of Chemical Physics*, 1986, **85**, No. 1, p. 519–526.

[37] Vlachy, V., and Prausnitz, J.M. Donnan equilibrium - hypernetted-chain study of one-component and multicomponent models for aqueous polyelectrolyte solutions. *Journal of Physical Chemistry*, 1992, **96**, No. 15, p. 6465–6469.

[38] Derjaguin, B.V., and Landau, L. Theory of stability of highly charged lyophobic sols and adhesion of highly charged particles in solutions of electrolytes. *Acta Phys. Chem. USSR*, 1941, **14**, p. 633–662.

[39] Verwey, E.J.W., and Overbeek, J.Th.G. (1948). *Theory of the Stability of Lyophobic Colloids*. New York: Elsevier.

[40] Hayter, J.B., and Penfold, J. An analytic structure factor for macroion solutions. *Molecular Physics*, 1981, **42**, No. 1, p. 109–118.

[41] Kalyuzhnyi, Yu.V., Reščič, J., and Vlachy, V. Analysis of osmotic pressure data for aqueous protein solutions via a one-component model. *Acta Chimica Slovenica*, 1998, **45**, No. 2, p. 194–208.

[42] Jiang, J.W., Blum, L., Bernard, O., and Prausnitz, J.M. Thermodynamic properties and phase equilibria of charged hard sphere chain model for polyelectrolyte solutions. *Molecular Physics*, 2001, **99**, p. 1121–1128.

[43] Jiang, J.W., Liu, H.L., Hu, Y., and Prausnitz, J.M. A molecular-thermodynamic model for polyelectrolyte solutions. *Journal of Chemical Physics*, 1998, **108**, No. 2, p. 780–784.

[44] Stevens, M.J., and Kremer, K. The nature of flexible linear polyelectrolytes in salt-free solution - a molecular-dynamics study. *Journal of Chemical Physics*, 1995, **103**, No. 4, p. 1669–1690.

[45] Collins, K. D. Charge density-dependent strength of hydration and biological structure. *Biophysical Journal*, 1997, **72**, No. 1, p. 65–76.

[46] Bernard, O., and Blum, L. Thermodynamics of a model for flexible polyelectrolytes in the binding mean spherical approximation. *Journal of Chemical Physics*, 2000, **112**, No. 16, p. 7227–7237.

[47] Blum, L., Kalyuzhnyi, Yu.V., Bernard, O., and Herrera-Pacheco, J.N. Sticky charged spheres in the mean-spherical approximation: A model for colloids and polyelectrolytes. *Journal of Physics – Condensed Matter*, 1996, **8**, No. 25A, p. A143–A167.

[48] Kalyuzhnyi, Yu.V. Thermodynamics of the polymer mean-spherical ideal chain approximation for a fluid of linear chain molecules. *Molecular Physics*, 1998, **94**, p. 735–742.

[49] Kalyuzhnyi, Yu.V., and Cummings, P.T. Multicomponent mixture of charged hard-sphere chain molecules in the polymer mean-spherical approximation. *Journal of Chemical Physics*, 2001, **115**, p. 540–551.

[50] Kalyuzhnyi, Yu.V., and Cummings, P.T. Multicomponent mixture of charged hard-sphere chain molecules in the polymer mean-spherical approximation (vol 115, pg 540, 2001). *Journal of Chemical Physics*, 2002, **116**, p. 8637–8637.

[51] Kalyuzhnyi, Yu.V., and Stell, G. Solution of the polymer msa for the polymerizing primitive model of electrolytes. *Chemical Physics Letters*, 1995, **240**, p. 157–164.

[52] Protsykevytch, I.A., Kalyuzhnyi, Yu.V., Holovko, M.F., and Blum, L. Solution of the polymer mean spherical approximation for the totally flexible sticky two-point electrolyte model. *Journal of Molecular Physics*, 1997, **73**, No. 4, p. 1–20.

[53] von Solms, N., and Chiew, Y.C. Analytical integral equation theory for a restricted primitive model of polyelectrolytes and counterions within the mean spherical approximation. i. thermodynamic properties. *Journal of Chemical Physics*, 1999, **111**, No. 10, p. 4839–4850.

[54] von Solms, N., and Chiew, Y.C. Analytical integral equation theory for a restricted primitive model of polyelectrolytes and counterions within the mean spherical approximation. ii. radial distribution functions. *Journal of Chemical Physics*, 2003, **118**, No. 9, p. 4321–4330.

[55] Wertheim, M. S. Fluids with highly directional attractive forces. 3. multiple attraction sites. *Journal of Statistical Physics*, 1986, **42**, No. 3–4, p. 459–476.

[56] Liao, Q., Dobrynin, A.V., and Rubinstein, M. Molecular dynamics simulations of polyelectrolyte solutions: Osmotic coefficient and counterion condensation. *Macromolecules*, 2003, **36**, No. 9, p. 3399–3410.

[57] Zhang, B., Yu, D.H., Liu, H.L., and Hu, Y. Osmotic coefficients of polyelectrolyte solutions, measurements and correlation. *Polymer*, 2002, **43**, No. 10, p. 2975–2980.

[58] Dymitrowska, M., and Belloni, L. Integral equation theory of flexible polyelectrolytes. ii. primitive model approach. *Journal of Chemical Physics*, 1999, **111**, No. 14, p. 6633–6642.

[59] Hofmann, T., Winkler, R.G., and Reineker, P. Integral equation theory approach to rodlike polyelectrolytes: Counterion condensation. *Journal of Chemical Physics*, 2001, **114**, No. 22, p. 10181–10188.

[60] Friedman, H. L. (1985). *A Course on Statistical Mechanics*. Prentice-Hall: Englewood Cliffs.

[61] Rasaiah, J. C. (1988). *The Liquid States and its Electrical Properties*. New York: Plenum.

[62] Vlachy, V., Ichiye, T., and Haymet, A.D.J. Symmetrical associating electrolytes – gcmc simulations and integral-equation theory. *Journal of the American Chemical Society*, 1991, **113**, No. 4, p. 1077–1082.

[63] Belloni, L. A hypernetted chain study of highly asymmetrical poly-electrolytes. *Chemical Physics*, 1985, **99**, No. 1, p. 43–54.

[64] Vlachy, V., Marshall, C.H., and Haymet, A.D.J. Highly asymmetric electrolytes – a comparison of Monte-Carlo simulations and the HNC integral-equation. *Journal of the American Chemical Society*, 1989, **111**, No. 12, p. 4160–4166.

[65] Blum, L. (1980). *Theoretical Chemistry; Advances and Perspectives*. New York: Academic Press.

[66] Blum, L., and Hoye, J.S. Mean spherical model for asymmetric electrolytes .2. Thermodynamic properties and pair correlation-function. *Journal of Physical Chemistry*, 1977, **81**, No. 13, p. 1311–1317.

[67] Harvey, A.H., Copeman, T.W., and Prausnitz, J.M. Explicit approximations to the mean spherical approximation for electrolyte systems with unequal ion sizes. *Journal of Physical Chemistry*, 1988, **92**, No. 22, p. 6432–6436.

[68] Sanchez-Castro, C., and Blum, L. Explicit approximation for the unrestricted mean spherical approximation for ionic-solutions. *Journal of Physical Chemistry*, 1989, **93**, No. 21, p. 7478–7482.

[69] Simonin, J.P., Bernard, O., and Blum, L. Real ionic solutions in the mean spherical approximation. 3. osmotic and activity coefficients for associating electrolytes in the primitive model. *Journal of Physical Chemistry B*, 1998, **102**, No. 22, p. 4411–4417.

[70] Kalyuzhnyi, Yu.V., and Vlachy, V. Integral-equation theory for highly asymmetric electrolyte-solutions. *Chemical Physics Letters*, 1993, **215**, No. 5, p. 518–522.

[71] Kalyuzhnyi, Yu.V., Vlachy, V., Holovko, M.F., and Stell, G. Multidensity integral-equation theory for highly asymmetric electrolyte-solutions. *Journal of Chemical Physics*, 1995, **102**, No. 14, p. 5770–5780.

[72] Holovko, M.F., and Kalyuzhnyi, Yu.V. On the effects of association in the statistical-theory of ionic systems - analytic solution of the PY-MSA version of the Wertheim theory. *Molecular Physics*, 1991, **73**, No. 5, p. 1145–1157.

[73] Kalyuzhnyi, Yu.V., Holovko, M.F., and Vlachy, V. Highly asymmetric electrolytes in the associative mean-spherical approximation. *Journal of Statistical Physics*, 2000, **100**, No. 1–2, p. 243–265.

[74] Kalyuzhnyi, Yu.V., Blum, L., Holovko, M.F., and Protsykevytch, I.A. Primitive model for highly asymmetric electrolytes. associative mean spherical approximation. *Physica A*, 1997, **236**, No. 1–2, p. 85–96.

[75] Kalyuzhnyi, Yu.V., and Cummings, P.T. Solution of the polymer Percus-Yevick approximation for the multicomponent totally flexible sticky 2-point model of polymerizing fluid. *Journal of Chemical Physics*, 1995, **103**, No. 8, p. 3265–3267.

[76] Baumgartner, E., and Fernandez-Prini, R. (1976). *Polyelectrolytes*, p. 1–33. Westport: Technomic, CT.

[77] Morawetz, H. Chemical reaction rates reflecting physical properties of polymer solutions. *Accounts of Chemical Research*, 1970, **3**, No. 10, p. 354–360.

[78] Morawetz, H. Revisiting some phenomena in polyelectrolyte solutions. *Journal of Polymer Science part B-Polymer Physics*, 2002, **40**, No. 11, p. 1080–1086.

[79] Morawetz, H., and Shaffer, J.A. Characterization of counterion distribution in polyelectrolyte solutions. ii. the effect of the distribution of electrostatic potential on the solvolysis of cationic esters in polymeric acid solution. *Journal of Physical Chemistry*, 1963, **67**, p. 1293–1297.

[80] Morawetz, H., and Vogel, B. Catalysis of ionic reactions by polyelectrolytes. reaction of pentaamminechlorocobalt(iii) ion and pentaamminebromocobalt(iii) ion with mercuric ion in poly(sulfonic acid) solution. *Journal of the American Chemical Society*, 1969, **91**, No. 3, p. 563–568.

[81] Rodenas, E., Dolcet, C., and Valiente, M. Simulations of micelle-catalyzed bimolecular reaction of hydroxide ion with a cationic substrate using the nonlinearized Poisson-Boltzmann equation. *Journal of Physical Chemistry*, 1990, **94**, No. 4, p. 1472–1477.

[82] Wensel, T.G., Meares, C.F., Vlachy, V., and Matthew, J.B. Distribution of ions around DNA, probed by energy-transfer. *Proceedings of the National Academy of Sciences of the United States of America*, 1986, **83**, No. 10, p. 3267–3271.

[83] Tapia, M.J., and Burrows, H.D. Cation polyelectrolyte interactions in aqueous sodium poly(vinyl sulfonate) as seen by Ce3+ to Tb3+ energy transfer. *Langmuir*, 2002, **18**, No. 5, p. 1872–1876.

[84] Tapia, M.J., Burrows, H.D., Azenha, M.E.D.G., Miguel, M.G., Pais, A.A.C.C., and Sarraguca, J.M.G. Cation association with sodium dodecyl sulfate micelles as seen by lanthanide luminescence. *Journal of Physical Chemistry B*, 2002, **106**, No. 27, p. 6966–6972.

[85] Keizer, J. Theory of rapid biomolecular reactions in solution and membranes. *Accounts of Chemical Research*, 1985, **18**, No. 8, p. 235–241.

[86] Reščič, J., and Vlachy, V. (1994). *Ion-ion correlation in the electrical double layer around a cylindrical polyion*, pages 24–33. Washington:ACS, DC.

[87] Reščič, J., Vlachy, V., Bhuiyan, B.L., and Outhwaite, C.W. Theoretical study of catalytic effects in micellar solutions. *submitted to Langmuir*, 2004.

[88] Ise, N. When does like like like? microscopic inhomogeneity in homogeneous ionic systems. *Proceedings of the Japan Academy Series B-Physical and Biological Sciences*, 2002, **78**, No. 6, p. 129–137.

[89] Ito, K., Yoshida, H., and Ise, N. Void structure in colloidal dispersions. *Science*, 1994, **263**(5143), No. 4, p. 66–68.

[90] Larsen, A.E., and Grier, D.G. Like-charge attractions in metastable colloidal crystallites. *Nature*, 1997, **385**, No. 6613, p. 230–233.

[91] Matsuoka, H., Harada, T., Kago, K., and Yamaoka, H. Exact evaluation of the salt concentration dependence of interparticle distance in colloidal crystals by ultra-small-angle x-ray scattering .2. the universality of the maximum in the interparticle distance salt concentration relationship. *Langmuir*, 1996, **12**, No. 23, p. 5588–5594.

[92] Ohshima, A., Konishi, T., Yamanaka, J., and Ise, N. "Ordered" structure in ionic dilute solutions: Dendrimers with univalent and bivalent counterions. *Physical Review E*, 2001, **6405**, No. 5, p. 051808.

[93] Zhang, B., Liu, H.L., and Hu, Y. Attraction between like-charge particles. *Progress in Chemistry*, 2001, **13**, No. 1, p. 1–9.

[94] Quesada-Perez, M., Gonzalez-Tovar, E., Martin-Molina, A., Lozada-Cassou, M., and Hidalgo-Alvarez, R. Overcharging in colloids: Beyond the Poisson-Boltzmann approach. *ChemPhysChem*, 2003, **4**, No. 3, p. 235–248.

[95] Spalla, O. Long-range attraction between surfaces: existence and amplitude? *Current Opinion in Colloid & Interface Science*, 2000, **5**, No. 1–2, p. 5–12.

[96] Hribar, B., and Vlachy, V. Evidence of electrostatic attraction between equally charged macroions induced by divalent counterions. *Journal of Physical chemistry B*, 1997, **101**, No. 18, p. 3457–3459.

[97] Hribar, B., and Vlachy, V. Clustering of macroions in solutions of highly asymmetric electrolytes. *Biophysical Journal*, 2000, **78**, No. 2, p. 694–698.

[98] Hribar, B., and Vlachy, V. Macroion-macroion correlations in presence of divalent counterions. *Journal of Physical chemistry B*, 2000, **104**, No. 17, p. 4218–4221.

[99] Hribar, B., and Vlachy, V. Macroion-macroion correlations in the presence of divalent counterions. effects of a simple electrolyte. *Acta Chimica Slovenica*, 2000, **47**, No. 2, p. 123–131.

[100] Hribar, B., and Vlachy, V. Properties of polyelectrolyte solutions as determined by the charge of counterions. *Revista de la Sociedad Quimica de Mexico*, 2000, **44**, p. 11–15.

[101] Hribar, B., and Vlachy, V. Monte Carlo study of micellar solutions with a mixture of mono- and trivalent counterions. *Langmuir*, 2001, **17**, No. 6, p. 2043–2046.

[102] Spohr, E., Hribar, B., and Vlachy, V. Mechanism of macroion-macroion clustering induced by the presence of trivalent counterions. *Journal of Physical chemistry B*, 2002, **106**, No. 9, p. 2343–2348.

[103] Spohr, E., Hribar, B., and Vlachy, V. unpublished results, 2004.

[104] Linse, P. Structure, phase stability, and thermodynamics in charged colloidal solutions. *Journal of Chemical Physics*, 2000, **113**, No. 10, p. 4359–4373.

[105] Reščič, J., and Linse, P. Gas-liquid phase separation in charged colloidal systems. *Journal of Chemical Physics*, 2001, **114**, No. 22, p. 10131–10136.

[106] Allen, M.P., and Tildesley, D.J. (1989). *Computer Simulation of Liquids*. Oxford: Clarendon Press.

[107] Blaul, J., Wittemann, M., Ballauff, M., and Rehahn, M. Osmotic coefficient of a synthetic rodlike polyelectrolyte in salt-free solution as a test of the Poisson-Boltzmann cell model. *Journal of Physical Chemistry B*, 2000, **104**, No. 30, p. 7077–7081.

[108] Deserno, M., Holm, C., Blaul, J., Ballauff, M., and Rehahn, M. The osmotic coefficient of rod-like polyelectrolytes: Computer simulation, analytical theory, and experiment. *European Physical Journal E*, 2001, **5**, No. 1, p. 97–103.

[109] George, A., and Wilson, W.W. Predicting protein crystallization from a dilute-solution property. *Acta Crystallographica Section D-Biological Crystallography*,1994, **50**, p. 361–365.

[110] Haynes, C.A., Tamura, K., Korfer, H. R., Blanch, H. W., and Prausnitz, J.M. Thermodynamic properties of aqueous alpha-chymotrypsin solutions from membrane osmometry measurements. *Journal of Physical Chemistry*, 1992, **96**, No. 2, p. 905–912.

[111] Reščič, J., Vlachy, V., Jamnik, A., and Glatter, O. Osmotic pressure, small-angle x-ray, and dynamic light scattering studies of human serum albumin in aqueous solutions. *Journal of Colloid and Interface Science*, 2001, **239**, No. 1, p. 49–57.

[112] Vilker, V.L., Colton, C.K., and Smith, K.A. The osmotic-pressure of concentrated protein solutions - effect of concentration and ph in saline solutions of bovine serum-albumin. *Journal of Colloid and Interface Science*, 1981, **79**, No. 2, p. 548–566.

[113] Allahyarov, E., Lowen, H., Hansen, J.P., and Louis, A.A. Nonmonotonic variation with salt concentration of the second virial coefficient in protein solutions. *Phys. Rev. E*, 2003, **67**, No. 5, p. 051404.

[114] Haas, C., Drenth, J., and Wilson, W.W. Relation between the solubility of proteins in aqueous solutions and the second virial coefficient of the solution. *Journal of Physical Chemistry B*, 1999, **103**, No. 14, p. 2808–2811.

[115] Kalyuzhnyi, Yu.V., and Vlachy, V. Study of a model polyelectrolyte solution with directional attractive forces between the macroions. *Journal of Chemical Physics*, 1998, **108**, No. 18, p. 7870–7875.

[116] Jimenez-Angeles, F., and Lozada-Cassou, M. Simple model for semipermeable membrane: Donnan equilibrium. *Journal of Physical Chemistry B*, 2004, **108**, No. 5, p. 1719–1730.

[117] Piazza, R. Protein interactions and association: an open challenge for colloid science. *Current Opinion in Colloid & Interface Science*, 2004, **8**, No. 6, p. 515–522.

[118] Vlachy, V., Hribar-Lee, B., Kalyuzhnyi, Yu.V., and Dill, K.A. Short–range interactions: from simple ions to polyelectrolyte solutions. *Current Opinion in Colloid & Interface Science*, 2004.

ANOMALOUS SMALL-ANGLE X-RAY SCATTERING IN ROD-LIKE POLYELECTROLYTES

A new tool to study the counterion distribution around rod-like macroions

M. Ballauff
Physikalische Chemie I, University of Bayreuth
95440 Bayreuth, Germany

Abstract   We present anomalous small-angle X-ray scattering (ASAXS) as a new tool for the study of rod-like polyelectrolyte. The polyelectrolyte consists of a stiff poly(p-phenylene) backbone with attached positive charged groups that are balanced by Bromine counterions. ASAXS uses the dependence of the scattering length of a given element if the energy of the incident X-ray beam is near the absorption edge of this element. The analysis of the ASAXS-data leads to three partial intensities. In particular, the scattering intensity that is solely due to the cloud of the counterions can be determined and compared to the prediction of the Poisson-Boltzmann cell model. Quantitative agreement is found. ASAXS is thus shown to be a new and highly effective tool for the analysis of polyelectrolytes.

Keywords: Polyelectrolytes, ASAXS, SAXS, counterion condensation

1.   Introduction

If salts of monovalent counterions are dissolved in water or other solvents of sufficiently high dielectric constant, full dissociation may take place [1]. The driving force of this process as well as the limits of solubility are well-understood by now. Moreover, the activity of the resulting ions can be treated in terms of the classical Debye-Hückel theory. In this theory infinite dilution is taken as the reference state. This is due to the fact that the ratio of the electrostatic energy to $k_B T$ is small for mono- or divalent salts. This means that counterions can be "diluted away" if the concentration is more and more lowered.

However, if a sufficiently high number of charges per unit length are affixed to a polymeric chain, a highly charged macroion results [2, 3]. As a consequence of this, the electrostatic energy of a counterion near to such a

D. Henderson et al. (eds.), Ionic Soft Matter: Modern Trends in Theory and Applications, 233–247.
© 2005 *Springer. Printed in the Netherlands.*

polyelectrolyte is not small anymore as compared to k_BT. Therefore there will be a strong correlation of a part of the counterions to the macroion and the definition of the reference state requires special care. For linear rod-like polyelectrolytes this problem has been treated already by Fuoss, Katchalski and their coworkers [4] and by Alfrey and coworkers [5] more than 50 years ago by introducing the Poisson-Boltzmann (PB) cell model [6]. Later Manning introduced the notion of counterion condensation in order to describe this marked correlation [7]. The main point of all these models is the fact that a certain fraction of the counterions is firmly correlated to the macroion. This correlation termed counterion condensation persist up to the highest dilution possible. In this way the high charge density leads to a new feature that cannot be captured properly by the conventional Debye-Hückel theory [2, 3].

In general, linear polyelectrolyte is characterized by its charge parameter ξ,

$$\xi = \frac{\lambda_B}{b}, \tag{1}$$

where b is the distance per unit charge and λ_B is the Bjerrum length given by $\lambda_B = e^2/4\pi\epsilon_0\epsilon k_BT$ where e is the unit charge, ϵ is the dielectric constant of the medium, while ϵ_0, k_B, and T have the usual meanings [2, 3, 6]. The charge parameter is the decisive quantity that compares the strength of electrostatic interaction to the thermal energy k_BT: If $\xi < 1$ the electrostatic attraction is weak as compared to the thermal energy and the Debye-Hückel approach is sufficient. However, if $\xi > 1$, counterion condensation sets in and the activity of the counterions will be strongly reduced. For rod-like polyelectrolytes the predictions of the PB cell model have been compared to computer simulations [8]. For monovalent salt a good agreement of simulated data with the predictions of the cell model was found. Divalent and higher-valent counterions, however, lead to much less satisfactory agreement.

Small-angle X-ray scattering (SAXS) or small-angle neutron scattering (SANS) are ideal tools for the investigation of the correlation of the counterions to rod-like macroions of length L [9]. For rods the measured intensity $I(q)$ (q: magnitude of scattering vector $\vec{q}$; $q = (4\pi/\lambda)\sin(\theta/2)$; λ: wavelength of radiation, θ: scattering angle) probes only particles whose long axis is approximately perpendicular to the direction of $\vec{q}$ if $2\pi/q > L$ [10]. Hence, the radial distribution of the counterions is directly related to the measured intensity $I(q)$ by a Hankel-transform. This is a very important point inasmuch as the loss of information in small-angle experiments due to an average over 4π is not taking place. Rod-like polyelectrolytes are therefore a class of model systems for which the prediction of theory can be tested best.

However, a full analysis of polyelectrolytes by SAXS or SANS requires that the scattering distribution of the counterions can be separated from the respective signal of the macroion. Hence, it should be possible to vary the contrast of the counterions in order to distinguish their signal and determine their partial scattering intensity. In case of SANS this can be done by using deuterated counterions [11]. SAXS, on the other hand, requires exchange of counterions as e.g. exchange of Chloride against Iodide ions to obtain information on the distribution of the counterion around a polymeric macroion [12–14]. However, both the use of bulky deuterated ions as well as the exchange of counterions may lead to artifacts. The interaction of the counterions with the macroion may entail specific effects as e.g. hydrophobic forces that depend on the very chemical nature of the counterion [2]. A method that could use conventional counterions but which could distinguish between the different contributions to the scattering would hence be a valuable tool for the study of polyelectrolytes in general.

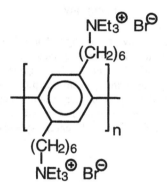

Figure 1 Chemical structure of the rodlike polyelectrolyte used for the ASAXS-study discussed here [19, 27].

Many years ago, Stuhrmann showed that anomalous dispersion can circumvent this problem in studies of polyelectrolytes by SAXS [15, 16]. This method utilizes the dependence of the scattering factor f if the energy of the incident radiation is near the absorption edge of the counterions [16]. Hence, he scattering factor f_{ion} becomes a complex function of the energy E of the incident radiation near the absorption edge of the ions [15, 16],

$$f_{ion} = f_0 + f'(E) + \tilde{i}f''(E). \tag{2}$$

The first term f_0 is the non-resonant term which equals the atomic number of the element [9]. The second and the third term in Eq. (2) are the real and the imaginary part due to the anomalous dispersion near the absorption edge and $\tilde{i}$ is the imaginary unit. The imaginary part $f''(E)$ is directly related to the absorption cross section for X-rays of energy E. Figure 2 displays the scattering factors thus defined as the function of the energy E of the incident radiation. Measurements of the small-angle X-ray scattering intensity near the absorption edge of the Bromine counterions allow therefore to change the scattering contribution of the Br^- ions in a systematic fashion while keeping all

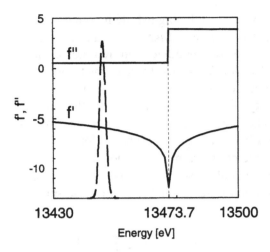

Figure 2.   Dependence of the effective scattering factors f'_{eff} and f''_{eff} of Bromine on the energy E of the incident X-ray beam. The scattering factor f' taken from [24, 25] is plotted against the energy of the incident beam. The energy of the edge is marked by a dashed line. The breadth of the energy distribution of the primary beam is approximated by a Gaussian and given by a dash-dotted line. The finite width of the primary beam imposes no problem unless if the experiment is done in the immediate neighborhood of the edge. Then the effective scattering factors f'_{eff} and f''_{eff} result from a convolution of the energy spread of the primary beam with f' or f'', respectively [19, 23].

other contributions constant. Hence, ASAXS can separately assess the spatial distribution of the macroion and the counterions which is not possible by the conventional SAXS-experiment.

Very recently, we [17–19] and others [20] have shown that anomalous small-angle X-ray scattering (ASAXS) can be used indeed for the study of polyelec-trolytes. In particular, it was shown that ASAXS leads in general to three partial intensities [17]:

  i) a non-resonant term that originates from non-resonant part of the scat-tering length density,

  ii) a cross term of the non-resonant and the resonant amplitude,

  iii) and a self-term related solely to the resonant part of the scattering length density.

A first experimental study was presented as well in which the anomalous dis-persion of Iodine counterions was used [17, 18]. However, the high energy necessary in order to reach the absorption edge (33 keV), rendered the data less secure. Recently, we have demonstrated that the three terms can be obtained in good accuracy from an ASAXS-experiment [19].

Here we review the application of ASAXS as applied to the analysis of stiff chain polyelectrolyte in solution. The data discussed here [19] have been obtained using the polyelectrolyte the chemical structure of which is shown in Fig. 1. This system has already been under scrutiny by conventional SAXS some time ago [14]. The paper is organized as follows: first we summarize the theory of ASAXS and its application to the problem at hand [18]. Moreover, we will briefly summarize the treatment of rod-like polyelectrolytes within the frame of the Poisson-Boltzmann cell model. An important point for the present analysis is the influence of mutual interaction of the dissolved polyelectrolytes. ASAXS-measurements need to be done at rather higher concentrations so that the interaction of the solute rods may come into play. Here it will be shown that this problem is negligible for the present system. Next possible difficulties encountered in an ASAXS experiment will be discussed and experimental results will be presented. A brief final section will conclude the present discussion.

2. Theory

2.1 *Theory of the anomalous small-angle X-ray scattering*

The scattering intensity originating from a solution of rod-like objects can be rendered as [10, 21],

$$I(q) = \frac{N}{V} I_0(q) S(q) \,. \tag{3}$$

Here N/V is the number of the dissolved polyelectrolyte molecules per volume whereas $I_0(q)$ denotes the scattering intensity of an isolated macromolecule. $S(q)$ is the effective structure factor that takes into account the effect of finite concentrations. As shown further below its influence can be disregarded for higher scattering angles.

For cylindrical objects in which the scattering length density varies only along the radial distance termed r_c, scattering intensity, $I_0(q)$, follows as [10],

$$I_0(q) = \int_0^1 F(q, \alpha) F^*(q, \alpha) d\alpha \,, \tag{4}$$

where α is the cosine of the angle between the $\vec{q}$ and $\vec{z}$. Here $\vec{z}$ denotes the unit vector along the long axis of the rod. For rods of length L it is given by [10],

$$F(q, \alpha) = L \frac{\sin(q\alpha L/2)}{q\alpha L/2} F_{cr}[\Delta\rho(r_c), q, \alpha] \,. \tag{5}$$

The amplitude of the cross section of the molecules $F_{cr}(\Delta\rho(r_c), q, \alpha)$ follows as,

$$F_{cr}(\Delta\rho(r_c), q, \alpha) = \int_0^\infty \Delta\rho(r_c) J_0 \left[q r_c \left(1 - \alpha^2\right)^{1/2} \right] 2\pi r_c dr_c, \qquad (6)$$

where $\Delta\rho(r_c)$ is the radial excess scattering length density and $J_0(x)$ is the Bessel-function of zeroth order. The scattering length densities can be obtained from the product of the local electron densities and the Thomson factor r_0, the scattering length of a single electron [10].

As mentioned above, for $q \gg 2\pi/L$ the rod gives only a contribution to the measured intensity if $\vec{q}$ is perpendicular to the long axis [10], i. e., if $\alpha \simeq 0$. Thus, the intensity measured at higher scattering angles is directly related to the Hankel-transform of the excess electron density $\Delta\rho(r_c)$ which means that [10],

$$I_0(q) \approx L \frac{\pi}{q} F_{cr}(\Delta\rho(r_c), q_\perp) F_{cr}^*(\Delta\rho(r_c), q_\perp), \qquad (7)$$

which for polydisperse samples can be re-written as, [18]

$$I_0(q) \approx L_n \left[\frac{\pi}{q} F_{cr} \left(\Delta\rho(r_c), q_\perp \right) F_{cr}^*(\Delta\rho(r_c), q_\perp) \right], \qquad (8)$$

where L_n is the number-averaged contour length.

The scattering of the macroion can be modelled in terms of a real excess electron density $\Delta\rho_{rod}$. With a being the minimum approach of the macroion and the counterions, it follows that $\Delta\rho(r_c) = \Delta\rho_{rod}$ for all $r_c \leqslant a$. For $r_c \geqslant a$ $\Delta\rho(r_c)$ is solely determined by the excess electron density of the counterions. Evidently, the integration in Eq. (6) must include all counterions, otherwise the condition of electroneutrality would be violated [17, 18].

Let $n(r_c)$ denote the radial number density of the counterions. It follows that $\Delta\rho(r_c) = \Delta f_{ion} n(r_c)$ for $r_c \geqslant a$ where Δf_{ion} is the number of excess electrons of a single ion. This quantity is given by [17, 18],

$$\Delta f_{ion} = f_0 - \rho_m V_{ion} + f' + \tilde{i} f'', \qquad (9)$$

where V_{ion} is the volume of the counterion, f_0 its non-resonant scattering factor, and f' and f'' are the real and the imaginary part of the energy-dependent scattering factor [15], respectively, while $\tilde{i}$ is the imaginary unit. The quantity ρ_m is the electron density of the aqueous solvent which is independent of energy. The respective scattering lengths are obtained by multiplication of these quantities by the Thomson factor r_0 [10, 15]. For the calculation of the SAXS-intensity measured far away from the absorption edge it suffices to take into account f_0. Only if measurements are conducted in the immediate neighborhood of the absorption edge of the counterions, contributions due to f' and f'' must be taken into account [15]. The excess electron density to be introduced into Eq. (6) follows as,

$$
\Delta\rho(r_c) = \begin{cases} \Delta\rho_{\text{rod}}, & 0 \leqslant r_c \leqslant a, \\ n(r_c)\Delta f_{\text{ion}}, & a \leqslant r_c \leqslant R_0, \\ 0, & r_c > R_0. \end{cases} \tag{10}
$$

Here R_0 denotes the cell radius introduced by the Poisson-Boltzmann cell model (see below). The term $\Delta\rho(r_c)$ may therefore be split into a non-resonant contribution $\Delta\rho_0(r_c)$ and the resonant contributions of the counterions according to [17],

$$
\Delta\rho(r_c) = \Delta\rho_0(r_c) + n(r_c)f' + \tilde{i}n(r_c)f'', \tag{11}
$$

where $\tilde{i}$ is the imaginary unit. Insertion of Eq. (11) into Eqs. (6) and (4) leads to three terms that are related to the Hankel-transforms of the terms $\Delta\rho_0(r_c)$, $n(r_c)f'$, and $n(r_c)f''$, respectively [17],

$$
\begin{aligned}
F(q,\alpha)F^*(q,\alpha) = & \left\{ L\frac{\sin(q\alpha L/2)}{q\alpha L/2} \right\}^2 \left\{ F_{\text{cr}}^2 \left[\Delta\rho_0(r_c), q, \alpha \right] \right. \\
& + 2f' F_{\text{cr}} \left[\Delta\rho_0(r_c), q, \alpha \right] F_{\text{cr}} \left[n(r_c), q, \alpha \right] \\
& \left. + \left(f'^2 + f''^2 \right) F_{\text{cr}}^2 \left[n(r_c), q, \alpha \right] \right\}. \tag{12}
\end{aligned}
$$

Equation (12) is the central result of the analysis. It demonstrates that ASAXS leads to three partial intensities. The first term $F_{\text{cr}}^2[\Delta\rho_0(r_c), q, \alpha]$ in the curved brackets leads to the usual SAXS-intensity measured far away from the absorption edge. The additional terms give the modification of $I_0(q)$ in the vicinity of the absorption edge. In many cases the cross term (second term of Eq. (12)) is dominating the ASAXS-terms. The self-term (third term of Eq. (12)), however, may be dominant if the contrast of the macroion is small (see [17] for an extended discussion of this point).

2.2   Poisson-Boltzmann cell model

An extensive discussion of the PB cell model has recently been presented by Deserno and coworkers [8] (see also preceding chapter by Vlachy et al). Hence, it suffices here to delineate the main features and how this model is compared to data deriving from a scattering experiment. The PB cell model treats the system as an assembly of N rods confined in cells of radius R_0.

The cell radius R_0 is determined by the condition $(N/V)\pi R_0^2 L = 1$. The distribution function $n(r_c)$ is then given by [22],

$$\frac{n(r_c)}{n(R_0)} = \left\{ \frac{2|\beta|}{\kappa r_c \cos[\beta \ln(r_c/R_M)]} \right\}^2. \tag{13}$$

The integration constant β can be calculated by use of the condition [22],

$$\arctan\left(\frac{\xi - 1}{\beta}\right) + \arctan\left(\frac{1}{\beta}\right) - \beta \ln\left(\frac{R_0}{a}\right) = 0, \tag{14}$$

for a set of parameters ξ, a, and R_0. The second constant R_M is given by,

$$R_M = a \exp\left\{ \frac{1}{\beta} \arctan\left(\frac{\xi - 1}{\beta}\right) \right\}. \tag{15}$$

The screening constant, $\kappa = 8\pi \lambda_B n(R_0) = 4(1 + \beta^2)/R_0^2$ [22]. The distribution function $n(r_c)$ thus obtained for a given number density N/V can be used to calculate the respective scattering intensity according to Eqs. (4) and (12).

Since the polyelectrolyte shown in Fig. 1 has already been studied by conventional SAXS, all parameters can be taken from this work [14]. Hence, the charge parameter ξ is 3.3. The cell radius R_0 is determined from the number density of the rod-like polyelectrolyte. Subsequently the integration constants β and R_M in Eqs. (14) and (15), are determined.

2.3   *Structure factor*

As mentioned above, ASAXS measurements require rather high concentrations of the dissolved polyelectrolytes. The cell model assumes a decoupling of the different polyelectrolytes. All calculations discussed in the Sec. 2.2 are referring explicitly to a hypothetical state in which the concentration of the polyelectrolyte is finite, i. e., the cell radius R_0 assumes a finite value but $S(q)$ (see Eq. (3)) is unity. Considering the influence of finite concentration it is important to note that the PB cell model predicts that most of the counterions are located in the immediate vicinity of the macroion (see the model calculations in [18]. Hence, an increase of concentration followed by a decrease of the distance between the rods is hardly seen in $I_0(q)$ calculated by this model. For the range of concentrations that are ranging from 1 to 20.0 g/L the cloud of counterions therefore does not change its spatial distribution in a profound manner. This prediction which is borne out of the cell model is well confirmed by experiments [18].

The foregoing consideration refers to the region of higher scattering angles in which the fine details of the clouds of the counterions become visible. Given the strong correlation of the counterions to the macroion it seems to be appropriate to consider the structure factor $S(q)$ (see Eq. (3)) in terms of a simplified model [18]. In the following we use the "decoupling approximation" [21, 26] which has been discussed and used frequently to describe $S(q)$ of anisometric objects. In this model an isotropic solution of rods of equal length is considered. The correlation of the centers of gravity of the rods is characterized by the pair correlation function $g(R)$. There is no correlation between the mutual orientation of the rods and the distance of their centers of gravity. In this approximation $S(q)$ follows as [21],

$$S(q) = 1 + \frac{\langle F(q,\alpha)\rangle_\Omega^2}{\langle F^2(q,\alpha)\rangle_\Omega} 4\pi \frac{N}{V} \int_0^\infty [g(R) - 1] \frac{\sin(qR)}{qR} R^2 dR, \qquad (16)$$

where $\langle \ldots \rangle_\Omega$ denotes the average over all orientations. This expression shows that the finite concentrations lead to a change of $S(q)$ because $g(R)$ deviates from unity. This is the usual correlation hole effect that is followed by a decrease of $S(q)$ at small q. On the other hand, the front factor of the integral in Eq. (16) is unity only at $q = 0$, for finite q-values it will diminish quickly and suppress the effect of the correlation hole. Hence, Eq. (16) predicts that $S(q)$ for suspensions of rod-like objects will deviate from unity only in the region of smallest scattering angles. For most of the q-range considered here it will be of marginal importance.

3. Experimental problems

In principle, ASAXS as a method has been known for a long time. Indeed, Stuhrmann was the first to apply this method to polyelectrolytes and to show the general feasibility [15]. By the time these experiments have been done, however, the experimental technique has not advanced enough to render ASAXS a tool having the necessary accuracy. In the following we shall enumerate briefly the main experimental problems of this method:

1. An important problem that need to be taken into account is the resolution of the energy of the primary beam. Figure 2 demonstrates this by showing the profile of the primary beam together with $f'(E)$ and $f''(E)$. It is obvious that the variation of f' which is most pronounced in the immediate neighborhood of the edge can only be used if the width of the primary beam is small enough. Moreover, a precise evaluation of the data requires that the finite width of the primary beam is taken into account by an appropriate average over f' and f'' [23]. This is done by weighing the respective scattering factors by the profile of the primary beam shifted to the respective energy. It leads to the effective scattering factors f'_{eff} and f''_{eff} which differ from f' and f'' in the immediate vicinity of the edge [19, 23].

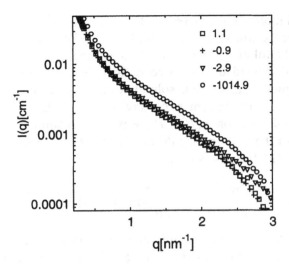

Figure 3. ASAXS intensities corrected for the fluorescence and the parasitic background by the solvent water. The difference of the energy of the incident beam to the edge is indicated in the graph.

2. The absorption edge must be localized with the highest precision possible. The ASAXS rests mainly on the variation of f' with the energy E of the incident beam. Hence, precise measurements of the absorption may be used to find the exact position of the edge and to detect small shifts of the monochromator of the instrument [19, 23].

3. The parasitic background caused by the scattering of water and by fluorescence must be subtracted properly. Fluorescence comes into play even below the edge because of the finite width of the primary beam (see Fig. 2). In the immediate neighborhood of the edge a part of the energy of the primary beam will be above the edge and hence cause fluorescence [23].

4. Absolute intensities must be determined with utmost precision. Equation (12) demonstrates that the entire ASAXS effect consists of a small decrease of the measured intensity when approaching the absorption edge. Any error in determining the absolute intensity would render the evaluation of the data impossible. Figure 3 demonstrates this problem by showing the net effect of ASAXS. Here the absolute intensities measured at four different energies are shown [19]. There is a small but measurable shift that can be evaluated if the calibration has been done accurately [19, 23].

These problems have been solved for the rod-like polyelectrolytes under consideration here [19] and for spherical polyelectrolytes [23]. For details of the procedures the reader is referred to these original papers.

4. Partial intensities

Equation (12) shows that ASAXS leads to three partial intensities, namely the first term which contains the non-resonant scattering, the cross term and the third term that is solely due to the resonantly scattering units. The latter partial intensity named self-term is the most interesting result because it is the scattering intensity of cloud of counterions only. All previous evaluations of ASAXS data proceeded by subtracting the first, non-resonant term from the experimental data. The non-resonant term could in principle be obtained through measurements far below the edge. Model calculations furthermore showed that the third partial intensity is small as compared to the cross term. Hence, this term was disregarded in previous ASAXS studies of polyelectrolytes [17, 20].

Subtracting two large terms in order to give a small difference is a numerically ill-posed problem. Here we proceed by a different scheme for the general treatment of the ASAXS data that is applicable to any system under consideration: Eq. (12) is a quadratic form in terms of the scattering factor f' if f'' is disregarded. This approximation is certainly justified for data below the edge

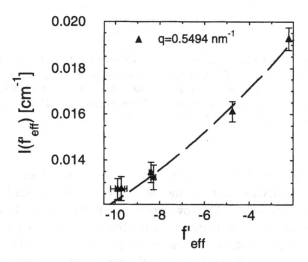

Figure 4. Decomposition of the ASAXS-intensities measured at different energies of the incident beam according to Eq. (12). The intensities measured at a q-value given in the graph are plotted against the effective real part f'_{eff} of the scattering factor (cf. the discussion of Fig. 2). The dashed line shows the fit according to Eq. (12) if f''_{eff} is disregarded. Taken from [19].

where f'' indeed is rather small (see Fig. 2). A detailed study of this problem has demonstrated that the variation of f'' is practically inconsequential even above the edge. This is due to the finite error in defining the position of the edge and other experimental uncertainties [19]. Hence, for each q-value the set of all scattering curves measured below and above the edge were plotted in Fig. 4 as the function of solely f'_{eff}. This plot shows the accuracy of the present decomposition. It can hence be used to determine the three partial intensities given in Eq. (12).

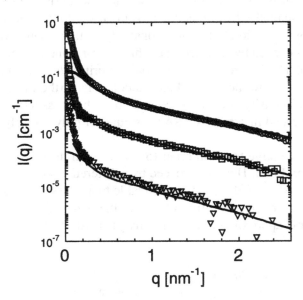

Figure 5.　　The partial intensities obtained by ASAXS are plotted against q and compared to the prediction of the cell model. The upper curve is the first term in Eq. (12) and refers to the intensity obtained far below the edge. The lowermost term is the self-term that solely refers to the scattering contribution of the counterions (third term of Eq. (12)). The curve in between is the cross-term (second term of Eq. (12)). The solid lines mark the prediction of the cell model. Taken from [19].

Figure 5 displays the three partial intensities. Here the upper curve (circles) correspond to the SAXS intensity measured by a conventional SAXS experiment far below the edge. The lowermost curve (triangles) is the self-term of Eq. (12) and the curve in between marks the cross-term (squares). As expected from previous model calculations, the intensities exhibit a very similar dependence on q [17]. Note that the self-term which is much smaller than the non-resonant term or the cross term can be obtained up to $q = 2.5 \text{ nm}^{-1}$. As mentioned above, this term provides the most valuable information of the ASAXS experiment. It refers to the scattering intensity that would result from a system in which the macroion is totally matched.

4.1   *Comparison with the Poisson-Boltzmann cell model*

The three partial intensities displayed in Fig. 5 can now serve for an unambiguous test of the cell model described in Sec. 2. The comparison with the cell model can be done as described in great detail recently [18]. With the distribution $n(r_c)$ the partial scattering intensities Eq. (12) can now be calculated and compared to the experimental data shown in Fig. 5.

The contrast parameters entering into this comparison have been determined recently for the system under consideration here (see Fig. 1). The contrast Δf_{ion} of the Br$^-$ counterion was determined from their respective crystallographic radii. Hence, as already discussed in previous papers [14, 17, 18] the hydration shell of the ion is treated as bulk water. This may induce a small error if Δf_{ion} is small. The value calculated in this way for Br$^-$ is $\Delta f_{ion} = 26$ e/ion. For the contrast of the macroion we used the value $\Delta \rho_{rod} = 25 \ e/nm^3$ [14].

The solid lines in Fig. 5 show this comparison [19]. Figure 5 shows that good agreement is reached for all three partial intensities. All ratios between the intensities as well as their dependence on q are captured by the cell model. Only the self-term is slightly underestimated but the small differences seen in Fig. 5 are hardly beyond the experimental uncertainty. All parameters are either fixed or have been taken from a previous analysis as $\Delta \rho_{rod}$. Moreover, as is obvious from Eq. (12), the self-term is not dependent on any contrast and is thus model-independent.

5.   Summary

The present survey has demonstrated the power of ASAXS study for the study of rod-like polyelectrolytes in aqueous solutions. Most importantly, ASAXS gives the intensity contribution that is solely due to the counterions (third term of Eq. (12)). In this way ASAXS furnishes data that are not available by a conventional scattering experiment. It is hence evident that ASAXS is the method of choice for the study of aqueous polyelectrolytes because it obviates the need of using labelled compounds or counterions.

Acknowledgements

Financial support by the Deutsche Forschungsgemeinschaft, Sonderforschungsbereich 481, Bayreuth, is gratefully acknowledged.

References

[1] Israelachvili, J.N. (1992). *Intermolecular and Surface Forces*, 2nd edition. London: Academic Press.

[2] Mandel, M. (1988). *Polyelectrolytes, Encyclopedia of Polymer Science and Engineering*, 2nd edition, vol. 11, p. 739, (Mark, F.H., Bikales, N.M., Overberger, C.G., Menges, G.). New York: Wiley.

[3] Schmitz, K.S. (1993). *Macroions in Solution and Colloid Suspension*. New York: VCH Publishers.

[4] Fuoss, R.M., Katchalsky, A. and Lifson, S. *Proc. Natl. Acad. Sci. U.S.A.*, 1951, **37**, p. 579.

[5] Alfrey, T., Berg, P.W. and Morawetz, H. *J. Polym. Sci.*, 1951, **7**, p. 543.

[6] Katchalsky, A. *Pure Appl. Chem.*, 1971, **26**, p. 327.

[7] Manning, G. *Ann. Rev. Phys. Chem.*, 1972, **23**, p. 117, (and further references cited therein).

[8] Deserno, M., Holm, Ch., Blaul, J., Ballauff, M. and Rehahn, M. *Eur. Phys. E J.*, 2001, **5**, 97.

[9] Glatter, O. and Kratky, O. (1982). *Small-Angle X-ray Scattering*. London: Academic Press.

[10] Porod, G. (1982). *Small-Angle X-ray Scattering*, (Glatter, O., Kratky, O.). London: Academic Press.

[11] Zakharova, S.S., Engelhaaf, S., Bhuiyan, L.B., Outhwaite, C.W., Bratko, D. and van der Maarel, J.R.C. *J. Chem. Phys.*, 1999, **111**, p. 10706, (and further citations given there).

[12] Wu, C.F., and Chen, S.H., Shih, L.B. and Lin, J.S. *Phys. Rev. Lett.*, 1988, **61**, p. 645.

[13] Chang, S.L., Chen, S.H., Rill, R.L. and Lin, J.S. *J. Phys. Chem.*, 1990, **94**, p. 8025.

[14] Guilleaume, B., Blaul, J., Wittemann, M., Rehahn, M. and Ballauff, M. *J. Cond Matter*, 2000, **12**, p. A245.

[15] Stuhrmann, H.B. *Ad. Polym. Sci.*,1985, **67**, p. 123.

[16] Stuhrmann, H.B., Goerigk, G. and Munk, B. (1991). *Handbook of Synchrotron Radiation*, (Ebashi, S., Koch, M. and Rubenstein, E.), **4**, chapter 17, p. 557. Elsevier.

[17] Guilleaume, B., Ballauff, M., Goerigk, G., Wittemann, M. and Rehahn, M. *Colloid Polym. Sci.*, 2001, **279**, p. 829.

[18] Guilleaume, B., Blaul, J., Ballauff, M., Wittemann, M., Rehahn, M. and Goerigk, G. *Eur. Phys. J. E*, 2002, **8**, p. 299.

[19] Patel, M., Rosenfeldt, S., Dingenouts, N., Pontoni, D., Narayanan, T. and Ballauff, M. *Phys. Chem. Chem. Phys.*, 2004, **6**, p. 2962.

[20] Das, R., Mills, T.T., Kwok, L.W., Maskel, G.S., Millett, I.S., Doniach, S., Finkelstein, K.D., Herschlag, D. and Pollack, L. *Phys. Rev. Lett.* 2003, **90**, p. 188103.

[21] Guinier, A. and Fournet, G. (1955). *Small-Angle Scattering of X-Rays*. New York: John Wiley.

[22] Le Bret, M. and Zimm, B. *Biopolymers*, 1984, **23**, p. 287.

[23] Dingenouts, N., Merkle, R., Guo, X., Narayanan, T., Goerigk, G. and Ballauff, M. *J. Appl. Cryst.*, 2003, **36**, p. 578.

[24] Henke, B.L., Gullikson, E.M. and Davis, J.C. *Atomic Data and Nuclear Data Table*, **54**, p. 181–342.

[25] Brennan, S. and Cowan, P.L. *Rev. Sci. Instr.*, 1992, **63**, p. 850.

[26] Weyerich, B., D'Aguanno, B., Canessa, E. and Klein, R. Faraday Discuss. Chem. Soc., 1990, **90**, p. 245.

[27] Wittemann, M., Kelch, S., Blaul, J., Hickl, P., Guilleaume, B., Brodowski, G., Horvath, A., Ballauff, M. and Rehahn, R., *Macromol. Symp.*, 1999, **142**, p. 43.

MACROIONS UNDER CONFINEMENT

Computer modelling of a layering phenomenon

A.D. Trokhymchuk [1,2], D. Henderson[2], D.T. Wasan[3], A. Nikolov[3]

[1]*Institute for Condensed Matter Physics*
National Academy of Sciences of Ukraine
79011 Lviv, Ukraine

[2]*Department of Chemistry and Biochemistry*
Brigham Young University
Provo, Utah 84602, USA

[3]*Department of Chemical Engineering*
Illinois Institute of Technology
Chicago, Illinois 60616, USA

Abstract   The layering of like-charged particles or macroions confined by two plane-parallel and two inclined surfaces is studied using a canonical Monte Carlo method combined with a simulation cell that contains both the confined and bath regions. The macroion solution is modelled within a one-component fluid approach that in an effective way incorporates a conventional double layer repulsion due to the ions of suspending electrolyte as well as an extra contribution due to the discrete nature of an aqueous solvent. The plane parallel and wedge-shaped geometries mimic the confinements that naturally occur in large number of systems widely known as colloidal dispersions. The effects of macroion charge, macroion and electrolyte concentrations on the particle layering and in-layer structuring are analyzed. The relation of obtained results to experiments on confined ionic micelle solutions and suspensions of charged polysterene spheres is discussed.

Keywords: Macroions, layering, thin colloidal films, solvent excluded volume effects, structural interactions, wedge confinement

1.   Introduction

Charged colloids, or macroions, are very important ingredients of a large number of chemicals and consumer products widely used in material science, electronics as well as in everyday life. Some other examples of matter comprised macroions, including biological systems, can be found in two preceding chapters presented by Vlachy et al. and Ballauff, where the theoretical and

D. Henderson et al. (eds.), Ionic Soft Matter: Modern Trends in Theory and Applications, 249–290.
© 2005 *Springer. Printed in the Netherlands.*

experimental aspects of the bulk polyelectrolyte solutions are discussed. We are interested in the confined macroions and more precisely in the macroion layer structuring next to a single macroscopic surface and/or between a pair of macrosurfaces.

The particle layering in the form of the step-wise stratification in thinning films formed from macroion solutions have been observed for some time [1–5]. The film stratification is a manifestation of the nontraditional structural forces acting between confining surfaces while the macroion films become a natural tool to probe these forces in colloidal dispersions [6–10]. From applied perspective, the phenomenon of macroion or nanocolloid layering has received considerable attention because it offers a new mechanism for stabilizing colloidal dispersions. The role of the confinement-induced structural forces in colloidal macrodispersions such as foams, emulsions, and particle suspensions has been reviewed by us recently [11]. From another side, special attention has been given to the layering phenomenon after the earlier experimental observations of ordering in some biological [12] and latex [13] suspensions. The presence of order-disorder transitions above a certain concentration in these systems was attributed exclusively to the fact that the macroions are enclosed in a restricted environment, i.e. are confined.

From the perspective of theoretical modelling, colloids dispersed in a macroion solution represent a complex material system governed by the interplay of structural forces and long-range electrostatic effects acting on different length scales. A powerful tool to study such systems is computer simulation. The purpose of this chapter is to discuss the application of computer simulations to the modelling of layering phenomenon in confined macroion solutions and to bring more understanding into nature of the processes that occur in complex colloidal systems. In doing this, besides the physical conclusions, we also summarize some methodological and modelling approaches that have been undertaken in our recent computer simulation studies regarding macroion structuring under the plane-parallel and wedge confinements [14–16]. First of all, we point out the correspondence between the thermodynamic parameters of the confined macroion suspension and those of the homogeneous or bulk reservoir. We deal with this issue by performing a canonical ensemble Monte Carlo (MC) calculations combined with a simulation cell that contains both the film region and bulk suspension.

Another important issue are the potential functions or potential model that are suitable to describe the macroion solution under confinement. These potential functions further are employed in a Metropolis algorithm to explore the configurational space during MC procedure to reach the thermodynamic equilibria in the entire system as well as to collect the ensemble averages. Because of the large size asymmetry of the species comprising a real macroion solution, the most widely used potential model is based on the one-component fluid (OCF) approximation. According to this approach, the macroion suspension is viewed as a one-component fluid of charged macroparticles that are interacting via an effective pairwise screened Coulomb potential [17, 18]; the molecular solvent and the ions of the supporting electrolyte usually are incorporated via a continuum approximation. One of the goals of this chapter is to study the role of excluded volume effects due to the molecular solvent. To fulfill this goal, we are using the OCF approximation that is based on an effective interaction which includes a conventional electrostatic repulsion between pair of macroions and an extra contribution that originates from the excluded volume effects due to the discrete nature of suspending fluid. In our treatment the excluded volume interactions are additive to the effective electrostatic potential and are evaluated within the hard-sphere model. Therefore, the proposed model is the OCF model with the screened Coulomb interaction plus excluded volume interactions due to the discrete molecular solvent. This model can be referred to as the solvent primitive model of the macroion (or polyelectrolyte) solution in an analogy to the solvent primitive model of electrolyte solution [19, 20] that takes into account the solvent excluded volume effects.

The chapter is organized as follows. We start in Sec. 2 with the brief description of the nature of the confinement for macroion solutions. The statistical mechanical description of colloidal suspensions and some other necessary details concerning modelling and simulation procedure are described in Sec. 3 and Sec. 4, respectively. The first results that we are discussing concern the macroion bulk structure and are presented in Sec. 5. The layering that occurs in macroion films formed in a plane-parallel slit is discussed in Sec. 6 while the results for macroion structuring under wedge confinement are presented in Sec. 7. The insight into macroion structure within the layer formations is outlined in Sec. 8. The links between theoretical predictions and experimental observation are discussed in Sec. 9. Finally, some conclusions are summarized in Sec. 10.

2. Confinement in macrodispersions

The systems of like-charged particles confined by plane-parallel or wedge-shaped slits that are the subject of present study can be prepared in a laboratory using, for example, a suspension of charged polysterene spheres confined between two flat glass plates [21] or a flat glass plate and a glass sphere [22]. However, our interest in these confinements originates from the fact that both are common and natural attributes of a large number of chemicals and consumer products, widely known as colloidal macro-dispersions.

2.1   *About macroion suspensions*

The sodium dodecyl sulphate (SDS) surfactant solutions is the prototype of a homogeneous macroion suspension in this study. For such surfactants, the ionic micelles, or macroions, are formed beyond the critical micelle concentration (CMC), which is 0.008 mol/L in the case of SDS. The micelles that are formed are fairly narrowly dispersed in size and are charged with the net charge being less than the aggregation number. Reiss-Hussen and Luzzati[23] used a low angle X-ray technique and have established that there exists a concentration range (0.03 to 0.10 mol/L) when the micelles are spherical (the micelle sphere-to-rod shape transition occurs at a concentration of 0.25 mol/L). The aggregation number of spherical micelles is 67, and the micelle (hard-core) diameter has a value of 4.8 nm and does not depend on the surfactant concentration (in this concentration range). The ionic micelles are our choice for the macroions in this study.

Table 1.   Some experimental data for SDS surfactant solution [1]

surfactant concn, C (mol/L)	Debye length, $1/\kappa$ (nm)	micellar volume fraction, η
0.03	3.20	0.013
0.10	2.45	0.051

The micelles are in equilibrium with SDS monomers, which are dissociated into Na^+ and DS^- ions in a solution. In this way, the ionic SDS monomers and an aqueous solvent serve as the simple electrolyte or suspending medium for micellar macroions. The micelles are partially dissociated and the degree of dissociation in a concentration range from 0.03 to 0.08 mole/l was measured and found to be around 27% [24], i.e. 18 charges per micelle. For this reason, the mean concentration of Na^+ ions in the solution is higher than the mean concentration of DS^- ions. However, to characterize the electrolyte, the Debye length, κ^{-1}, which is calculated from the definition:

$$\kappa^2 = \frac{4\pi e^2 N_A}{10^3 \epsilon k T} \left(C_{Na^+} + C_{DS^-} \right),   \tag{1}$$

can be used. In Eq. (1), N_A is Avogadro's number, k is the Boltzmann constant, T is the temperature, ϵ is the solvent dielectric constant, e is the charge of an electron (in CGS units), and C_{Na^+} and C_{DS^-} are the ionic concentrations (mol/L). The concentration C_{Na^+} accounts for both the Na^+ ions dissociated from the SDS monomers and from the micelles. The resulting concentration of micelle particles has been calculated using the Hartley model [25] assuming a spherical micelle shape and an aggregation number 67. From the concentration of micelles one can calculate the volume fraction, η, occupied by micellar macroions. Table 1 presents the set of experimentally-derived data [1] which are using in present study as an input for modelling of the bulk macroion solution. The data that correspond to surfactant concentration 0.03 mol/L will be referred to as a low electrolyte concentration while those associated with surfactant concentration 0.10 mol/L will be referred to as a high electrolyte concentration.

2.2   *Scaling in macro-dispersions*

An important common feature of macroion solutions is that they are characterized by at least two distinct length scales determined by the size of macroions (an order up to 10nm in the case of ionic micellar solutions) and size of the species of primary solvent (water molecules and salt ions, i.e. few Angstroms). Considering practical colloidal macro-dispersions, like foams, gels, emulsions, etc., usually we are dealing with as many as four distinct length scales: *molecular* scale (up to 1nm) that characterizes the species of the primary solvent (water or simple electrolytes); *submicroscopic* or *nano* scale (up to 100nm) that characterizes nanoparticles or surfactant aggregates called micelles; *microscopic* or *mesoscopic* scale (up to 100μm) that encompasses liquid droplets or bubbles in emulsion and foam systems as well as other colloidal suspensions, and *macroscopic* scale (the walls of container etc).

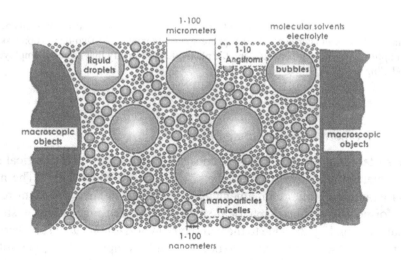

Figure 1.   Length scales in colloidal dispersions.

The different grades of confinement that occur in such a complex fluid system must be distinguished. On the lower, nanoscopic level confinement is realized by a pair of colloids that approaching one another form the annular planar and annular wedge geometries with respect to the primary (molecular or electrolyte) solvent. At the same time, on the mesoscopic level similar annular plane parallel and annular wedge confinements are formed with respect to the colloidal particles themselves due to a pair of droplets that are approaching one another. The latter case of confined colloidal suspensions is a typical to occur in concentrated dispersions. Taking into account such a hierarchy of the confinements is quite important because it shows the way how structural forces that operate at the lower nanoscopic scale between colloidal particles may influence the interactions on the higher mesoscopic scale that govern the observed properties of both micro- and macrodispersions.

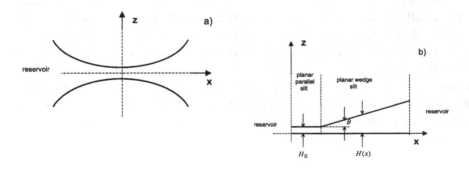

Figure 2.   (a) Two-dimensional sketch of the most general wedge-shaped confinement that occurs in dispersions and corresponds to a sphere-sphere geometry. (b) Two-dimensional sketch of the simplified wedge confinement using the plane-plane geometry that has been employed in MC modelling.

2.3   Geometry of the confinement

The system that we are interested consists of two *uncharged* spherical surfaces immersed into suspension of the *charged* colloidal particles. The most general example of the confinement that is realized in such a system refers to that formed by a pair of spherical surfaces (e.g., see Fig. 1) and is shown schematically on Fig. 2a. Due to the symmetry in horizontal (x, y) and vertical (z) directions, the sphere-sphere geometry can be simplified to a plane-plane geometry, preserving the main features of the confinement formed by two large spheres. Such a non-trivial model of the confining that occurs in dispersions is shown on part b of Fig. 1. It consists of both the plane parallel slit and planar wedge slit. The main characteristic of the plane parallel confinement is the separation between confining surfaces or slit thickness, $h = H_0$. An important characteristic of the wedge confinement is the wedge angle θ, such that the value of $\tan \theta$ determines how the local separation between surfaces $h(x)$ varies with the distance x along the wedge. In the case when plane parallel and wedge-shaped films are connected as it is shown in Fig. 2b, the local wedge thickness is $h(x) = H_0 + x \tan \theta$.

3. Statistical mechanical description

The properties of any physical system are determined to a good extent by the pair forces acting between the constituents. The interaction between two macroions in an electrolyte solution is dominated by electrostatic and solvation forces [26]. The electrostatic forces are determined by the number of charges carried by the macroions and by the concentration of supporting electrolyte solution. The solvation forces are promoted by the molecular solvent. The electrostatic effects in macroion solutions are described remarkably well within the so-called *restricted primitive model* (RPM) for supporting electrolyte where the ions are treated explicitly while the polar solvent is considered as a continuum dielectric background. To progress, we must include into the modelling a discrete or molecular nature of the solvent.

3.1 *Solvent primitive model for macroion solutions*

A particularly simple model of the solvent is obtained using a system of hard spheres together with the dielectric background. This model has been called the *solvent primitive model* (SPM). At first glance, this would seem to be an unpromising model of a polar solvent. However, this model does recognize that the solvent molecules occupy space. With regard to the subject of the present study, the system we are interesting, i.e. macroion solution, can be modelled as a homogeneous (or bulk) three-component fluid mixture that represents a low-charge electrolyte solution consisting of neutral hard spheres (solvent molecules) with number density ρ_s, diameter d_s, plus two species of charged hard spheres (cations and anions) with charges eZ_+, eZ_-, number densities ρ_+, ρ_-, and diameters d_+ and d_-. A macroscopic dielectric permittivity of the continuum dielectric background is ϵ. Because the number density of the solvent molecules dominates that of the ions, for simplicity we assume: $d_+ = d_- = d_s \equiv d$, i.e. all discrete entities that form the supporting electrolyte solution have the same geometrical dimensions. Additionally, there are the macroparticles (m) that represent macroions bearing a fixed charge eZ and whose diameter is D and number density, ρ. Some time we will refer to macroions as colloidal particles, that are identical within the present study. The system as whole can be called a *suspension* with the macroparticles referred to as the species belong to the *suspended phase*, while the three-component fluid mixture forms the *suspending fluid*.

The properties of a suspending fluid near one or two giant particles can be obtained by computer simulation, using either the MC or molecular dynamics techniques, or by means of the analytical theories based on the Poisson-Boltzmann or Ornstein-Zernike equations, or density functional theories. However, due to the large size asymmetry of the species comprising macroion solutions, the application any of these techniques in its original version is problematic. At the same time, combining some of these techniques allows to exploit their advantages and seems to be very productive. The combination of integral equation theory and computer simulations by means of the one-component fluid approximation forms a theoretical basis of the approach outlined in the present study.

3.2   One-component fluid approach

According to the one-component fluid (OCF) approximation (see, for example, paragraph 2.3 in preceding chapter by Vlachy et al.), the complex macroion solution is viewed as a single component fluid of like-charged colloids (or macroions) interacting via the effective forces. A crucial aspect in the OCF approach is the definition and calculation of an effective pair potential. Different routes to derive the expression for the effective interaction between two macroions in the presence of simple electrolyte have been considered since the classical DLVO potential was proposed by Derjaguin and Landau [17] and Verwey and Overbeek [18] in the 1940s. The following recent attempts such as those of Sogami and Ise [27], Spalla and Belloni [28] as well as others have been stimulated by experimental observations for colloidal suspensions [29–32] that seem to be inconsistent with the DLVO theory. All of these calculations of the effective interaction between a pair of macroions are consistent at the level of Poisson-Boltzmann (PB) approximation but differ quantitatively and conceptually in the case when the size of the electrolyte ions is not negligible. Furthermore, in the present study we are aiming at an even more sophisticated case in which the size of the solvent molecules has to be taken into account. This task requires that the correlations of the solvent molecules and electrolyte ions around the two fixed macroions must be accounted for. A route that directly utilizes this concept is based on the evaluation of the effective average interaction, $W(r)$, via the macroion radial distribution function, $g_{mm}(r)$, using the exact relation

$$W(r) = -k_B T \ln g_{mm}^o(r) \,. \tag{2}$$

The superscript "^{o}" indicates that only two macroions are involved, i.e., statistically it means that the theoretical macroions should be in the limit of infinite dilution. The latter point is very important as otherwise the correlations between a pair of fixed macroions are affected by many-body effects and $W(r)$ will not be a true pair interaction. Moreover, the degree of success of Eq. (2) is concerned now with the modeling of suspending fluid that is used to calculate the radial distribution function, $g^o_{mm}(r)$, and with the approximations involved in these calculations. The function $W(r)$ also is known as the *potential of mean force*.

3.3   *Ornstein-Zernike equation and effective interaction*

Different techniques can be applied to calculate the effective interaction in a form of the potential of mean force. Because we are looking for $W(r)$ which in the following will be used in computer simulations, the mostly suitable is the integral equation theory (IET) approach. The advantage of IET lies in a fact that in many most important applications this approach leads to an analytical equation for $W(r)$. The IET technique is based on the Ornstein-Zernike (OZ) equation [33]

$$h_{\lambda\mu}(r_{12}) = c_{\lambda\mu}(r_{12}) + \sum_{\nu} \rho_{\nu} \int h_{\lambda\nu}(r_{13})c_{\nu\mu}(r_{32})d\mathbf{r}_3, \qquad (3)$$

where the Greek subscripts range over all the suspending fluid components $+, -, s$, as well as macroparticle (m (, and $\mathbf{r}_{\alpha}$ is the position of particle α. Finally, $r_{\lambda\mu} = |\mathbf{r}_{\lambda} - \mathbf{r}_{\mu}|$. The function $h_{\lambda\mu}(r) = g_{\lambda\mu}(r) - 1$ is the *total correlation functions* for a pair of particles of component λ and μ that are separated by the distance r. The function $g_{\lambda\mu}(r)$ is the *pair* or *radial distribution functions* (RDF). The function $c_{\lambda\mu}(r)$ is the *direct correlation function*.

Let us assume that there are only one or two macroparticles in the mixture. Thus, the macroparticle is present in zero concentration, $\rho = 0$. Further, we assume $\rho D^3 = 0$; otherwise, the macroparticle would affect the entire fluid. In other words, the macroparticles are part of the bulk fluid at vanishingly small density, i.e. $\rho \to 0$. Using this fact, we obtain a system of three equations

$$h_{ij}(r_{12}) = c_{ij}(r_{12}) + \sum_{k} \rho_k \int h_{ik}(r_{13})c_{kj}(r_{32})d\mathbf{r}_3, \qquad (4)$$

$$h_{im}(r_{12}) = c_{im}(r_{12}) + \sum_{k} \rho_k \int h_{mk}(r_{13})c_{ki}(r_{32})d\mathbf{r}_3, \qquad (5)$$

and

$$h^o_{\text{mm}}(r_{12}) = c^o_{\text{mm}}(r_{12}) + \sum_k \rho_k \int h_{\text{m}k}(r_{13})c_{k\text{m}}(r_{32})d\mathbf{r}_3 , \tag{6}$$

where now subscripts are range over the suspending fluid components only, i.e., $i, j, k = +, -, s$. The first equation, (4), is just the OZ equation for the homogeneous three-component suspending fluid. Equations (5) and (6) involve the giant particles and, due to this, play an important role in the statistical mechanics of inhomogeneous fluid systems, i.e., systems involving diluted mesoscopic objects. Namely, Eq. (5) describes the fluid inhomogeneity by means of the local density profiles $\rho_i(r) = \rho_i g_{i\text{m}}(r)$ of the suspending fluid species i near the macroparticle or surface. In turn, Eq. (6) describes the correlations between only two macroparticles mediated by the suspending fluid, and consequently yields information about the effective pair interaction between a pair of macroparticles via the relation (2).

Equations (5) and (6) are applicable to any size D of the macroparticle. In the case of macrodispersions $D \gg d_i$, and for practical purposes it is more convenient to rewrite Eq. (6) introducing new distance variables: $x = r_{13} - D$ that is the normal distance or gap width between the surfaces of two macrospheres, as well as $s = r_{13} - D/2$ and $t = r_{23} - D/2$. Thus, Eq. (6) has the form

$$h^o_{\text{mm}}(x) = c^o_{\text{mm}}(x) + \pi D \sum_k \rho_k \int_{-\infty}^{\infty} h_{\text{m}k}(s)ds \int_{x-s}^{\infty} c_{k\text{m}}(t)dt , \tag{7}$$

a result obtained by Attard et al [34] and Henderson [35]. Applying the hypernetted chain (HNC) approximation [36, 37] to describe the correlations between a pair of suspended macrospheres, Eq. (7) becomes

$$\ln g^o_{\text{mm}}(x) = \beta u_{\text{mm}}(x) + \pi D \sum_k \rho_k \int_{-\infty}^{\infty} h_{\text{m}k}(s)ds \int_{x-s}^{\infty} c_{k\text{m}}(t)dt , \tag{8}$$

where $u_{\text{mm}}(x)$ is the "bare" interaction energy of the two macroion spheres at the gap width x. Taking into account relation (2), the effective interaction energy between two macrospheres mediated by a suspending fluid, has form

$$W(x) = u_{\text{mm}}(x) + k_B T \pi D \sum_k \rho_k \int_{-\infty}^{\infty} h_{\text{m}k}(s)ds \int_{x-s}^{\infty} c_{k\text{m}}(t)dt . \tag{9}$$

In Eq. (9), the correlation functions for macroparticles do not appear on the RHS of this equation. Thus, applying a particular closure for the macroparticle correlations need not relate to the closures used for the correlation functions of the suspending fluid on the RHS of Eq. (9). This means, that using the Percus-Yevick (PY) approximation to describe the correlations between macrospheres and comparing obtained result with Eqs. (8) and (9), we obtain

$$W(x) = -k_B T \ln g_{\text{mm}}^{HNC,o}(x) = -k_B T h_{\text{mm}}^{PY,o}(x), \quad \text{for } x \geq 0. \quad (10)$$

The result (10) is consistent with the observation that energy of interaction is proportional to the diameter D of the macrospheres. Thus, the HNC approximation predicts correctly that $W(x)$ is proportional to D whereas both the PY approximation and mean spherical approximation (MSA) both lead to the incorrect prediction that $W(x)$ is proportional to the logarithm of D. This is because both the PY approximation and MSA are linearized versions of the HNC approximation. If the PY approximation or the MSA are used for the macroparticle correlations then the ansatz,

$$W(x) = -k_B T h_{\text{mm}}^o(x), \quad \text{for } x \geq 0, \quad (11)$$

should be applied. This ansatz is consistent with Eq. (10) and is equivalent to using the HNC approximation for the correlation between two macrospheres.

3.4   *Solvent excluded volume effects*

Following the OZ/MSA solution of the Eqs. (4)-(6), the pair correlation function $h_{\text{mm}}^o(r)$, between two macroions suspended in a SPM electrolyte, is a sum,

$$h_{\text{mm}}^o(r) = h_{\text{hs}}^o(r) + \Delta h_{\text{el}}^o(r), \quad (12)$$

where the hard-sphere term, $h_{\text{hs}}^o(r)$, is obtained within the PY approximation [39] and depends on the total volume fraction occupied by electrolyte ions and solvent molecules, $\eta_0 = (\pi/6)(\rho_+ + \rho_- + \rho_s)d^3$; the second term, $\Delta h_{\text{el}}^o(r)$, that controls the electrical response of the system, is unaffected by solvent density and depends on electrostatic parameters like, the inverse Debye length, κ, and dielectric permittivity, ϵ. As a result, the effective potential between a pair of macroions, $W(r)$, can be presented in a form:

$$W(r) = u_{\text{mm}}(r) + u_{\text{mm}}^{\text{ex}}(r) + u_{\text{mm}}^{\text{el}}(r), \quad (13)$$

where

$$u_{\text{mm}}(r) = \begin{cases} \infty, & r \leq D, \\ 0, & r > D, \end{cases} \quad (14)$$

refers to the direct interaction due to the macroion hard cores of a diameter D and represents the direct (independent from the medium) part of excluded volume forces, while the terms with superscripts (ex) and (el) represent the effective contributions to the energy from excluded volume forces and electrostatic forces, respectively. The effective excluded volume interaction, $u_{mm}^{ex}(r)$, arises in colloidal suspensions because the suspending fluid is composed of the small discrete species (of a diameter d in our case) that represent the molecular solvent and the non-zero diameter electrolyte ions; $u_{mm}^{ex}(r)$ consists of two terms in the form[40]:

$$u_{mm}^{ex}(r) = \begin{cases} \phi^{dep}(r), & D < r \leq r^\star \\ (r/d)^{-1}\phi^{str}(r), & r > r^\star, \end{cases} \tag{15}$$

where $\phi^{dep}(r)$ and $\phi^{str}(r)$ are the so-called depletion (the analog of that found by Asakura and Oosawa (AO) [41]) and structural parts of the excluded volume interactions, respectively. At a low volume fraction, η_0, of the species of suspending fluid, the depletion contribution, ϕ^{dep}, is well described by the Asakura and Oosawa result [41] while the structural component, ϕ^{str}, vanishes. However, with an increase in the fine species packing fraction, the purely attractive AO-like depletion part is followed by a repulsive barrier and to a good approximation, ϕ^{dep} can be described by expression [40]:

$$\frac{1}{kT}\phi^{dep}(x) = \left(\frac{D+d}{2d}\right)\left(a_0 + a_1 x + a_2 x^2 + a_3 x^3\right), \tag{16}$$

with coefficients $a_{n=0,\dots,3}$ that depend on the volume fraction η_0. The structural part, ϕ^{str}, at higher volume fractions of the fine species is of oscillatory nature, i.e. structural forces can be either repulsive or attractive, and their contribution to the excluded volume interaction is described well by the equation [40, 42, 43]:

$$\frac{1}{kT}\phi^{str}(x) = \left(\frac{D+d}{2d}\right) A\cos(\omega x + \varphi)\exp(-\mu x), \tag{17}$$

where again the coefficients A, ω, φ and μ depend on the volume fraction η_0 occupied by the species of suspending fluid. The distance x in Eqs. (16) and (17), is the gap width between the outer surfaces of two macroions; it is related to the distance r between the centers of two macrospheres by $x = r - D$. The distance $r = r^\star$ in Eq. (15) defines the separation between two macroion surfaces where the components of the depletion and structural forces, $\phi^{dep}(r)$ and $\phi^{str}(r)$, are merging. This position lies within the gap width of a one diameter, d, of the suspending fluid species and is sensitive to their volume fraction η_0 [40]. In a particular case when the density of small particles is fixed at value, $\eta_0 = 0.3$, the parameter $r^\star = 1.0789D$, i.e., the depletion and structural forces are merged at the gap width, $x \approx 0.08D/d$.

3.5   *Electrolyte contribution to the effective interaction*

The electrostatic contribution, $u_{mm}^{el}(r)$, to the effective interaction between two macroions can be evaluated in a simplified form suggested by Blum and Høye [38]

$$\frac{1}{kT} u_{mm}^{el}(r) = \frac{Z^2 \lambda_B}{r(1 + \kappa D)^2} \exp(-\kappa[r - D]), \quad r > D, \qquad (18)$$

where $\lambda_B = e^2/(4\pi\epsilon_r\epsilon_0 kT)$, is the Bjerrum length and κ is the inverse Debye screening length defined by Eq. (1). In general, we could utilize a more accurate expression for electrostatic contribution incorporating, for example, dipole alignment and ion correlation effects[44]; but we felt that the level of the screened Coulomb potential is capable of capturing the physical aspects we wish to address in our study. Besides, the appropriate analytical equations for electrostatic contribution to the effective interaction between macroions, that is an important attribute of computer simulation studies, are not available for the moment.

3.6   *Two macroions in the solvent primitive model electrolyte*

Let us discuss briefly the interaction between two macroion particles. This is a hypothetical case for both theory and experiment. However, the forces acting between a pair of isolated particles is a key ingredient in the statistical mechanical description of the properties of a macroscopic system. If a pair of macroions would be in a vacuum, then the forces acting between them will be determined by the individual features of each macroion that, in our case, are limited to the hard-core diameter, D, and macroion charge number, Z. In contrast, within the OCF approach the pair interaction is effective and depends on the physical properties of a suspending fluid. In the present study such properties include the strength of the electrolyte characterized by the Debye screening length, κ, the dielectric constant of the model solvent, ϵ, and the total (solvent plus electrolyte) volume fraction, η_0, occupied by a suspending fluid. The first two parameters, κ and ϵ are the same as in the classical DLVO theory and are responsible for the electrostatic forces. The third parameter, η_0, is not accounted for within the DLVO theory and is responsible for the excluded volume forces. Assuming the volume fraction of the suspending fluid is equal to zero (i.e. neglecting the discrete nature of suspending medium) eliminates the contribution of excluded volume forces and our model reduces to a DLVO-like model with effective electrostatic forces only.

Figure 3a shows separately the electrostatic (DLVO-like) and excluded volume contributions to the effective interaction potential between two macroions, $W(r)$. The electrostatic contribution is all the way repulsive and decays mono-

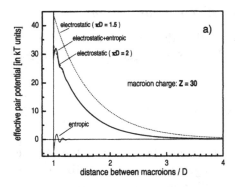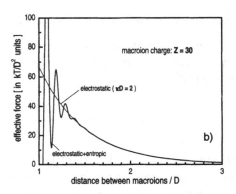

Figure 3. (a) The effective pair interaction between two isolated macroions. The thick solid line represents the total interaction $W(r)$ for the case of high electrolyte concentration. The two dashed lines correspond to an electrostatic contribution at both the low and high electrolyte concentrations. (b) The total effective force, $dW(r)/dr$ between two macroions that corresponds to the total effective interaction shown on part a).

tonically with distance, extending to a few macroion hard-core diameters. The intensity of the electrostatic repulsion depends strongly on the macroion charge number, Z. The range of electrostatic interaction is determined by the suspending electrolyte solution: an increase in the electrolyte concentration (increase in κD) leads to a shortening of the range of the repulsion but at the same time affects (slightly raises up) the strength of repulsion. In contrast, the excluded volume contribution exhibits an oscillatory decaying profile. It has repulsive and attractive parts with the periodicity of oscillations scaled by the geometrical dimensions of the species of suspending medium. The excluded volume contribution vanishes when the separation between two macroions exceeds the distance of four-to-five diameters of the suspending fluid entities (approximately 2 nm, or less than half of a micellar hard-core diameter in the case of SDS surfactant micellar solution). Since the number density of the solvent molecules dominates that of the electrolyte ions, we suggest that the excluded volume forces are not sensitive to the variations of the electrolyte concentration. In what follows, the contribution of excluded volume forces to the effective interaction between two macroions has been evaluated in all our calculations at the conditions that correspond to a fixed volume fraction of suspended fluid, namely, $\eta_0 = 0.3$ or 30% that is of the order of typical volume fraction of aqueous electrolytes.

An example of the effective force (derivative of the pair potential with respect to the separation) experienced by the two approaching macroions is shown in Fig. 3b. The oscillatory decay part of the force profile reflects in an effective way the impact of the discrete nature of the solvent on interparticle forces. The structural component of excluded volume forces modulates the repulsive electrostatic interactions so that it becomes oscillatory; at short separations (less than the solvent diameter) the attractive depletion force dominates the electrostatic repulsion. At a distances larger than a few solvent diameters, only the electrostatic repulsive forces govern the macroion-macroion interactions.

4. Monte Carlo simulations

In the following, we employed the effective pair interaction $W(r)$ between two macroions in a Metropolis scheme of MC simulations to study the macroion structuring under plane-parallel and wedge-shaped confinements.

4.1 *Description of the system used in simulations*

The system that we studied by computer simulations consists of two *uncharged* macrosurfaces to mimic the plane-parallel or wedge slit. The surfaces are immersed into *effective* macroion solution at a fixed macroion bulk concentration, ρ. According to the OCF approach, the macroions that represent a suspended phase, are considered explicitly; we model them as the charged hard spheres with a hard-core diameter D, bearing a fixed charge eZ. The electrolyte solution, that serves as a suspending fluid, is treated in an effective way being present in computer modelling by means of the Debye radius κ^{-1}, the dielectric permittivity of an aqueous solvent ϵ, and the total (ions plus water molecules) volume concentration η_0.

Two systems discussed in the present study are determined by two different volume fraction $\eta = (\pi/6)\rho D^3$ occupied by the micellar macroions in a bulk suspension, namely, $\eta = 0.01$ and 0.05. According to Table 1, these two volume fractions are associated with the SDS micellar solutions characterized by two different surfactant concentrations of 0.03 and 0.10 mol/L and two different values of the Debye radius κ^{-1}. To be close to the real macroion suspensions, the value of the Debye parameter, used in computer modelling, was fixed at $\kappa D = 1.5$ when macroion volume fraction is $\eta = 0.01$, and $\kappa D = 2$ in the case of $\eta = 0.05$. Afterwards, two simulated macroion system that are characterized by these two sets of parameters are referred to as the macroion suspension of low surfactant concentration and high surfactant concentrations, respectively.

In addition, we note that all results presented in this study mimic the aqueous solutions at a room temperature, $T = 298K$. Consequently, the corresponding value of the Bjerrum length, λ_B, in Eq. (18) is 0.714 nm. The macroion particle to suspending fluid species size ratio, $D : d$, that is necessary input to effective modelling of excluded volume forces, is chosen to be $10 : 1$. The volume fraction occupied by the suspending fluid is fixed at $\eta_0 = 0.30$. The main results are presented for the macroion charge number $Z = 30$ although some simulations with lower values of Z, namely, 20 and 10 have been performed as well.

4.2   *Simulation cell*

The basic cell employed in a canonical ensemble MC simulations has been chosen in form shown in Fig. 4. The cell has the shape of a rectangular box of dimensions $l_x \times l_y \times l_z = 30 \times 10 \times 30$ (in the units of macroion diameter, D) and contains of both bulk and confined regions. There are two (left and right) bulk regions which are the parts of the same reservoir connected by means of periodic boundary conditions in x direction, i.e., particles that left the cell from the left side enter on the right side and vice versa.

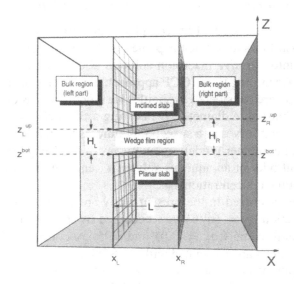

Figure 4.   Sketch of the basic cell employed in canonical Monte Carlo simulations.

The horizontal confining slit is formed by two (upper and bottom) hard-wall slabs placed in the center of the cell. Each slab is bounded by three planes. Two, left and right, vertical planes YZ of each slab at $X = x_L$ and $X = x_R$, are necessary to prevent the species of a bulk solution from entering into the slab's interior; these two planes are separated by the distance $l = x_R - x_L$ that defines the lateral extension or length of the film region in x direction; in the other lateral direction (y) the film is periodically repeated, i.e. is infinite. The horizontal XY plane at $Z = z^{bot}$ forms a bottom surface of the slit. The wedge geometry is obtained by inclining of the XY plane of the upper slab so that it intersects the left vertical YZ plane at $Z = z_L^{up}$ while an intersection with the right vertical YZ plane occurs at $Z = z_R^{up}$. Thus, the thickness of the wedge on left side is $h(x_\mathrm{L}) \equiv H_\mathrm{L} = z_\mathrm{L}^\mathrm{up} - z^\mathrm{bot}$ and on right side $h(x_\mathrm{R}) \equiv H_\mathrm{R} = z_\mathrm{R}^\mathrm{up} - z^\mathrm{bot}$. The wedge angle θ can be fixed using $\tan\theta = (H_\mathrm{R} - H_\mathrm{L})/l$. If the thickness of the wedge on left side is the same as on right side, the wedge is transformed to a planar slit of thickness $h \equiv H_\mathrm{L} = H_\mathrm{R}$.

The main advantage of using the simulation cell designed here is that the suspension confined to the slit between two slabs in the center of the cell is in equilibrium with the reservoir. We can observe directly this coexistence as well as the transition between a film and bulk suspensions as the thickness of the slit progressively increases.

4.3   System energy

The system energy, E, is the sum of macroion-macroion and macroion-wall interactions in the form,

$$E = \sum_{i,=1}^{N-1} \sum_{j=i+1}^{N} W(r_{ij}) + \sum_{i=1}^{N} u_\mathrm{mw}(z_i), \qquad (19)$$

where r_{ij} is the distance between the centers of two macroions while z_i is the normal distance from the macroion center to confining surface. The macroion-macroion interaction $W(r)$ is given by Eqs. (13)-(18). Since the confining surfaces are uncharged hard walls, there is no electrostatic contribution to the macroion-wall interaction, $u_\mathrm{mw}(z)$; the excluded volume interaction between a macroion sphere and a planar wall is similar to that between two macroions, but with a factor of two, according to the Derjaguin approximation [26], which relates the interaction between two spheres and that between a sphere and a flat wall.

4.4   Simulation details

At the beginning of the simulation runs, the number N of macroion particles are randomly placed into the both bulk regions of the simulation cell, keeping the slit region free of particles. As the equilibration run starts, the macroions fill the slit space by forming a macroion film. As the filling process proceeds, the additional macroions are injected to the bath to maintain the macroion bulk volume fraction, η, constant. The ensemble averages are calculated after the equilibrium between bath and film is reached. A constancy of the system potential energy and a constancy of the particle bulk volume concentration are used as the criteria for having reached an equilibrium between the film region and reservoir. Typically, 10^6 MC steps are needed for reaching the equilibration and are discarded; 10^7 steps are used for the averaging.

   The simulations have been performed for N macroions with N varying from 500 up to 700, depend on the slit thickness, to maintain a constant value of the macroion bulk volume fraction, η, when the volume of the slit region and the number of macroions in the slit region were changing.

5.   Macroion bulk structure

   Although our main interest is in the properties of confined macroions, our first goal in each simulation run was to establish the equilibrium between macroions confined to the slit and those in a bulk region. After equilibrium was reached, we monitored the trends in macroion structuring in a slit region as well as in a bulk region. The structuring in a homogeneous bulk is characterized by means of the radial distribution function $g(r)$.

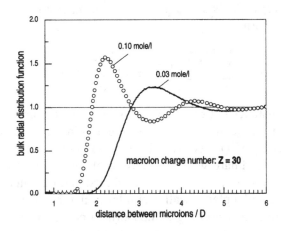

Figure 5.   MC data for the radial distribution function between macroions in a bulk solution. The low and high surfactant concentration conditions have been simulated. These correspond to macroion volume fractions, $\eta = 0.01$ (open circles) and 0.05 (solid line), respectively. The macroion charge number is fixed at $Z = 30$ in both cases.

5.1   *Effect of surfactant concentration on macroion RDF*

   Typical radial distribution functions of the bulk macroion solution at the conditions that correspond to both low and high surfactant concentrations are shown in Fig. 5. The main differences between two curves for high and low surfactant concentrations are concerned the positions of the maxima and min-ima and the magnitude of the oscillations. In the case of high surfactant con-centration the location of the first maximum is found at about two times the micelle hard-core diameter ($r/D \approx 2$) while for the low surfactant concentra-tion the first maximum is shifted towards a large distances and found to be at more than three times the hard-core micelle diameter ($r/D \approx 3.3$). This is expected result and usually is interpreted as an increase of the effective macroion diameter (a sum of the macroion hard-core diameter and the Debye atmosphere about the macroion) with decrease in a surfactant concentration: at low surfactant concentration the total amount of simple ions in the solution is less and the Debye length according to Eq. (1) is larger. The amplitude of the oscillations is related to the macroion bulk concentration itself: with an increase in surfactant concentration more macroions are present in solution. As a result of the increased macroion bulk volume fraction the first maximum of the radial distribution function at surfactant concentration $C = 0.10$ mole/l is higher than corresponding maximum at lower surfactant concentration, $C = 0.03$ mole/L.

Lowering the surfactant concentration decreases the micellar volume fraction and moderates the electrostatic interaction. This results in a lower probability for the two micellar particles to approach each other and in a weakening of long-range ordering imposed by electrostatic forces (see Fig. 5).

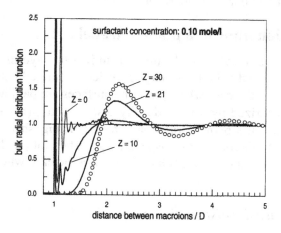

Figure 6. MC data for the radial distribution function between macroions in a bulk solution at high surfactant concentration and different macroion charge numbers as depicted in the figure. The thin solid line shows MC data obtained with excluded volume forces only, i.e. $Z = 0$.

5.2 Effect of macroion charge on macroion RDF

To illustrate how the distribution of the macroion particles in bulk region is affected by the number of charges carried by particles, Fig. 6 shows the MC data for two other values of macroion charge number, namely, $Z = 10$ and 21 maintaining in both cases the same value of κ that corresponds to a high surfactant concentration (see Table 1). For comparison on the same figure we show (by the thin solid line) the macroion radial distribution functions that have been generated from MC simulations using excluded volume interactions only that corresponds to the limiting case $Z \to 0$. We observe that excluded volume and electrostatic forces are acting on different distance scales: excluded volume forces - in the close neighborhood of the macroions while electrostatic forces - on the intermediate to large distances between macroions. The impact of these forces on the radial distribution function is different and depends on the volume fraction of micellar particles and the strength of electrostatic interaction between them. In practice both of these factors are controlled by surfactant concentration. In the case where the number of charges carried by micellar macroions is larger, e.g. $Z = 30$, the excluded volume contribution to the effective pair interaction becomes irrelevant for the bulk radial distribution functions (at least for considered range of the surfactant concentration): flocculation of micellar macroions in a bulk solution becomes energetically unfavorable and only electrostatic forces operate. As a consequence, even at low surfactant concentration, the micellar macroions show an ordering (see Fig. 5) with the nearest neighbors positioned at $r/D \simeq 3$.

6. Macroion films in a plane-parallel slit

The subject of main interest in the present study is the layering phenomenon in films that are formed from like-charged particles confined between two uncharged surfaces. In the case when confining surfaces are parallel, the particle layering is characterized by a local density distribution $\rho(z)$ across the slit. Two kinds of films in a plane-parallel slit can be distingushed. The first one is that formed from the macroion suspension adsorbed into a slit of the fixed thickness. The other kind of film can be formed in the case when slit surfaces are movable.

6.1 *Local density distribution across the slit*

Figure 7 shows MC data for the macroion local density distribution across the slit between two uncharged surfaces fixed at a distance of ten macroion hard-core diameters, i.e. $H/D = 10$. The macroion film that is formed in such a slit is in equilibrium with the resorvoir at a macroion bulk volume fraction, $\eta = 0.05$, that corresponds to a high surfactant concentration. The average number density of macroions in the film region (a number of macroions divided over the volume of film region) is about 30% higher than that in the bulk suspension. We can observe that the distribution of macroion density in the film shows a tendency for macroions to form the layers parallel to the confining surfaces. However, as we can see there is very evident difference in the macroion layering next to the surface and that in the middle zone of the film.

There are two well-defined layers next to each of the slit surfaces that in the following we will refer to as the surface layers. The surface layers are highly

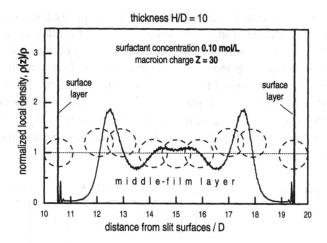

Figure 7. MC data for the normalized local density distribution of the macroions in a slit formed by two uncharged surfaces fixed on a distance $H/D = 10$. The bulk volume fraction of macroions is $\eta = 0.05$ (high surfactant concentration) and macroion charge number is fixed at $Z = 30$.

ordered with respect to the confining surfaces. The shape of the macroion density profile in this part of the film is rather narrow and shows a high concentration of macroions condensed directly on the film surfaces; this density steeply decreases going to the surface layer boundaries. As a result, each surface layer can be treated as a quasi-2D monolayer with a δ-like density profile, indicating that the thickness of the surface layers is close to that of one diameter of the macroion particles.

The macroions in the middle of the film region could be considered as belonging to a thick middle-film layer or superlayer which show a tendency to be segregated from the surface layers. In contrast to the surface layers, the macroions in the middle zone of the film are less structured with respect to the film surfaces; the midlle-film layer is broaden and has the thickness that exceeds six macroion hard-core diameters.

6.2   *Macroion layering in the films*

To obtain the film that is stabilized by itself, we have performed a number of MC simulation runs for the plane-parallel slit but of variable thickness (starting from the thickness, $H/D = 10$, shown in Fig. 7) in order to find the local minima in the configurational potential energy E per film particle. The transformation of the macroion layer structuring when the distance between confining surfaces, i.e. slit thickness becomes smaller, has been monitored during simulations with the thickness step $\Delta H/D$ equals to 1/10 of macroion hard-core diameter. We found that each case of the energy local minimum results in a slit thickness H that allows for a macroion film that is composed of an integral number of the middle-film particle layers. The first such thickness is $H/D = 9.8$ and corresponds to the three macroion layers in the middle of the film. Further, we observed the macroion films with two and one layer in the middle zone for the slit thicknesses $H/D = 7.5$ and 5, respectively and, with no layers or macroion "vacuum" for the slit thickness $H/D = 3.5$. The MC data for the macroion local density distribution $\rho(z)$ across the last three films are shown on Fig. 8.

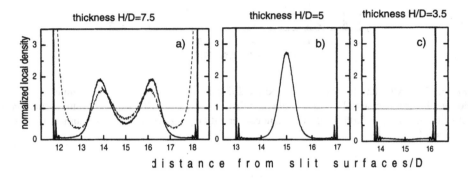

Figure 8.   MC data for the normalized local density distribution of the macroions in a film formed at macroion bulk volume fraction $\eta = 0.05$. The macroion charge is fixed at $Z = 30$. (a) The film has thickness $H/D = 7.5$ and contains four particle layers: two surface monolayers and two middle-film layers. The dashed line shows MC data obtained from the simulation without excluded volume forces. (b) The film has thickness $H/D = 5$ and contains three particle layers: two surface monolayers and one middle-film layer. (c) The film has thickness $H/D = 3.5$ and contains two surface monolayers only.

The data we are going to discuss are drawn by the solid lines on Fig. 8. The first part, Fig. 8a, shows the results for macroion suspension confined to a slit that has a thickness of around $H/D = 7.5$; the macroion film is organized in total into four layers: two surface monolayers and two layers in a middle of the film. The film shown in Fig. 8b is formed by a slit that has thickness $H/D = 5$; the film consists of two surface monolayers and a single layer of macroions in the center of the film. Finally, we can observe only two surface monolayers for the film shown on Fig. 8c that is formed in a slit that has the thickness $H/D = 3.5$. These three films of different thicknesses have been formed under the same thermodynamic conditions ensured by the equilibrium with the bulk reservoir maintained in all three case at a macroion bulk volume fractions $\eta \approx 0.05$ that corresponds to a high surfactant concentration, as discussed in Table 1. We note, that despite of the differences in the slit thickness, the surface monolayers for all three films on Fig. 8 are identical and similar to those for the film discussed on Fig. 7. At the same time, the structuring of the middle part of the film in between two surface monolayers, experiences some crucial transformations when slit is shrinking.

6.3   *Effect of surfactant concentration on the layering*

The films discussed on Figs. 7 and 8 are formed from the macroion suspension at a high surfactant concentration, i.e macroion volume fraction is $\eta = 0.05$ and the Debye parameter is $\kappa D = 2$. As the surfactant concentration decreases, the macroion suspension, according to Table 1, decreases the macroion volume fraction and value of the Debye parameter becomes smaller too. How the decreasing in surfactant concentration affects the macroion layering can be seen from the comparison of the films shown on Figs. 8b and 9a. Although for both surfactant concentrations the qualitative features of the macroion layering are quite similar, e.g. in both cases there is a single layer in the middle region of the film, some quite evident trends with decrease in surfactant concentration can be observed. First of all, the macroion layering in the middle region of the film become less pronounced and the middle-film layer grows in its thickness. The features of the surface layers are less sensitive to the surfactant concentration, however, the particle concentration in the surface monolayers, as well as overall in the film, decreases. As a result, the total thickness of the equilibrium macroion film with the same number of particle layers increases with the surfactant concentration getting lower. In particular, the thickness of the film composed of three particle layers at a high surfactant concentration (Fig. 8b) is $H/D = 5$, whereas for the low surfactant concentration (Fig. 9a) similar thickness is $H/D = 7$.

6.4   *Effect of macroion charge on the layering*

The macroion charge for MC data presented in Figs. 7 and 8 was fixed at $Z = 30$. To reveal the effect of macroion charge on the macroion layering, Fig. 9b shows MC data identical to those on Fig. 8a but assuming that macroion charge number is smaller, namely, $Z = 10$. The tendencies in the local density distribution of macroions across the film when the number of charges carried by each macroion is lowering can be summarized as follows. First of all, the slit thickness that corresponds to the energy local minimum in both cases is the same, i.e. $H/D = 7.5$. However, we can see that when the electrostatic repulsion between macroions becomes weaker the layer structuring in the middle-film region becomes less pronounced. Secondly, the distance of the middle-film superlayer from the surface layers decreases and entire layer in the middle-film region becomes more thick. The comparison these two data from Figs. 8a and 9b underlines the leading role of electrostatic forces in the layering observed in the thin macroion films.

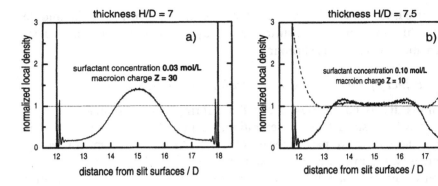

Figure 9. MC data for the normalized local density distribution of the macroions in a plane-parallel film. (a) The macroion bulk volume fraction, $\eta = 0.010$ (low surfactant concentration) and macroion charge, $Z = 30$. The film contains two surface monolayers and one middle-film layer, and has the thickness, $H/D = 7$. (b) The macroion bulk volume fraction $\eta = 0.045$ (high surfactant concentration) and macroion charge, $Z = 10$. The film contains two surface monolayers and two weak middle-film layers, and has the thickness, $H/D = 7.5$. The dashed line shows MC data obtained from the simulation without excluded volume forces.

6.5   Role of the solvent excluded volume in the layering

As we learned from Fig. 6, the role playing by excluded volume forces in the macroion structuring in a bulk increases and becomes essential by reducing the macroion charge number. To reveal the role of excluded volume forces in the modelling of confined macroion suspensions, Figs. 8a and 9b show (by a dashed lines) a comparison between the macroion local density as predicted by MC simulations of the popular DLVO-based continuum model when macroions interact only via a screened Coulomb potential in the form of Eq. (18). Two tendencies, when the excluded volume forces are superimposed on the electrostatic interaction, are clearly seen: (i) the moderation of particle layering both next to the surface and in the middle of the film, and (ii) the progressive segregation of the middle-film layer from the surface layers. However, similar to the bulk suspension, we can see that the results are sensitive to the macroion charge number Z.

In the case of higher charge, $Z = 30$ on Fig. 8a, both models result that the like-charged particles being confined to a film that has a thickness around $H/D = 7.5$ tend to be organized into four particle layers. For the middle-film layers formed with and without excluded volume forces, only some quantitative differences in the particle local density distribution are observed. The main difference introduced by excluded volume forces is found in the surface layers. Taking into account the discrete nature of the solvent results that the surface layers themselves show a structuring with respect to the film surfaces.

This results in each surface layer consisting of a well-defined sublayers in an immediate vicinity of the film surface; the shape of the density profiles of the surface sublayers has a δ-like form indicating that surface sublayers are the quasi-two-dimensional monolayers. The surface layers formed within the DLVO-like model being thinner than the middle-film layers still are far from to be monolayers. As a result, the segregation of the middle-film layers from the surface layers is not so evident in this case. As expected, the difference between models with and without excluded volume forces increases when the macroion charge becomes smaller (Fig. 9b).

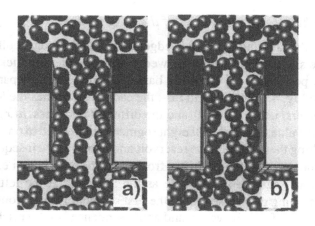

Figure 10. Snapshots of the representative macroion configurations from a MC runs for a film shown in Fig. 8b. (a) The case when excluded volume forces are taken into account (b) The case without excluded volume forces.

A snapshot of the representative configurations of the macroions in a three-layer film obtained from the simulations with and without excluded volume forces is presented on Fig. 10. The simulations without excluded volume forces (Fig. 10b) serve as a methodological example which illustrates that an adequate modelling of complex colloidal suspension should necessarily take into account the discrete nature of a primary suspending fluid. In the case of a three-layer film, the excluded volume forces play an important role in the organization of both the surface and middle-film layers. In general, the excluded volume forces become more important with a decrease of the interparticle distances; this is the case for particle layers, both near the film surfaces and in the middle of the film.

7.   Layering in the wedge slit

From what we learned in the preceding section, the macroion layering in
a slit is highly sensitive to the distance between confining surfaces. This distance, or slit thickness, is constant in the case of a plane-parallel slit. According
to Fig. 2, the wedge geometry allows for a continuous change of the slit thickness within the same film. Due to this, the structuring in a wedge-shaped slit is
characterized by a pair of the local density profiles, $\rho(x)$ and $\rho(z)$, i.e. along
and across the slit, respectively. The local density profile along the slit, $\rho(x)$,
provides an insight into the filling of the wedge by macroions while the features of macroion layering can be obtained from the profile, $\rho(z)$, similarly to
the plane-parallel slit.

7.1   *Looking for a modelling of the long wedge*

To analyze the layering in a long wedge, we merged the data collected from
eight separate simulation runs for the wedge slits of different thickness. Each
run has been performed by using the basic cell introduced in paragraph 4.2
and shown on Fig. 4. According to this, the each segment has the length of 10
particle diameters, i.e. $l/D = 10$ and the different thickness $h(x) = x \tan \theta$
but at the same value of $\tan \theta$. All eight segments are related to one continuous
wedge by linking them to the same reservoir and choosing their sequence (from
I to VIII) in the way that the thickness from the right-hand side, i.e., at the end
of each previous segment is the same as the thickness at the left-hand side,
i.e., at the beginning of next segment; the first segment I starts from the wedge
distance $x/D = 5$ from the vertex and entire sequence is shown schematically
on Fig. 11.

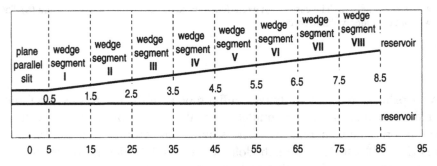

Figure 11.   The scheme of the wedge segments. The each single wedge segment was simulated separately using a basic simulation cell shown in Fig. 3. The numbers inside the wedge
indicate local wedge thickness $h(x)$ in units of particle diameter D.

The wedge segment I is characterized by the thickness $H_L/D = 0.5$ at the beginning and $H_R/D = 1.5$ at the end, so that the tangent of the wedge angle is fixed at $\tan \theta = 0.1$, i.e., the wedge angle is about $5°$. The remaining seven wedge segments are assigned in the way as it is shown on Fig. 11.

7.2 Snapshots of the wedge film

Figure 12 shows the set of six snapshots of the representative macroion configurations for the first six wedge film segments sketched in Fig. 11. With increasing wedge thickness, i.e. going from wedge segment I to segment VI, some characteristic regions within a wedge film can be even visually distinguished from Fig. 12.

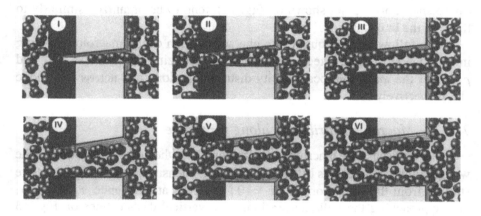

Figure 12. Snapshots of MC generated representative configurations of the macroions in wedge segments from I through VI that are the parts of the same wedge as shown in Fig. 11.

The first wedge segment I consists of a region that has no macroions because of wedge wall geometrical restriction that prohibits macroions from penetrating into a wedge if the local thickness is less than particle hard-core diameter. In reality this geometrical or hard-core "vacuum" with regard to the macroions presumably will be filled by the species comprising the suspending fluid that is coarse-grained in present study. The actual macroion film starts with a single-layer of particles that are spread roughly till the middle of the next wedge segment II. At the end of this segment II, one can note a tendency for a single-layer particle formation to split into two horizontal layers. Such a partition finally occurs on the wedge segment III where a pair of well-defined monolayers formed by macroions tight to the each of two confining walls, are observed. Interestingly, according to the sketch on Fig. 11, the local wedge thickness, $h(x)$, at the end of the segment III is larger than three particle diameters ($\approx 3.5D$), i.e. the separation between two surface layers allows for a three-layer particle formation. However, the electrostatic repulsion from the surface macroions prevents the macroions of the bulk solution from entering the space between the surface monolayers. This electrostatic "vacuum" phase extends roughly till the middle of the next segment IV, i.e. until a wedge thickness of around four particle diameters (see schematic drawing on Fig. 11). At this wedge thickness a novel single layer of particles starts to form in the middle zone of the slit between two surface layers. This three-layer region extends to the entire following segment V and partially to the next segment VI. Only close to the end of the segment VI, do the macroions between two surface layers show less layering features. The macroions in the two remaining segments VII and VIII (these snapshots are not shown on Fig. 12), tend to be organized similarly to those of the bulk solution.

In the following, this qualitative characterization of the macroion layering in a wedge slit is supported and detailed by analyzing the profiles $\rho(x)$ and $\rho_x(z)$ of the macroion local density distribution along and across the wedge slit, respectively.

7.3 *Macroion distribution along the wedge*

First we discuss the macroion distribution along the angular direction of the wedge. Figure 13 presents MC data for the local density $\rho(x)$ along the wedge starting from the wedge origin ($x = 0$) and ends at a distance $x/D = 85$, i.e. approaching the bulk suspension. The vertical dashed lines on Fig. 13 mark the sequence of wedge segments that have been simulated in separate runs so that the results presented on Fig. 13 can be compared directly with the snapshots discussed on Fig. 12. We note, to avoid the so-called edge effects in the estimate of the film density, two slabs of the thickness of one macroion diameter (one slab next to the left and one next to the right slit ends) have been discarded from the procedure of ensemble averaging.

According to its definition as the total number of macroions over an occupied volume, the density $\rho(x)$ represents the local distribution of the total particle number density along the wedge slit. By integrating over the wedge length, we find that, on average and for the system considered in this study, the overall macroion density in the wedge film, $\langle\rho\rangle$, is higher than its reservoir counterpart, ρ. Additionally, we can see from Fig. 13 that profile $\rho(x)$ exhibits an oscillatory shape indicating an inhomogeneity of macroion density along the wedge. In general, the magnitude of oscillations of $\rho(x)$ decays as both the distance x from the wedge corner and the wedge thickness $h(x)$, are progressing. At the same time, the magnitude and periodicity of the oscillations of $\rho(x)$ both show quite distinct features that depend on the range

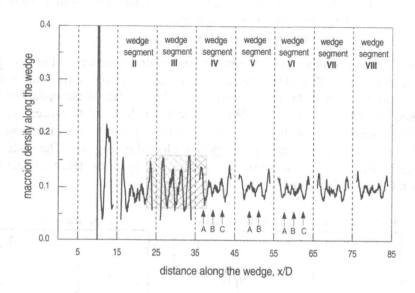

Figure 13.   Total local density $\rho(x)$ of the macroions along the wedge angular direction in wedge segments from I through VIII. The rectangular pattern indicates wedge region with an electrostatic "vacuum" phase. The letters and arrows mark various positions along the wedge that are of interest.

of distances and on the local thickness of the wedge. In particular, for the first wedge segment I, density $\rho(x) = 0$ for distances $x/D < 10$ or for thicknesses $h/D < 1$. When distance x increases, the local density $\rho(x)$ exhibits a sharp, narrow and high (δ-like) peak localized right after the distance $x/D = 10$. At the same time, the following second peak of $\rho(x)$ is broadened, extending over a distance interval of a few particle diameters starting from $x/D > 11$; it covers the remaining part of the wedge segment I and the initial part (up to $x/D \approx 17$) of the following wedge segment II. The magnitude of the second peak of $\rho(x)$ is almost one order lower than the magnitude of the preceding first peak. The third distinct part in the shape of density profile $\rho(x)$ is associated with the distance range $17 \leq x/D \leq 23$; it is characterized by a low average density (the same as in a bulk or even lower) and very weak density oscillations. In contrast, the following portion of the density profile $\rho(x)$ shows well-defined oscillatory features including both magnitude and periodicity; it extends through the entire wedge segment III and includes the final and initial parts of the left-hand and right-hand neighboring segments II and IV, respectively. As the distance from the wedge vertex progressing further beyond $x/D > 40$ and corresponding wedge thickness beyond $h/D > 4$, the density profile $\rho(x)$ shows less variety. A common feature for the remaining wedge segments from IV to VIII is that the magnitude of oscillations of the total local density $\rho(x)$ along the wedge decreases and their periodicity becomes less pronounced.

7.4 *Macroion distribution across the wedge*

The density profiles $\rho_x(z)$ across the wedge are presented on Fig. 14. The subscript indicates that local density $\rho_x(z)$ was evaluated across the wedge slabs centered at a distance x from the wedge corner; these distances are marked by arrows on Fig. 13. We note that for different distances x, the local thickness $h(x)$ of the wedge slab is different; due to this, the range of z values is different for each $\rho_x(z)$ profile plotted on Fig. 14. The each part on Fig. 14 displays a set of the density profiles $\rho_x(z)$ across three consecutive wedge segments IV, V and VI.

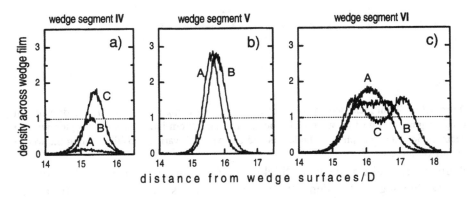

Figure 14.   Normalized local density of the macroions across the wedge in three consecutive wedge segments IV, V and VI. Only the density distribution at the middle of the wedge is shown while the surface layers are discarded; they are the same as in the case of the film across a plane-parallel slit on Figs. 7 and 8.

To save the space, we do not show on Fig. 14 the δ-like peaks that correspond to the surface monolayers of macroions on the each wedge walls; these peaks are pretty the same as those that were established in the plane-parallel slit and discussed on Figs. 7 and 8. As we can see from the first fragment, Fig. 14a, there are still no macroions in the wedge middle zone for the initial part of the wedge segment IV (the density profile marked by letter A). However, as the wedge thickness increases, around a half length of this segment at a wedge thickness $h/D \approx 4$, a new macroion monolayer starts to develop in the middle zone of the wedge film (curves B and C). The monolayer becomes well-defined through out the forthcoming wedge segment V (see two profiles on Fig. 14b), where the wedge thickness varies in a range from

$4.5 \leq h/D \leq 5.5$. The density distributions $\rho_x(z)$ presented on Fig. 14c indicate that macroion monolayer in the middle zone of the wedge extends partially even to the beginning of the following wedge segment VI (curve A), where the wedge thickness lies in the range $5.5 \leq h/D \leq 6.5$. At the same time, we can see that further increasing of the wedge thickness in this segment is accompanied by the broadening of monolayer (curve B) and, finally, at the thickness $h/D \approx 6$ this monolayer starts to split into two layers (curve C). The bilayer film in the middle zone of the wedge extends through the following wedge segment VII that is not shown on Fig. 14.

A further increase of the wedge thickness (the MC data are not presented here) is accompanied by the transformation of the macroion film (in between two wedge-shaped surface monolayers) from bilayer to threelayer, from threelayer to fourlayer and so on, with the each new layer being less and less defined up to approaching the structuring similar to that in a bulk solution.

8. In-layer macroion structuring

Insight into the features of macroion structuring within the layers can be obtained by analyzing the MC data for the in-layer radial distribution functions, extracted from both the surface and middle-film layers.

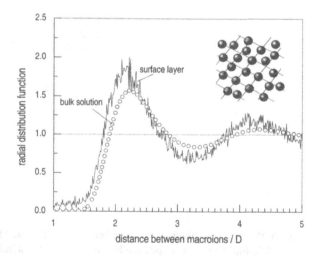

Figure 15. MC data for the radial distribution functions between macroions. The thin solid line represents the data for a surface layer while open circles are MC data for a bulk suspension that is in equilibrium with the film. The inset shows the snapshot of a representative configuration of the macroions within a surface layer.

8.1   The surface layers

The macroion-macroion radial distribution function (rdf) that corresponds to the surface layer is shown on Fig. 15. It exhibits the shape that is typical for a two-dimensional (2D) radial distribution function reflecting a tendency for surface macroions to form a hexagonal ordering along the confining wall (see also the inset with the snapshot of the representative configuration of the surface macroions during simulation). The 2D hexagonal ordering within the surface monolayers can be enhanced by an increase either in the macroion volume concentration η in a bulk region or by an increase of the macroion charge number Z. The high roughness of MC data for the radial distribution within surface layer is not the result of statistical errors but reflect the fact that the surface macroions are tightly bound (this fact also is known as a structural arrest [45] or particle halos [46]) to the neutral confining surface by a combining effect of the strong electrostatic repulsion between the surface macroions and the middle-film macroions and the short-range attraction due to the discrete nature of suspending fluid or solvent excluded volume forces.

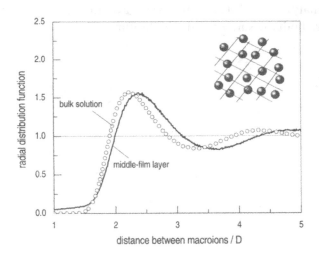

Figure 16.   MC data for the radial distribution functions between macroions. The thick solid line corresponds to a middle-film layer while open circles are MC data for a bulk suspension that is in equilibrium with a film. The inset shows the snapshot of a representative configuration of the macroions within a middle-film layer.

8.2   The middle-film layer

The macroion-macroion radial distribution function within a middle-film layer also differs from that of the bulk suspension, i.e., from a typical three-dimensional (3D) radial distribution function, that can bee seen from the data presented on Fig. 16. In contrast to the surface layer, the middle-film layer radial distribution function exhibits almost perfectly smooth and continuous behavior. Another important difference between the surface and middle-film in-layer radial distribution functions results in the change (an increase) of the nearest neighbor distance between particles that is signaled by the shift of the first maximum in the in-layer radial distribution function from $r/D \simeq 2.2$ in the case of microions within a surface monolayer, to $r/D \simeq 2.5$ when the macroions form a layer in the middle of the film.

8.3   *In-layer structuring and the film thinning*

As we pointed out at the beginning of this section, the macroion density in the film region differs from that in the bulk suspension. Consequently, each pair of the radial distribution functions shown on Figs. 15 and 16 correspond to slightly different density conditions; this may rise the question when two figures are discussing. Although the conditions of the relevancy of the comparison between 2D and 3D structures might be questionable, we stress that radial distribution functions shown in Figs. 15 and 16 correspond to the 2D and 3D configurations that are in thermodynamic equilibria and/but not necessary at the same densities. In general, the main feature of the 2D radial distribution function against its 3D counterpart concerns the shift of the position of the first maximum to smaller distances, that reflects a shortening (on average) of the nearest neighbor distances in 2D system when compared with 3D configuration. In the case of a middle-film layer radial distribution shown in Fig. 16, the position of first maximum, $r/D \simeq 2.5$, is encountered to be larger than in the case of a corresponding bulk, i.e. 3D radial distribution. In the context of the process of the thinning of macroion film [3], the middle-film region represents (see also the snapshots in Fig. 10a) an intermediate phase between the surface monolayers (i.e., a truly 2D structuring) and a bulk phase (i.e., 3D structuring). If we assume that the transition from 2D to 3D structuring is continuous, we would expect that the position of the first maximum in a radial distribution for an intermediate (middle-film) region will be in the range between the corresponding values for 2D and 3D systems; then the difference of the middle-film layer radial distribution function from that of the bulk suspension can be attributed to a partial break of the 3D symmetry in the middle-film layer. The fact that the data for the first maximum of the middle-film layer radial distribution function are out of range between 2D and 3D configurations can be interpreted as the indication that such a middle-film layer is unstable and on the stage to squeeze out of the middle-film region.

9.   Relation to the experiment

The main results presented in this chapter are obtained for macroion suspension defined by the bulk volume fraction of the macroion particles, $\eta = 0.05$, the macroion charge number, $Z = 30$, the Debye length of the supporting electrolyte, $\kappa D = 2$. Such a choice of the parameters makes this model system a suitable prototype of some aqueous ionic micellar solutions such as cetyltrimethylammonium bromide (CTAB) and sodium dodecyl sulfate anionic surfactant (SDS).

9.1   *Ionic micelle films*

The overall thickness of the film in Fig. 10 that contains two surface monolayers and a single particle layer in between, is $5D$. Using that the micelle hard-core diameter in SDS or CTAB surfactant solution $D \approx 4.8$ nm [23], the thickness of the film presented on Fig. 10 is around 24 nm, that is in close agreement with the position of the secondary minima in the measured force between crossed mica cylinders immersed in a micellar solution of CTAB [47] and with the measured thickness of the film with one micellar layer in stratifying film from 0.1 mol/L SDS [3]. Further, providing that D is independent of micellar concentration, it is obvious that our modelling predicts the thickness of the surface layers to be independent of micellar concentration that is in direct agreement with observations [48], where the surface layers of micelar suspension are identified as the adsorbed surfactant bilayers. In turn, the electrostatic repulsion between charged CTAB bilayers stabilize the mica surfaces immersed in a micellar solution [47]; this again is in agreement with our simulation results in Fig. 8c, where the stable film was found to be comprised of the two surface layers only. At this point, we wish to emphasize that these findings can be revealed exclusively by modelling when the excluded volume forces due to the discrete nature of the suspending fluid are taken into consideration.

9.2   *Film thinning*

Let us assume that during the film thinning process observed in [1], the film changes its thickness by squeezing out one layer of micelles. Then the height of the step-wise layer-by-layer thinning will depend on the effective thickness of the squeezed layer. To verify this assumption, the effective thicknesses of the squeezed layers in simulations have been estimated as the difference between the metastable or local thickness of the films containing three and two, two and one, one and no middle-film particle layers and compared with observation for the heights of the thinning steps (see Table 2). The film containing no middle-film layers, i.e. just two surface monolayers (Fig. 8c) is mostly stable and its thickness corresponds to the final film thickness in the film thinning process at the surfactant concentrations considered in the present study. These also agree qualitatively well with the force measurements of Richetti and Kékicheff [9]. These authors measured the force between two cross mica cylinders immersed into micellar CTAB solutions. The zeroes of the measured force can be compared with the metastable film thicknesses in our modeling. From the highest micellar concentration ($\eta = 0.073$) studied by Richetti and Kékicheff we can see that the zeroes of the force (see Fig. 3 of the Ref. [9]) are at $\sim$ 15nm; $\sim$ 25nm and $\sim$ 35nm. These results correlate well with our findings for both metastable film thickness revealed from the film thinning experiments as well as from the computer modelling (see Table 2).

Table 2. Values of the film thickness, h_n at the metastable states determined from the interferogram of thinning SDS film at the high surfactant concentration, $C = 0.10$ mol/L [1] and the computer simulation data for the metastable or local film thicknesses containing an integral number m of particle layers. MC simulations are performed at the macroion bulk volume fraction, $\eta = 0.05$ and macroion charge number, $Z = 30$.

observation		modelling		
no. of metastable states, n	film thickness, nm	no. of middle-film layers, m	film thickness, nm	D
0	16.2	0	15.84	3.3
1	26.3	1	24.48	5.1
2	35.9	2	36.00	7.5
3	-	3	47.04	9.8

For low surfactant concentration case, the difference in the film thickness between the films containing three and two micellar layers is about 15.8 nm. The experimentally observed height of the step-wise thickness transition is about 15.3 nm. For high surfactant concentration, the difference in the film thickness for the films with three and two micellar layers is about 8.6 nm (see Table 2) while the experimentally observed height of the step-wise thickness transition is about 10.1 nm. The increased deviation between simulation predicted film-thickness difference and the measured value of the step-wise thickness transition at a high surfactant concentration is probably due to the fact that our present modelling does not account for the contributions of the finite size of the small electrolyte ions as well as the solvent molecules to the electrostatic part of an effective interaction $W(r)$ between the pair of micellar macroions. Our present model accounts for the contributions of the finite size of the electrolyte ions and the solvent molecules to the excluded volume interaction only.

9.3   *Layering of the charged latex particles*

The layering of charged particle has been studied experimentally by Pieransky et al [22] using polystyrene spheres confined between two wedge-shaped glass boundaries. The polystyrene particles in that study experienced a mutual Coulomb repulsion (similar to that in our study) as well as a repulsion by the wedge walls due to both image charges and a surface charge on the glass plates (in contrast to our study where the wedge walls are uncharged hard walls). The layer formations that have been observed include particle monolayer, particle bilayer, three-layer, etc., up to the film composed of seven particles layers. However, due to the differences in the wedge wall set up, the authors of experimental study [22] did not observe an abnormal increase in particle density close to the wedge vertex, as we did in computer simulation study (see the main peak in $\rho(x)$ on Fig. 13). Indeed, charged particles try to stay away from the vertex formed by like charged walls because of electrostatic repulsion. Furthermore, for the same reason Pieransky et al [22] did not observe a 1D particle configuration along the wedge vertex (wedge segment I on Fig. 12). Additionally, the surface monolayers formed by macroions tight to the wedge walls that are predicted by computer simulations as the separation between walls exceeds two particles diameters (wedge segments II and III on Fig. 12), have not been observed experimentally. In contrast, experimentally observed particle monolayer starts to form at the wall separations of around two particles diameters. This monolayer region in observation is preceded by a free of particles region that has been recognized by Pieransky and his colleagues as a "vacuum" phase. Since the experimental wedge thickness associated with the boundary of the "vacuum" phase is larger than one particle diameter, the absence of polystyrene spheres in this part of the wedge is not due to the geometrical restriction, that is a case of the wedge segment I on Fig. 4. Most probably that it is due to the electrostatic repulsion from the wedge-shaped charged plates. In this case, the so-called "vacuum" phase observed by Pieransky et al [22] can be referred to as electrostatic "vacuum" in contrast to a hard-core "vacuum" for a uncharged wedge on Fig. 12 (wedge segment I). A careful examination of the experimental and simulation data guides us to conclude that particle monolayer that starts to form between two surface monolayers (see wedge segment IV on Fig. 12) is the analog of that observed experimentally by Pieransky et al [22]. In fact this layer in the computer simulations has been formed by solution macroions that are repelled by surface macroions similar to the experiment where polystyrene spheres have been repelled by charged glass boundaries. Interestingly, the local thickness h where the monolayer starts to form in experiment ($\approx 2D$) correlates well with that in computer modelling (middle of the wedge segment IV has thickness $h/D \approx 4$) if we assume that two surface monolayers (of the total thickness $2D$) are the part of the "charged" wedge walls. In this case, the space between two surface monolayers that is free of macroions on the entire segment III and partially on the segments II and IV on Fig. 12 (region marked by pattern on Fig. 13), is nothing but the electrostatic "vacuum" phase that is similar to that in the observation of Pieransky et al [22].

9.4 *Segregation of the charged particles to uncharged surfaces*

An important difference between charged and uncharged particles under confinement is related with the magnitude and shape of the main peak of $\rho(x)$ on Fig. 13 as well as with the magnitude and shape of the wall peaks of the density profiles $\rho_x(z)$ on Figs. 7-9. The δ-like shape of these peaks indicates the formation of a 1D macroion chain along the wedge corner in the first case, and the formation of the macroion surface monolayers in the second case. Both of these formations are the result of macroion segregation on the wedge walls. In contrast, the corresponding density profiles for uncharged particles under wedge confinement (e.g., see Fig. 1 of Ref. [49]) do not show evidence for neither a 1D particle configuration at the vertex or a 2D particle assembly on the surface of the wedge walls. This suggests that particle segregation is driven predominatingly by an electrostatic repulsive forces between macroions in suspension and occurs simply because the region next to the wedge surfaces represent a big space without (depleted of) electrostatic repulsion. This is an important difference between suspensions of charged and uncharged colloids that can affect significantly the processes that carry out in practical colloidal suspensions and, in particular, their stability. The segregation of highly charged nanoparticles to the regions near large-sized uncharged colloids has been predicted also by Garibay-Alonso et al [50] in their study of the binary mixtures of charged particles with high size asymmetry. Recently, Tohver et al [46] refer to this process as nanoparticle halos and suggested exploiting the haloing mechanism for regulating the effective charge and, hence, stability of mesoscopic colloids.

10. Summary

The layering and layer ordering of like-charged particles or macroions, confined to a plane-parallel and a wedge slit formed by two uncharged hard walls has been investigated by canonical Monte Carlo method combined with a simulation cell containing both the bath and film regions. The macroion suspension is modelled using an effective one-component approach with potential functions that include both electrostatic and excluded volume forces. The effective electrostatic forces are considered in a conventional fashion by employing a screened Coulomb interaction that incorporates as parameters the inverse Debye length to characterize the contribution of the salt ions, and the dielectric permittivity to characterize the contribution of aqueous solvent. The excluded volume forces in the present study incorporate some new features in comparison to those used previously; in particular, the extra term that originates from a discrete nature of suspending fluid has been added to the effective macroion-macroion and macroion-wall interactions. The excluded volume forces are characterized by the volume fraction associated with suspending fluid.

The computer simulation data discussed in this chapter have confirmed the suggestion raised from observations that as a result of the confinement, the macroions form layers parallel to the confining surfaces. Two distinct types of particle layering have been revealed when the suspension of macroion particles is confined to the gap formed by uncharged macroscopic hard walls: surface monolayers and middle-film layers. The surface monolayers are more dense and well structured representing a quasi-2D assembly of macroions adsorbed and tightly bound to each of the confining surfaces. The middle-film layers are less packed, diffuse and show a tendency to be segregated from the surface layers. The layer structuring is improved with an increase in macroion concentration.

In general, the phenomenon of particle structuring in the confined macroion suspensions is a combined effect of the electrostatic repulsion between macroions and the excluded volume interaction between macroions and confining walls. The strong electrostatic repulsion between macroions practically diminishes the role of the excluded volume forces in a bulk suspension. Although being not so decisive in the modelling of homogeneous colloidal suspensions, the excluded volume forces originate from the discrete nature of the primary solvent become a necessary attribute in the modelling of confined colloidal suspensions. Therefore, for complex colloidal suspensions or colloidal dispersions, one must distinguish between the excluded volume contribution into the interaction between smaller nanocolloidal particles (micelles, etc.) and that of the interaction between nanocolloids and larger mesocolloids (oil droplets, etc.) or/and macroscopic confining surfaces. The simulations with and without excluded volume forces discussed in this chapter, serve as a methodological example which illustrates that an adequate modelling of complex colloidal suspension should necessarily take into account the discrete nature of a primary suspending fluid.

On the other hand, the recent studies of the uncharged hard-sphere colloidal suspensions [51, 52] predict the formation of surface monolayers of the neutral colloidal particles similar to that in the charged suspension, however, the segregation of the remaining film from the surface layers is not so evident. This suggest that former is promoted exclusively by the electrostatic forces and results from the strong long-range repulsion of the macroions in the middle-film region and short-range excluded volume attraction between the nearby macroions and film surfaces. Summarizing, the electrostatic forces enhance the macroion layering while the excluded volume forces contribute to the formation of the surface monolayers and to the in-layer structuring.

The macroion structuring in between two inclined surfaces has been analyzed by carrying out a set of separate simulation runs for a set of consecutive wedge segments designed to represent a single long wedge. As the wedge thickness progressively increases, the sequence of characteristic regions along the wedge film has been established. The wedge film starts with no macroion region for the wedge thicknesses less than macroion diameter; this part of the wedge is referred to as a geometrical or hard-core "vacuum" because of the hard-wall geometrical restriction that prevents the macroions from being present. As soon as the wedge thickness becomes the same as macroion diameter, a 1D ordered macroion chain parallel to the wedge vertex is formed by macroions segregated on the uncharged wedge walls because of the electrostatic repulsion in a bulk solution. This macroion chain within the wedge film is followed by a film region that extends till a wedge thickness of up to slightly above two macroion diameters and represents a macroion monolayer because of the wall geometrical restriction that does not allow the fit of two lateral particle layers at the thickness less than two particle diameters. Shortly, after the wedge thickness exceeds two macroion diameters, the wedge film starts to split into two well-defined quasi-2D monolayers formed by macroions bound to each wedge walls by mutual electrostatic repulsion. Once formed, the surface monolayers of macroions persist for all remaining wedge thicknesses up to the bulk. At small separations between surface monolayers the electrical repulsion prevents bulk macroions to enter in between, forming a one more region within the wedge that is free of macroions. In contrast to one next to the vertex, the former is of electrostatic origin and was identified as the so-called "vacuum" phase that has been observed experimentally [22]. The estimates of the wedge thickness boundary of the "vacuum" phase observed in simulations correlate well with the reported experimental data.

In relation with the experimental studies of ionic surfactant micellar solutions, the simulation data discussed in this chapter partially supports the "squeezing layer" mechanism [3]. The middle-film layer was found to be poorly organized; the average macroion-to-macroion distance inside this layer is larger than the same distance inside the surface layer. The information extracted from Monte Carlo data on the macroion in-layer structuring is particularly helpful in understanding the mechanism of film thinning in colloidal suspensions. The radial distribution functions obtained during simulations, one for the bulk suspension and second for the particles within surface monolayers, each represents the most preferable 3D and 2D, respectively, ordering that exist in nature. The radial distribution function of the macroions within the middle-film layer represents some virtual realization of the particle structuring that occurs in the interspace between 2D and 3D. By increasing the distance between confining surfaces, the middle-film layer transforms into the bulk suspension, i.e. the 3D system. In contrast, when the distance between film surfaces decreases, there are two possibilities for the macroions comprising the middle-film layers: to be transformed into the thin quasi-2D monolayer, beginning surface crystallization, or to squeeze out the film region causing the film thinning process.

Finally, we wish to point out that the effective interaction $W(r)$ used in present study, is still primitive, with the electrostatic contribution being treated at a lower level than the excluded volume contribution. An expression for the effective interaction that accounts for contributions from all types of electrostatic interactions in the system, i.e. between electrolyte ions and solvent molecules in a more consistent manner is preferable. A more accurate approach could be based on the complete treatment of the electrostatic contribution for both ions and polar solvent molecules.

Acknowledgements

This work was supported in part by DOE and NSF grants.

References

[1] Nikolov, A.D., Wasan, D.T. *J. Colloid Interface Sci.*, 1989, **133**, p. 1.

[2] Nikolov, A.D., Wasan, D.T., Denkov, N.D., Kralchevsky, P.A., Ivanov, I.B. *Progr. Colloid Polym. Sci.*, 1990, **82**, p. 87.

[3] Nikolov, A.D., Wasan, D.T. *Langmuir*, 1992, **8**, p. 2985.

[4] Wasan, D.T., Nikolov, A.D., Lobo, L.A., Koczo, K., Edwards, D.A. *Progr. Surface Sci.*, 1992, **39**, p. 119.

[5] Koczo, K., Nikolov, A.D., Wasan, D.T., Borwankar, R.P., Gonsalves, A. Layering of Sodium Caseinate Submicelles in Thin Liquid Films – A New Stability Mechanism for Food Dispersions. *J. Colloid Interface Sci.*, 1996, **178**, p. 694.

[6] Nikolov, A.D., Kralchevsky, P.A., Ivanov, I.B., Wasan, D.T. *J. Colloid Interface Sci.*, 1989, **133**, p. 13.

[7] Kralchevsky, P.A., Nikolov, A.D., Wasan, D.T., Ivanov, I.B. *Langmuir*, 1990, **6**, p. 1180.

[8] Bergeron, V., Radke, C.J. *Langmuir*, 1992, **8**, p. 3020.

[9] Richetti, P., Kékicheff, P. *Phys. Rev. Lett.*, 1992, **68**, p. 1951.

[10] Wasan, D.T., Nikolov, A.D. In: *Supramolecular in Confined Geometries*, S. Manne and G. Warr, Eds., 1999, ACS Symposium Series No. 736, p. 40.

[11] Wasan, D.T., Nikolov, A.D., Henderson, D., Trokhymchuk, A. (2002). *Encyclopedia of Surface and Colloid Science*, p. 1181. New York: Marcel Dekker.

[12] Parsegian, V.A., Brenner, S.L. The Role of Longe Range Forces in Ordered Arrays of Tobacco Mosaic Virus. *Nature*, 1976, **259**, p. 632.

[13] Furusawa, K., Tomotsu, N. Direct Observation Studies for the Structure of the Electrical Double Layer of Concentrated Monodisperse Latices. *J. Colloid Interface Sci.*, 1983, **92**, p. 504.

[14] Trokhymchuk, A., Henderson, D., Nikolov, A., Wasan, D.T. *J. Phys. Chem.*, 2003, **107**, p. 3927.

[15] Henderson, D., Trokhymchuk, A., Nikolov, A., Wasan, D.T. *Ind. Eng. Chem. Res.*, 2005, **44**, p. 1175.

[16] Trokhymchuk, A., Henderson, D., Nikolov, A. and Wasan, D.T. *Langmuir*, 2005, in press.

[17] Derjaguin, B., Landau, L. *Acta. Phys. Chim. URSS*, 1941, **14**, p. 633.

[18] Verwey, E.J.W., Overbeek, J.Th.G. (1948). *Theory of the Stability of Lyophobic Colloids*, Amsterdam: Elsevier.

[19] Henderson, D., Lozada-Cassou, M. *J. Colloid Interface Sci.*, 1986, **114**, p. 180.

[20] Zhang, L., Davis, H.T. and White, H.S. *J. Chem. Phys.*, 1993, **98**, p. 5793.

[21] Murray, C.A., Van Winkle, D.H. *Phys. Rev. Lett.*, 1987, **58**, p. 1200.

[22] Pieransky, P., Strzelecki, L. and Pansu, B. *Phys. Rev. Lett.*, 1983, **50**, p. 900.

[23] Reiss-Husson, F., Luzzati, V. *J. Phys. Chem.*, 1964, **68**, p. 3504.

[24] Sasaki, T.; Hattori, M., Sasaki, J., Nukina, K. *Bulletin of the Chem. Soc. of Japan* 1975, **48**, p. 1397.

[25] Hartley, G.S. (1936). *Aqueous Solutions of Paraffin-Chain Salts*. Paris: Hermann.

[26] Israelachvili, J.N. (1992). *Intermolecular and Surface Forces*, 2nd edition. London: Academic Press.

[27] Sogami, I., Ise, N. *J. Chem. Phys.*, 1984, **81**, p. 6329.

[28] Spalla, O., Belloni, L. *Phys. Rev. Lett.*, 1995, **74**, p. 2515.

[29] Ise, N., Smalley, M.V. *Phys. Rev. B*, 1994, **50**, p. 16722.

[30] Ito, K., Yoshida, H., Ise, N. *Science*, 1994, **263**, p. 66.

[31] Larsen, A.E., Grier, D.G. *Nature*, 1997, **385**, p. 230.

[32] Crocker, J.C.; Grier, D.G. *Phys. Rev. Lett.*, 1996, **77**, p. 1897.

[33] Ornstein, L.S. and Zernike, F. *Proc. Acad. Sci. Amsterdam*, 1914, **17**, p. 793.

[34] Attard, P., Bernard, D., Ursenbach, C., Patey, G.N. *Phys. Rev. A*, 1991, **44**, p. 8224.

[35] Henderson, D. (1991). *11th Symposium on Thermophysical Properties*, Boulder CO, June 23-27; *Fluid Phase Equil.*, 1992, **76**, p. 1; *J. Chem. Phys.*, 1992, **97**, p. 1266.

[36] Hansen, J.P., and McDonald, I.R. (1986). *Theory of Simple Liquids*. New York: Academic Press.

[37] Barker, J.A. and Henderson, D. *Rev. Mod. Phys.*, 1976, **48**, p. 587.

[38] Blum, L., Hye, J.S. *J. Chem. Phys.*, 1977, **81**, p. 1311.

[39] Henderson, D. *J. Colloid Interface Sci.*, 1988, **121**, p. 486.

[40] Roth, R., Evans, R., Dietrich, S. *Phys. Rev. E*, 2000, **62**, p. 5360.

[41] Asakura, S., Oosawa, J. *J. Chem. Phys.*, 1954, **22** p. 1255.

[42] Trokhymchuk, A., Henderson, D., Wasan, D. *J. Colloid Interface Sci.*, 1999, **210**, 320.

[43] Trokhymchuk, A., Henderson, D., Nikolov, A., Wasan, D.T. *Langmuir*, 2001, **17**, p. 4940.

[44] Trokhymchuk, A., Holovko, M.F., Henderson, D. *Molec. Phys.*, 1993, **80**, 1009.

[45] Imhof, A., Dhont, J.K.G. *Phys. Rev. E*, 1995, **52**, p. 6344.

[46] Tohver, V., Smay, J.E., Braem, A., Braun, P.V., Lewis, J.A. *Proc. Natl. Acad. Sci.*, 2001, **98**, p. 8950.

[47] Pollard, M.L., Radke, C.J. *J. Chem. Phys.*, 1994, **101**, p. 6979.

[48] Pashley, R.M., Ninham, B.W. *J. Phys. Chem.*, 1987, **91**, p. 2902.

[49] Boda, D., Chan, K-Y., Henderson, D., Wasan, D.T., Nikolov, A.D. *Langmuir*, 1999, **15**, p. 4311.

[50] Garibay-Alonso, R., Mendez-Alcaraz, J.M., Klein, R. *Physica A*, 1997, **235**, p. 159.

[51] Trokhymchuk, A., Henderson, D., Nikolov, A., Wasan, D.T. *Phys. Rev. E*, 2001, **64**, 012401.

[52] Trokhymchuk, A., Henderson, D., Nikolov, A., Wasan, D.T. *J. Colloid Interface Sci.*, 2001, **243**, p. 116.

GRAIN INTERACTION AND ORDERING IN A DUSTY PLASMAS

Phenomenon of nonlinear screening in colloidal plasmas

O. Bystrenko, T. Bystrenko, A. Zagorodny
Bogolyubov Institute for Theoretical Physics
03143 Kiev, Ukraine

Abstract   Recent results concerning the grain interactions and ordering in colloidal plasmas are presented. The chapter focuses on the phenomenon of nonlinear screening and its effects on the structure of colloidal plasmas, the role of trapped ions in grain screening, and the effects of strong collisions in the plasma background. It is shown that the above effects may strongly modify the properties of the grain screening giving rise to considerable deviation from the conventional Debye-Hückel theory as dependent on the physical processes in the plasma background.

Keywords: High-Z impurity, charged colloidal suspension, dusty plasma, nonlinear screening, bound ionic states, strong collisions

1. Introduction

During last decades, the properties of charged macroparticles embedded in plasmas have been the focus of attention of researchers. Our interest in this issue is connected, first of all, with its implications in spatial ordering phenomena in colloidal plasmas (CP) such as dusty plasmas (DP) or charged colloidal suspensions (CCS). Colloidal plasmas/inxxcolloidal plasmas consist of a large number of highly charged ($Z \simeq 10^3 - 10^5$) colloidal particles of submicron size immersed in a plasma background. Experiments have revealed a number of collective effects in colloidal plasmas, in particular, the formation of Coulomb liquids or crystals associated with the strong Coulomb coupling in the colloidal subsystem [1–5]. It is clear that the properties of grain screening play therewith an important role, since the effective screened potentials produce the most significant contributions to the grain-grain interactions and thus determine collective properties of the colloidal component in colloidal plasmas.

D. Henderson et al. (eds.), Ionic Soft Matter: Modern Trends in Theory and Applications, 291–314.
© 2005 *Springer. Printed in the Netherlands.*

The simplest approach conventionally employed to describe the grain screening in colloidal plasmas is the Debye-Hückel (DH) approximation, or, its modification for the case of the grain of finite size, the DLVO theory [6, 7]. The DH approximation represents the version of Poisson-Boltzmann (PB) approach linearized with respect to the effective potential based on the assumption that the system is in the state of thermodynamical equilibrium. The DH theory yields the effective interparticle interaction in the form of the so-called Yukawa potential which constitutes the basis for the Yukawa model.

Extensive molecular dynamics and Monte Carlo (MC) computer simulations performed for the Yukawa system (YS) [8–10] indicate that the latter provides a possibility for qualitative explanation of formation of condensed state in colloidal plasmas. However, it is clear that the accurate description of grain screening in colloidal plasmas requires more accurate approaches. Let us point out some important issues which should be first of all taken into account.

In the first place, these are nonlinear effects in the grain screening. Simple estimates evidence that the magnitude of the ratio $e\phi/k_BT$ (where e is the electron charge, ϕ is the potential, k_B and T are the Boltzmann constant and the temperature, respectively), which determines the significance of nonlinear effects, is, near the grain surface, on the order of $\simeq 10$ in real experiments on dusty plasmas and charged colloidal suspensions. This means that the linear approximation may fail in this case.

In the second place, a distinguishing feature of DP is that the charge of dust grains is maintained by plasma currents to the grain surface. Thus, DP are far from thermodynamical equilibrium even in the steady state. In these conditions, the Boltzmann distribution for plasma particles does not hold, which makes the equilibrium PB as well as the DH theory inapplicable. In other words, the kinetic description of grain screening is needed. Note that in this case the properties of grain screening may essentially depend on the presence of collisions in the plasma background.

It should be pointed out that the above questions have been the subject of large number of studies, where a number of important results has been obtained. The effects of nonlinear screening in the thermodynamically equilibrium case of charged colloidal suspension were studied in [11–13]. It was found that in the presence of the nonlinear effects the effective potential at distances can be described by the linear Debye-Hückel theory, however, with some effective charge being less than the bare grain charge.

The basic reference model for the case of collisionless plasma background with regard to the absorption of plasma particles by the grain, the OML theory, has been initiated by the paper of Bernstein and Rabinovitz [14]. As mentioned in this work, the asymptotical behavior of the screened potential for collisionless case is inversely proportional to the square distance; the authors also formulate here the question about the role of the bound ionic states in the grain screening. Later the OML theory and closely related collisionless approaches have been developed in numerous works [15–19, 22]. The particular interest to the collisionless case is due to its industrial implications and the fact that the laboratory and astrophysical DP may be considered in most of the cases with a good accuracy collisionless. It was found that within the range of plasma parameters and grain sizes typical for experimental observations, the effective potentials in the vicinity of the grain approach the predictions of DH theory, i.e., the allowance for charging by plasma currents does not affect considerably the properties of screening.

Let us say a few words about the role of bound ionic states. In most of the literature, the effects related to the ions trapped by negatively charged grains are neglected. Nevertheless, it is *a priori* unclear, to what extent the properties of screening can be affected by the presence of the bound states. An attempt to give some insight onto this problem is made in [14, 19–22].

Strictly speaking, the relative contribution of bound states within the collisionless models is in principle indeterminate, which is, eventually, the consequence of the time-reversibility of Vlasov equation. The matter is that the stationary solutions of the Vlasov equation are dependent on the way of formation of the steady state of the system. Thus, to tackle this problem, one has to employ additional considerations or principles for evaluating the number of trapped ions.

As mentioned above, for the first time this problem was pointed out in the work [14]. The authors related the generation of bound states to the ion collisions. Recently, this idea was used to take into consideration the presence of trapped ions in both analytical [21, 22] and numerical [23] studies. These papers give answers to a number of important questions, but many aspects of the problem still remain open. In particular, the bound ion distributions found in [21, 22] in the approximation of small collision frequency on the basis of calculations of free and bound ion balance, do not exhaust the variety of many other distributions which could exist in the absence of collisions.

The opposite case of strongly collisional plasma background is much less examined. In [24–26] the grain screening has been studied on the basis of the continuous drift-diffusion (DD) approximation. The effects of grain charging by plasma currents are essential in this case and the effective screened field considerably deviates from the predictions of DH theory. The main conclusion of the authors is that the the effective potentials can be still fitted by DH theory, however, with effective parameters, and the screening length is, typically, longer than the Debye radius.

The goal of this chapter is to give a concise review of further important results on the above issues obtained recently in [27–31].

2. Nonlinear phenomena in the grain screening and plasmas structure

In the case of thermodynamical equilibrium, an accurate description of non-linear effects in grain screening can be obtained within the Poisson-Boltzmann (PB) approach describing the plasma as a two-component gas with Boltzmann distribution. The relevant equation for the case of a single spherical high-Z grain of a radius a in a plasma background reads,

$$\Delta\varphi(r) = -4\pi en \left\{ \exp\left[-\frac{e\varphi(r)}{kT} \right] - \exp\left[\frac{e\varphi(r)}{kT} \right] \right\}, \qquad (1)$$

with the boundary conditions for the effective self-consistent potential φ,

$$\varphi'(a) = Ze/a^2,$$
$$\varphi(\infty) = 0,$$

specifying the electric field on the grain surface and the potential at infinity. Here e is the charge of a positively charged plasma particle and n is the plasma concentration at infinity.

The conventional treatment based on the assumption,

$$\frac{e\varphi}{kT} \ll 1, \qquad (2)$$

yields, after linearization with respect to φ, the well-known DLVO solution [6, 7],

$$\varphi(r) = \frac{Z'e}{r} \exp\left(-\frac{r}{r_{\mathrm{D}}} \right), \qquad (3)$$

with the effective charge,

$$Z' = Z\frac{\exp\left(\frac{a}{r_{\mathrm{D}}} \right)}{1 + \frac{a}{r_{\mathrm{D}}}}, \qquad (4)$$

where r_{D} denotes the Debye screening length.

It is clear that at short distances the condition (2) is definitely violated, thus, the transition $a \to 0$ with the DH limit,

$$\varphi_{\mathrm{D}}(r) = \frac{Ze}{r} \exp\left(-\frac{r}{r_{\mathrm{D}}} \right), \qquad (5)$$

is incorrect. In other words, in the case of a grain of a small size the nonlinear effects in screening may be significant and the applicability of Eqs. (3) and (4) would break down.

To estimate the validity of the linear approximation, it is convenient to introduce the plasma-grain coupling,

$$\chi = \frac{Ze^2}{kTa},$$

giving the potential-to-kinetic energy ratio for a plasma particle on the grain surface. As mentioned above, its magnitude for dusty plasmas and charged colloidal suspension with high-Z impurities is on the order of $\simeq 10$, which casts doubt on the validity of the linear DH theory for the description of screening.

Below we consider the problem of screening of a finite-size charge Z in a plasma background for the range $\chi \simeq 1 - 50$ in two ways. The first one is the accurate numerical solution of the above PB equations. The other one is the method of MC computer simulations providing a microscopic description of screening.

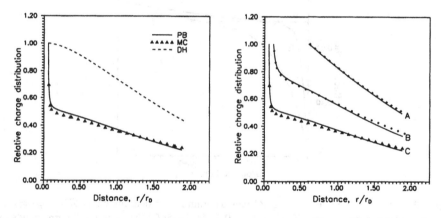

Figure 1. Comparison of the relative charge distributions near the charged grain obtained within the linear DH approximation (dashed line), nonlinear PB theory (solid line), and MC simulations (symbols) for for $Z = 25, \Gamma = 0.1$. The plasma-grain coupling is $\chi = 20$ (left), and $\chi = 2(A), 10(B), 20(C)$ (right).

The PB boundary problem (1) has been solved numerically, by using the shooting method and the second-order Runge-Cutta numerical algorithm [32]. The MC simulations of the screening were performed for the NVT-ensemble using the conventional Metropolis algorithm [33], within the finite model with the microscopic two-component plasma background, represented by a large number of charged hard spheres confined in a spherical volume with the grain fixed in the center. The goal of computations was the effective screened potential $\varphi(r)$ and the charge distribution function $Q(r)$ defined as the ratio of the total charge residing within a sphere of a radius r to the grain charge.

Let us say a few words about the choice of parameters. The PB theory is based on the notion of the mean field, which loses its sense for strongly coupled plasma background, in the case that,

$$\Gamma = \frac{e^2}{kTd} \geqslant 1,$$

as the plasma correlations become significant. Here $d = (4\pi n)^{-1/3}$ is the average distance between plasma particles. Typical for colloidal plasmas are the values $Z \gg 1$ and $\Gamma \ll 1$. We give below the comparison of the results obtained within the two above approaches for $Z = 25$, $\Gamma = 0.1$ and 0.05, $\chi = 2, 10$ and 20.

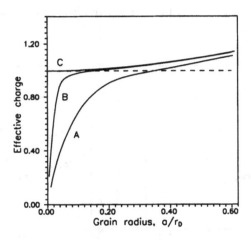

Figure 2. Relative effective charge Z^* versus grain radius determined as $Z^* = \varphi/\varphi_D$ at $r \gg a$. Other parameters are: $Z = 25$; $\Gamma = 0.1(A)$; $0.05(B)$. The curve (C) corresponds to the linear DLVO approximation; the dashed line is the DH theory.

As follows from our computations, both approaches evidence at strong plasma-grain coupling, in a distinct contrast with the linear DH theory, the existence of an interesting effect associated with the accumulation of plasma charge on the grain surface or "plasma condensation", which sharply affects the characteristics of screening, Figs. 1 and 2. While the asymptotical behavior of the screened potential at long distances retains its Yukawa-like form given by Eq. (5), the magnitude of the effective charge Z^* can be well described by the DLVO theory only for small values of χ. For stronger plasma-grain coupling, in a sharp contrast with the predictions of linear screening theory, the effective charge approaches zero, which evidences a pronounced enhancement of screening in this case.

An important point is the existence of a critical magnitude of this parameter, $\chi \simeq 4$, that weakly depends on the other plasma parameters. It is interesting to note that this critical magnitude is much larger than unity, which means that the linear screening approximation is well accurate even for the the expansion parameter (2) ranging somewhere between 4 and 5.

It is clear that the above phenomenon of nonlinear screening is of importance for structural properties of colloidal plasmas. Its effect on the phase diagram for charged colloidal suspension can be illustrated on the basis of the model of effective intergrain forces in the following way.

As mentioned above, the basic reference systems of colloidal plasmas based on the notion of effective interaction is the Yukawa system with the interparticle effective potential given by

$$\frac{V(x)}{k_B T} = \frac{\Gamma}{x} \exp\left(-\frac{x}{\Delta}\right), \qquad (6)$$

with two dimensionless parameters: the coupling Γ and the screening length Δ; x is the dimensionless distance. Our study employs the connection between the dimensionless parameters Γ and Δ of a Yukawa system and the microscopic parameters of colloidal plasma, that can be established as follows.

Let us consider a two-component asymmetric strongly coupled plasma, which is a simplest example of colloidal plasma. A good microscopic model for it is a system of charged hard spheres interacting by Coulomb forces. In the case that the size of a plasma particle is negligibly small (in agreement with physical situation in colloidal plasma), such a system can be described by three dimensionless parameters: the packing fractions for the colloidal component, $v = n\pi\sigma^3/6$, the charge asymmetry Z, and the plasma-grain coupling χ. Under the assumption that the screening of grains is produced purely by plasma background and that the screening can be described in terms of the linear DH theory for point charges, one easily gets the effective interaction in the form of Eq. (6), with the parameters of the Yukawa system determined by

$$\Gamma = \frac{Z^2 e^2}{k_B T d}, \qquad (7)$$

$$\Delta = \frac{r_D}{d}, \qquad (8)$$

and the dimensionless distance specified as $x = r/d$. Here $d = (4\pi n_c)^{-1/3}$ is the average distance between colloidal plasma particles; $r_D = (4\pi n_{bg} e^2 / k_B T)^{-1/2}$ is the Debye screening length produced by the single plasma component.

Thus, the parameters entering the effective Yukawa interaction are expressed here via the microscopic plasma parameters. Our further consideration is based upon the idea, that the properties of colloidal plasma can be described by effective pair interaction in the form of Eq. (6) even in the case that the nonlinear screening is significant. As mentioned above and shown in [13], the effective screened potential retains in this case the Yukawa-like form at long distances. The effective charge, however, should be found from the exact solution of the relevant PB equation; the background density n_{bg}, which determines the Debye length, is assumed to be equal to the average plasma background concentration. Within such an approach, the account of nonlinear effects reduces to re-scaling the well known melting curve for Yukawa system [9] with the use of the relevant *effective charge* Z^* instead of the *bare charge* Z. This latter can be evaluated by numerically solving the nonlinear PB equation for a single grain in one-component plasma background in a spherical cell with relevant parameters. An important point is that due to the connection between the microscopic parameters of two-component asymmetric plasmas and the parameters of the Yukawa model one can obtain important qualitative conclusions about a *minimal charge asymmetry* Z_{min} needed for formation of Coulomb lattices in colloidal plasma. Namely, there is a connection between the parameters Z, Γ and Δ in a form,

$$Z = \Gamma \Delta^2, \tag{9}$$

which is the consequence of the relations (7) and (8) and the global charge neutrality condition $Zn = n_{bg}$. Therefore, the parameter Z specified via Eq. (9), which has the physical meaning of charge asymmetry, can be used for the description of Yukawa system instead of coupling parameter Γ. In other words, the relation (2.9) makes it possible to transfer the melting curves onto the $Z - \Delta$ plane.

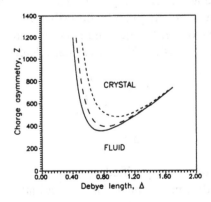

Figure 3. Melting curves for colloidal plasmas in $Z - \Delta$ plane. Solid line: $v = 5 \cdot 10^{-2}$; long dashes: $v = 5 \cdot 10^{-3}$; short dashes: $v = 5 \cdot 10^{-4}$. Nonlinear screening effects in shifting the melting curves to higher values of asymmetry Z at small packing fractions.

The results of computations of the melting curves are given in Fig. 3. As can be seen from this figure, there exists a minimal charge asymmetry $Z_{min} = 355$ needed to obtain a crystal state. The same conclusion and a close value of $Z_{min} = 360$ was obtained in [34] based on the Lindemann melting criterion for the case of specific effective grain-grain forces. The lower melting curve in the Fig. 3 is close to that given in that work. However, we see that the allowance for the nonlinear screening results in shifting melting curves to higher values of charge asymmetry Z at small packing fractions of colloidal component.

It should be noted, that our consideration is based on the effective Yukawa interaction in the form given by Eq. (6), which is expected to work in the case of diluted charged colloidal suspensions with high charge asymmetry and weakly coupled plasma background [35]. The effects of nonlinear background screening are commonly accepted to be associated with induced many-body forces between colloidal particles and are expected to become relevant for moderate packing fractions. The present example shows that the nonlinear screening may be important in the case of small packing fractions as well.

A more accurate description of the structure of colloidal plasmas can be obtained by means of MC computer simulations based on the microscopic model of asymmetric two-component plasmas (TCP). As shown above, the nonlinear grain screening obtained within PB theory has the direct analogue in the MC simulations with the microscopic description of plasma background, the phenomenon of "plasma condensation" near grain surface. This suggests that the above phenomenon should manifest itself in MC simulations of asymmetric two-component plasmas affecting its structural properties as well.

Below we give our results of MC simulations of strongly coupled two-component plasmas with the charge asymmetry up to $Z = 100$ based on the "primitive model" aimed at the elucidation of the nonlinear effects on the structural properties of two-component plasmas. Within the "primitive" model, a two-component plasmas is considered as an overall charge neutral mixture of charged spherical grains in a compensating plasma background. In all the simulations we assume the size of a plasma particle to be negligibly small, in accord with the physical situation in colloidal plasmas. We performed MC simulations of such a system for canonical ensemble by using the conventional Metropolis algorithm and periodic boundary conditions [33]. An accurate account of long-range Coulomb forces was achieved due to the Ewald summation procedure [36].

The idea of simulations was to study radial grain-grain and plasma-plasma distributions near the critical point defined by the value $\chi \simeq 4$. In particular, we performed a number of simulations for different values of plasma-grain coupling χ but for a fixed value of coupling $\Gamma_c = Z^2 e^2 / k_B T d_c$ in the colloidal component (we use here a slightly different definition for the average interparticle distance $d_c = (4\pi n_c/3)^{-1/3}$). The range of parameters was: the charge asymmetry from $Z = 10$ to 100, volume fractions of colloidal component from $v_c = 0.001$ to 0.1, the coupling from $\chi = 1$ to 50. Note that these parameters are connected with the coupling Γ_c by the relation,

$$\Gamma_c = Z\chi v_c^{1/3} . \tag{10}$$

Therefore, by varying the charge asymmetry Z of a two-component plasmas, one can change the parameter χ while holding the above coupling Γ_c constant.

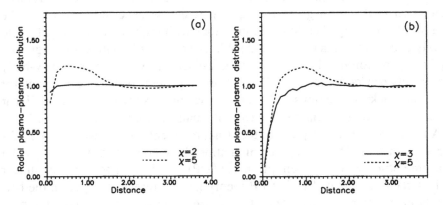

Figure 4.   Radial plasma-plasma distribution functions for infinite two-component plasmas. a) $\Gamma_c = 26$, $v_c = 0.01$. Solid line: $\chi = 2$, $Z = 60$. Dashed line: $\chi = 5$, $Z = 24$. b) $Z = 10$, $v_c = 0.01$. Solid line: $\chi = 3$, $\Gamma_c = 6.5$. Dashed line: $\chi = 5$, $\Gamma_c = 10.8$. The unit of distance is σ_c.

The results for different values of parameters Z, v_c and χ are presented on Figs. 4-6. The most remarkable result consists in a pronounced change in the behavior of the system near the point $\chi \simeq 4$. If the coupling χ between the components is less than 4, the grain-grain distribution in a two-component plasmas exhibits an oscillatory behavior characteristic of a liquid phase. It means that the effects of screening produced by the plasma component, do not change qualitatively the properties of colloidal component, and the latter behaves like one-component plasma. In Fig. 4, we can see that in this case (for $\chi = 2$ and 3) the plasma-plasma distributions are characteristic of a gas phase. In the case of strong plasma-grain coupling $\chi > 4$, the reduction in grain-grain correlations and the appearance of correlations (on the length of the order of the grain diameter σ_c) in plasma-plasma distributions are observed, Figs. 4 and 5. These indicate a pronounced enhancement of grain screening and the accumulation of plasma particles near grain surfaces.
Remarkably, it was observed in all the simulations near the same threshold value $\chi = 4$ within a wide range of other parameters of two-component

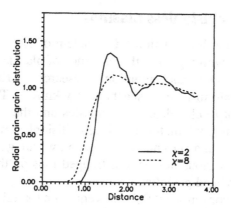

Figure 5.   Radial grain-grain distribution functions for infinite two-component plasmas for the same grain-grain coupling $\Gamma = 26$ and packing fraction $v_c = 0.01$ (liquid state). Solid line: $\chi = 2$; Dashed line: $\chi = 8$; the charge asymmetry $Z = 60$ and 15. The unit of distance is d_c.

plasma regardless of the way of simulations. Direct visual observations of equilibrium configurations for strong plasma-grain coupling also evidence the accumulation of plasma particles near grains, Fig. 6.

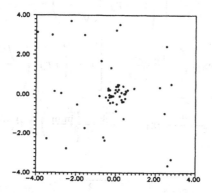

Figure 6.   Equilibrium configuration for the plasma component near a single grain, $Z = 100$; the coupling in the plasma background is $\Gamma_p = 0.05$; $\chi = 20$. The unit of distance is σ_c.

Thus, we see that the qualitative change in the structural properties near the point $\chi = 4$ is rather general feature of asymmetric two-component plasma and the threshold value obtained in MC simulations of this system is in a good agreement with the studies of nonlinear screening of a single grain based on the continuous PB theory.

3.   Grain in collisionless plasmas

As mentioned in the Introduction, the problem of grain screening in colli-sionless background with regard to the influence of plasma particle loss at the grain surface has attracted much attention of researchers. However, the effects of trapped ions remain in many respects poorly known. The purpose of this section is an attempt to elucidate the properties and the role of bound ionic states in grain screening within the nonlinear collisionless model in the case of a grain charged by plasma currents. In particular, we are going to focus on the effects produced by various numbers of trapped ions on the charge densities and the effective screened potentials.

We start from the conventional Poisson's equation for a single charged spher-ical grain of a radius a immersed in a plasma background,

$$\Delta\phi(r) = -4\pi e\left[Z_i n_i(r) - n_e(r)\right],\tag{11}$$

with the ion and electron densities $n_i(r)$ and $n_e(r)$ being specified as

$$n_i(r) = n_{ib}(r) + n_{if}(r),$$

where

$$n_{if}(r) = \frac{n_{0i}}{2}\exp\left[-\frac{Z_i e\phi(r)}{k_B T_i}\right]\left\{1 - erf\left(\frac{v_{ib}}{\sqrt{2}s_i}\right) + \frac{2}{\sqrt{\pi}}\frac{v_{ib}}{\sqrt{2}s_i}\exp\left(-\frac{v_{ib}^2}{2s_i^2}\right)\right.$$

$$+ \sqrt{1 - \frac{a^2}{r^2}}\exp\left(-\frac{v_{imin}^2}{2s_i^2}\right) \times \left[1 - \left(erf\left(\frac{\sqrt{v_{ib}^2 - v_{imin}^2}}{\sqrt{2}s_i}\right)\right.\right.$$

$$\left.\left.\left.- \frac{2}{\sqrt{\pi}}\frac{\sqrt{v_{ib}^2 - v_{imin}^2}}{\sqrt{2}s_i}\exp\left(-\frac{v_{ib}^2 - v_{imin}^2}{2s_i^2}\right)\right)\theta(v_{ib}^2 - v_{imin}^2)\right]\right\},\tag{12}$$

$$n_{ib}(r) = \mathcal{A}n_{0i}\exp\left[-\frac{Z_i e\phi(r)}{k_B T_i}\right]\sqrt{1 - \frac{a^2}{r^2}\exp\left(-\frac{v_{imin}^2}{2s_i^2}\right)}$$

$$\times \left[erf\left(\frac{\sqrt{v_{ib}^2 - v_{imin}^2}}{\sqrt{2}s_i}\right) - \frac{2}{\sqrt{\pi}}\frac{\sqrt{v_{ib}^2 - v_{imin}^2}}{\sqrt{2}s_i}\exp\left(-\frac{v_{ib}^2 - v_{imin}^2}{2s_i^2}\right)\right]$$

$$\times \theta(v_{ib}^2 - v_{imin}^2),\tag{13}$$

and

$$
[n_e(r) = \frac{n_{0e}}{2} \exp\left[\frac{e\phi(r)}{k_B T_e}\right] \left\{ 1 + erf\left(\frac{v_{e0}}{\sqrt{2}s_e}\right) \right.
$$

$$
- \sqrt{\frac{2}{\pi}} \frac{v_{e0}}{s_e} \exp\left(-\frac{v_{e0}^2}{2s_e^2}\right) + \sqrt{1 - \frac{a^2}{r^2}} \exp\left(\frac{v_{e0}^2}{2s_e^2}\frac{a^2}{r^2 - a^2}\right)
$$

$$
\times \left[1 - erf\left(\frac{v_{e0}}{\sqrt{2}s_e}\sqrt{\frac{r^2}{r^2 - a^2}}\right)\right.
$$

$$
\left.\left. + \sqrt{\frac{2}{\pi}} \frac{v_{e0}}{s_e}\sqrt{\frac{r^2}{r^2 - a^2}} \exp\left(-\frac{v_{e0}^2}{2s_e^2}\frac{r^2}{r^2 - a^2}\right)\right]\right\}. \quad (14)
$$

Here $n_{ib/if}(r)$ is the density of bound/free ions, $n_{0i/0e}$ is the ion/electron density at infinity; $\phi(r)$ is the self-consistent effective potential; e is the absolute value of the electron charge; $T_{i/e}$ is the ion/electron temperature; $s_{i/e} = \sqrt{k_B T_{i/e}/m_{i/e}}$ is the thermal ion/electron velocity; $m_{i/e}$ is the ion/electron mass, and Z_i is the ion charge number. Also, we introduced here the notations,

$$
v_{i0}^2 = \frac{2eZ_i}{m_i}\left[|\phi(a)| + \phi(r)\right], \qquad v_{ib}^2 = \frac{2eZ_i}{m_i}|\phi(r)|,
$$

$$
v_{e0}^2 = \frac{2e}{m_e}\left[|\phi(a)| + \phi(r)\right], \qquad v_{imin}^2 = \frac{a^2 v_{i0}^2}{r^2 - a^2}.
$$

The relations (12)–(14) can be obtained by integrating the Maxwellian distributions over velocities taking into account the energy and angular momentum conservation laws and the limitations imposed by the presence of the absorbing grain. I.e.,

i) we take into account all the ion and electron trajectories which do not touch the grain,

ii) we exclude from the phase space all the finite ion trajectories which intersect the grain surface,

iii) and we exclude the outgoing free ion or electron trajectories, which previously met the grain.

Notice that the above equations also follow from the stationary solution of the Vlasov equation with the appropriate boundary conditions (Maxwellian distributions at the infinity and zero value of distribution functions with positive radial velocity at the grain surface). It should be noted, that in the derivation of the density for bound ions, we also start from the Maxwellian distribution, though the finite trajectories do not reach the infinity and, therefore, cannot be coupled to the heatbath. Thus, we employ an additional assumption that the bound states are initially created with equilibrium distribution.

The relation (13) contains a free parameter, the amplitude $\mathcal{A}$, which determines the relative contribution of bound ionic states to the charge density. As mentioned in the Introduction, its value is indeterminate within the collisionless model, because the concentration of bound states cannot be related in any way to the ion concentration at infinity. However, some reasonable estimates for the magnitude of $\mathcal{A}$ can be obtained as follows. Consider the limit $a \to 0$ in Eqs. (12)–(14). It can be verified that the value $\mathcal{A} = 1$ can be found from the additional requirement for the distributions to be Boltzmannian, which corresponds to the case of thermodynamical equilibrium. It is natural to assume that, in general case, the value of $\mathcal{A}$, though being dependent on various physical situations (i.e., on how the system reaches its steady state), would have a magnitude of the same order.

In order to solve the problem (11)–(14) within the interval $a \leqslant r \leqslant r_{\max}$, we have to formulate the boundary conditions for the effective potential $\phi(r)$,

$$\phi(a) = \phi_0 , \tag{15}$$

$$\phi(r_{\max}) = \phi_{\mathrm{as}} . \tag{16}$$

Here the right boundary $r_{\max}$ has to be chosen at sufficiently long distance, $r_{\max} \gg r_{\mathrm{D}}$, so that the potential is described by its asymptotical value ϕ_{as}. The latter is known [17, 18], it reads,

$$\begin{aligned}
\phi_{\mathrm{as}}(r) &\simeq -\pi e n_{0i} a^2 \left(1 + \frac{2e Z_i |\phi_0|}{k_{\mathrm{B}} T_i} \right) \frac{r_{\mathrm{D}}^2}{r^2} \\
&= -\frac{T_e}{2 Z_i (T_e + T_i)} \left(Z_i + \frac{k_{\mathrm{B}} T_i}{2e|\phi_0|} \right) \frac{a^2}{r^2} |\phi_0| .
\end{aligned} \tag{17}$$

The boundary value of the potential ϕ_0 at the grain surface is determined by the balance of plasma currents to the grain surface. In order to find it, we use the well-known equation [37],

$$\frac{\omega_{pe}^2}{s_e} e^{-u} = \frac{\omega_{pi}^2}{s_i} (t + u) . \tag{18}$$

Here $\omega_{p\sigma}^2 = 4\pi e_\sigma^2 n_\sigma / m_\sigma$, $s_\sigma = (k_{\mathrm{B}} T_\sigma / m_\sigma)^{1/2}$, $t = T_i / T_e Z_i$; the particle density of σ species at infinity is n_σ, and $u = e|\phi_0|/k_{\mathrm{B}} T_e$ is the sought-for dimensionless potential at the grain surface.

The two-point boundary value problem for the effective potential (11)–(16) was solved numerically, by using the shooting methods [32]. The computations were performed for the following range of parameters: $\tau = T_i / T_e = 0.08 -$

1.0; $\rho = a/r_{\mathrm{D}} = 0.015 - 3.0$; $\mathcal{A} = 0 - 10$; $Z_{\mathrm{i}} = 1$ and $\mu = m_{\mathrm{i}}/m_{\mathrm{e}} = 10^4$. The results of computations are shown on the figures from 7 through 9.

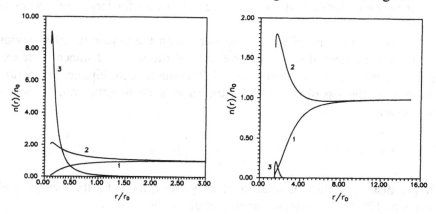

Figure 7. Charge densities for electrons (1), free (2) and bound (3) ionic states vs. distance for $\tau = 0.4$, $\mathcal{A} = 1.0$, $\rho = 0.1$ (left), and $\rho = 1.5$ (right).

In Fig. 7 the behavior of plasma charge densities associated with the calculated effective potentials are displayed within a typical range of parameters. As is seen from them, the bound ionic states tend to concentrate in the vicinity of the grain surface. The most remarkable feature in the behavior of bound ion states is that, beginning with some critical distance r_{c}, the density of bound states is strictly equal to zero.

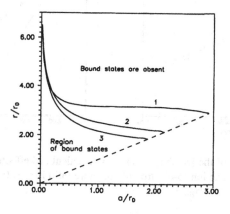

Figure 8. Critical distance r_{c} (dividing the region with bound ionic states and that where they are absent) vs. the grain radius for $\tau = 0.08$ (1); 0.4 (2); 1.0 (3). The amplitude $\mathcal{A} = 1.0$.

In Fig. 8 we give the relevant dependencies for r_{c} obtained in our calculations. Notice that, with increasing the grain size, the value r_{c} diminishes. As a result, in the case of very large grains, for $a \simeq 2 - 3r_D$, the bound states cannot exist at all. This conclusion is in agreement with the results of [14], where it was mentioned that the role of bound states for large grain sizes is insignificant.

Let us show that this effect is associated with the asymptotical behavior of the effective potential inversely proportional the square distance. The expression for the density of bound states (13) contains a multiplier, θ-function accounting for the loss of ions with the trajectories meeting the grain. Its argument is given by

$$\eta = v_{\text{ib}}^2 - v_{\text{imin}}^2 = \frac{2eZ_i}{m_i\left(1 - a^2/r^2\right)} \left\{ |\phi(r)| - \frac{a^2}{r^2}|\phi(a)| \right\}.$$

At larger distances, the potential $\phi(r)$ may be replaced by its asymptotical expression (17). In this case, the argument of the θ-function,

$$\eta(r \to \infty) = -\left[1 + 2\tau\left(1 - \frac{1}{4Z_i u}\right)\right] \frac{a^2|\phi_0|}{r^2} \frac{eZ_i}{m_i(1 - a^2/r^2)(1 + \tau)}$$

is always negative, $\eta(\infty) < 0$, since $Z_i \geqslant 1$ and $u = e|\phi_0|/k_B T_e \geqslant 1$. It means that the density of bound states is equal to zero (i.e. they are absent) at larger distances, where the potential assumes its asymptotical form.

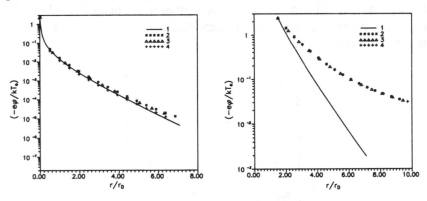

Figure 9.   Comparison of the DH theory (1) with the calculated effective potentials for $A = 0$ (2); 0.1 (3); 1.0 (4). The ion-to-electron temperature ratio is $\tau = 0.08$; the grain radius is $\rho = 0.015$ (left) and $\rho = 1.5$ (right).

In Fig. 9 the calculated effective potentials are displayed. As can be seen from them, the allowance for the bound ionic states for the amplitudes $A = 0 - 1$ results in rather insignificant changes in effective potentials, suggesting that the densities of free and bound ions adjust themselves self-consistently to produce very close potentials for different values of A.

Remarkably, for smaller grain sizes, $a \simeq 0.01 r_{\rm D}$, the critical radius $r_{\rm c}$ tends to increase indicating that the region where the potential takes its asymptotical form moves to larger distances. In this case, the effective potentials within the region $r < r_{\rm c}$ are very close to those predicted by DH theory. This conclusion is in agreement with the theoretical results of the papers [16, 20], as well as with the recent experiments [38], where the Yukawa type of the effective grain-grain interactions was demonstrated in the direct measurements.

4.   Grain charged by plasma currents

In this section we consider the screening of a spherical grain charged by plasma currents in a weakly ionized high pressure gas. As will be shown below, the properties of grain screening in this case substantially depend on the type of boundary conditions. In contrast to the works [24–26], were a complicated semirealistic multigrain systems with relevant specific boundary conditions are considered, we are going to examine the simplest case of a *single* grain with the emphasis on the basic features of this problem.

Thus, we examine a single spherical grain of a radius a imbedded in a weakly ionized high pressure gas. In this case, it is natural to use the drift-diffusion (DD) approach, because collisions of plasma particles with neutrals play here a dominant role. Assuming two types of plasma particles (ions and electrons) only , we write the general time-dependent equations for the unknown ion/electron densities $n_{\rm i,e}$ and self-consistent potential ϕ in the form,

$$\frac{\partial n_{\rm i,e}}{\partial t} = -{\rm div}\mathbf{j}_{\rm i,e} + I_0 - \alpha n_{\rm i} n_{\rm e}, \qquad (19)$$

$$\Delta \phi = -4\pi e(n_{\rm i} - n_{\rm e}). \qquad (20)$$

Here α is the coefficient of recombination, I_0 is the intensity of plasma ionization (we examine the case of uniformly distributed plasma sources). The expression for the current densities $\mathbf{j}_{\rm i,e}$ has the form,

$$\mathbf{j}_{\rm i,e} = -\mu_{\rm i,e} n_{\rm i,e} \nabla \phi - D_{\rm i,e} \nabla n_{\rm i,e},$$

where $\mu_{\rm i,e}$ and $D_{\rm i,e}$ are the ionic/electronic mobility and diffusivity, respectively. These latter are assumed to be related due to the Einstein's equation $\mu_{\rm i,e} = z_{\rm i,e} e_{\rm i,e} D_{\rm i,e} / k_{\rm B} T$ (here $z_{\rm i,e} = \pm 1$ is the ion/electron charge number). In a weakly ionized gas with dominating plasma-neutrals collisions, it is reasonable to assume that the ion and electron temperatures are equal. Thus, we consider below only the case that $T_{\rm i} = T_{\rm e} = T$. The grain charge emerges as a result of plasma currents due to the difference in electron and ion diffusivities. With regard for spherical symmetry, the relevant equation for the grain charge number Z reads,

$$\frac{\mathrm{d}Z}{\mathrm{d}t} = -4\pi a^2 \left(j_{(r)i} - j_{(r)e}\right) , \tag{21}$$

where the subscript (r) denotes the radial component of a current.

In order to formulate the boundary conditions, we admit that the system is confined in a spherical volume of sufficiently large radius $R \simeq 50 - 500 r_\mathrm{D}$ (where r_D is the Debye screening length) with the grain placed at the center. The boundary conditions are specified at the surface of this sphere and at the surface of the grain. In our simulations, we consider the two basic cases and two types of boundary conditions, respectively. In the first case (I), the sources of plasma ionization, which compensate the loss of plasma particles due to the absorption on the grain surface, are assumed to be far from the grain (outside the spherical volume). The action of these sources is modelled by maintaining constant electron and ion densities on the surface of the sphere, $n_i = n_e = n_0$. According to this, we write the boundary conditions for the densities $n_{i,e}$,

$$n_{i,e} = n_0 \qquad \text{at} \qquad r = R,$$

and assume the rates of plasma ionization and recombination over the volume I_0 and α to be equal to zero. In the second case (II), we examine the problem with uniformly distributed plasma sources ($I_0 \neq 0$) with allowance for the plasma recombination over the volume ($\alpha \neq 0$). Note, that in this case the quantities I_0 and α are related to the unperturbed bulk plasma density n_0 by the equation $I_0 = \alpha n_0^2$ valid in the absence of the grain. The relevant boundary conditions read,

$$\frac{\partial n_e}{\partial r} = \frac{\partial n_i}{\partial r} = 0 \qquad \text{at} \qquad r = R.$$

The boundary conditions for the potential at the grain surface have the form,

$$\frac{\partial \phi}{\partial r} = -\frac{Z(t)e}{a^2} \qquad \text{at} \qquad r = a,$$

and for the densities $n_{i,e}$ we use the boundary conditions [24],

$$n_{i,e} = 0 \qquad \text{at} \qquad r = a$$

appropriate for the case of strongly collisional background.

We solved the above system of equations (19)–(21) by means of the method of lines and Gear's method. In addition, we performed a limited number of Brownian dynamics (BD) simulations based on particle-in-cell (PIC) method [39] with spherically symmetric concentric cells and the boundary conditions corresponding to the case (I). In these simulations, the plasma background is modelled by finite numbers of particles of two sorts representing the ion and electron components. The dynamics of the system is governed by the reduced Langevin equations of overdamped motion,

$$h\frac{d\mathbf{x}_k}{dt} = -\nabla_k U + \mathbf{F}_k(t).$$

Here $\mathbf{x}_k$ is the radius vector of the k-th particle, and U is the potential energy of the configuration. The friction coefficient h and the random force $\mathbf{F}_k(t)$ are determined by the properties of the heatbath (in our case the role of the heatbath plays the high pressure neutral gas). Random force acting on k-th particle is specified by the Gaussian distribution,

$$P(\mathbf{B}_k(\Delta t)) = \frac{1}{(4\pi h^2 D \Delta t)^{3/2}} \exp\left[-\frac{|\mathbf{B}_k(\Delta t)|^2}{4h^2 D \Delta t}\right],$$

which determines the probability for the momentum,

$$\mathbf{B}_k(\Delta t) = \int_t^{t+\Delta t} \mathbf{F}_k(t)dt,$$

to be transferred to the k-th plasma particle during the time span Δt. The random forces, which act on different plasma particles are uncorrelated. It is clear that the quantities h and D related to the ion and electron components are different. In the above expressions, we dropped the subscripts for simplicity. Note that the friction coefficient h can be expressed via diffusivity and temperature, $hD = k_B T$, which enables one to establish the correspondence with the continuous DD approach. A detailed presentation of the issues concerning Brownian dynamics and its relation to the continuous probabilistic approaches, such as Fokker-Planck and Smolukhovsky equations, can be found in [40, 41]. Here we would like to point out that the overdamped BD represents the direct microscopic analogue to the DD approach, since the latter can be derived from the Smolukhovsky equations for one-particle distributions (i.e., within the additional mean field approximation). The aim of BD simulations was to test the results of DD approximation.

The range of parameters is typical for the dusty plasmas experiments in high pressure weakly ionized noble gases like Ne or Ar: plasma background coupling $\Gamma \simeq 10^{-3}$; plasma density $n_0 \simeq 10^{10}$ cm^{-3}; the density of the neutrals $n \simeq 10^{18}$ cm^{-3}; radius of the grain $a \simeq 10^{-3}$ cm; electron-ion recombination coefficient $\alpha \simeq 10^{-7}$cm$^3/sec$; the ratio of the Debye length to the grain radius $r_D/a \simeq 0.1 - 50$. The ratio of diffusitivities in all computations was fixed, $A = D_e/D_i = 10^3$ (with the exception for the BD simulations). The goal of simulations was the final time-independent density and charge distributions which establish themselves after sufficiently long period of relaxation.

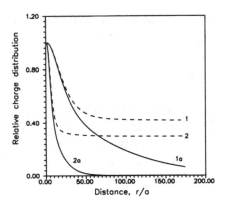

Figure 10. Comparison of charge distributions for different types of boundary conditions for the same bulk plasma parameters. The Debye length is $r_D/a = 10$ for (1) and (1a), and $r_D/a = 2$ for (2) and (2a). Dashed and solid lines relate to the BC (I) and (II), respectively.

The results of computations are given in the figures. In Fig. 10, we give the relative charge distributions for the different types of boundary conditions. Remarkably, in the case of boundary conditions (I), we observe the Coulomb-type asymptotical behavior of the screened field with the effective charge determined by the asymptotical value of the charge distribution. Note that such asymptotical behavior of the screened field may be viewed as a consequence of the Ohm's law for the problem under consideration. In contrast to the case (I), the screening in the case of ionization over the volume has a finite screening length $\simeq 10 - 50 r_D$. The computations performed for the same plasma parameters, in particular, for the same steady bulk density n_0 at long distances for the cases (I) and (II) indicate that there exist a sheath ranging up to $10 r_D$ independent on the type of boundary conditions (provided that the ionization rate is relatively low). At longer distances, a distinct difference in the asymptotical behavior is observed. The stationary grain charges acquired by the grain in both cases are nearly equal.

Figure 11 illustrates the behavior of the relative charge distributions as a function of the rate of ionization. The bulk plasma density is held constant therewith due to simultaneous appropriate change of the recombination coefficients. The approximate straightness of the lines outside the sheath (on log scale) suggests the exponential type of the screening at distances. Different rates of ionization (and recombination) correspond to the different slopes and the screening lengths, respectively. The higher intensity of ionization, the shorter is the length of screening. At higher rates, the relative indifference of the sheath is likely to break down, and the properties of screening approach the predictions of the DH theory. We see that, typically, the charging plasma cur-

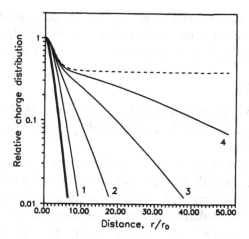

Figure 11. Relative charge distributions versus ionization rates for BC (II) at a fixed bulk plasma density. The dimensionless intensity of plasma sources, $i_0 = I_0 a^5 / D_i$ is (1) $1.25 \cdot 10^{-2}$; (2) $2.5 \cdot 10^{-3}$; (3) $5 \cdot 10^{-4}$; (4) 10^{-4}. The bold line is the linear DH theory; dashed line is DD approach for BC (I). The grain radius a/r_D is 0.158.

rents in the presence of collisions result in increase in the length of screening, as compared to the equilibrium DH theory. These results correlate qualitatively with those of [26] dealing with more complicated case of non-isothermic nitrogen plasma.

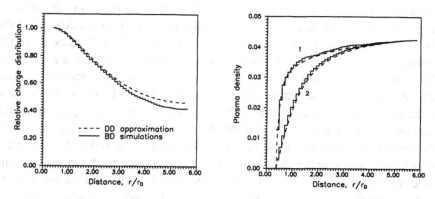

Figure 12. Comparison of of the results of DD and BD simulations for the same parameters, $a/r_D = 0.373$; $A = 10.0$. Left: relative charge distributions. Right: comparison of ion (1) and electron (2) densities obtained in DD approximation (dashed lines) and in BD simulations (solid lines).

Comparison of the continuous DD approach and the microscopic Brownian dynamics simulations shows a qualitative agreement between both cases,

Fig. 12. Some discrepancy (DD approach yields approximately 10% higher absolute value of the stationary grain charge) is, apparently, the result of microscopic effects in the plasma background in BD simulations.

5.   Summary

Thus, we see that the properties of screening of high-Z impurities in colloidal plasmas may considerably vary depending on the physical processes in the plasma background.

The nonlinear effects in screening in the thermodynamically equilibrium case of a high-Z grain with a fixed charge (e.g., the case of colloidal suspensions) are essential for strong plasma-grain coupling. The nonlinearity is associated with the accumulation of plasma particles on the grain surface and results in a sharp decrease in the effective charge as compared the linear theory. The linear DLVO theory works well only for weak plasma-grain coupling, $\chi < 4$. The nonlinear effects have a number of consequences for the structural properties of strongly coupled colloidal plasmas. In particular, they give rise to qualitative changes in pair distribution functions and result in shifting melting curves to larger magnitudes of charge asymmetry.

The grain screening in the collisionless background, with regard to the absorbtion of plasma particles by the grain, is close to the predictions of the DH theory (in the vicinity of grains) for the range of plasma parameters typical for dusty plasmas and for small grain sizes ($a \ll r_{\text{Deb}}$). At longer distances, the asymptotical behavior of the effective potentials inversely proportional to the squared distance is observed. The bound ionic states result in considerable changes in the plasma densities near the grain, however, they weakly affect the effective potentials in these conditions. The presence of bound states is limited by some critical distance ($\simeq 2 - 3r_{\text{Deb}}$), beyond which they cannot exist at all.

The processes of grain charging in strongly collisional plasma background result in considerable deviation from the equilibrium DH theory. In the case that the plasma sources are placed at infinity, we observe at long distances the Coulomb field with a certain effective charge. The effect of screening manifests itself in the decrease of this effective charge as compared to the stationary grain charge. The smaller the ratio of the Debye length to the grain size, the smaller effective charge is observed. In the case that the plasma sources are distributed uniformly over the volume, there exist a finite screening length depending on the rate of ionization. Typically, this screening length in the presence of plasma currents and strong collisions considerably exceeds the Debye radius. The stationary grain charge as well as the field within the sheath around the grain ($\simeq 10r_{\text{D}}$) does not depend on the type of boundary conditions and the ionization rate provided that this latter is relatively low. At higher ionization rates, the properties of screening approach the predictions of DH theory.

In conclusion, we would like to mention that an important question, which still remains poorly examined, is the properties of grain screening in weakly collisional and intermediate case. It would be interesting to study this issue within the Bhatnagar-Gross-Krook model, or on the basis of Fokker-Planck equations. Of particular interest is collisionless limit obtained within these approaches, which could be compared to the results of the paper [21]. Further valuable information on the above issues could be obtained by means of microscopic computer simulations in the spirit of [23].

References

[1] Thomas, H., Morfill, G.E., Demmel, V., Goree, J., Feuerbacher, B., and Möhlmann, D. *Phys. Rev. Lett.*, 1994, **73**, p. 652.

[2] Chu, J.H., and Lin, I. *Physica*, 1994 **A205**, p. 183.

[3] Tsytovich, V.N. *Phys. Usp.*, 1997, **40**, p. 53.

[4] Pieransky, P. *Contemp. Phys.*, 1983, **24**, p. 25.

[5] Löwen, H. *Phys. Rep.*, 1994, **237**, No. 5, p. 249.

[6] Derjaguin, B.V., and Landau, L. *Acta Physicochimica (USSR)*, 1941, **14**, p. 633.

[7] Verwey, E.J., and Overbeek J.Th.G. (1948). *Theory of the Stability of Lyophobic Colloids.* Amsterdam: Elsevier.

[8] Meijer, E.J., and Frenkel, D. *J. Chem. Phys.*, 1991, **94**, p. 2269.

[9] Robbins, M.O., Kremer, K., and Grest, G.S. *J. Chem. Phys.*, 1988, **88**, p. 286.

[10] Dupont, G. et al. *Mol. Phys.*, 1993, **79**, p. 453.

[11] Fortov, V.E., and Yakubov, I.T. (1992). *Physics of Nonideal Plasmas.* New York: Hemisphere.

[12] Xu, Y., and Chen, Y.-P. *Phys. Scripta*, 1999, **60**, p. 176.

[13] Alexander, S., Chaikin, P.M., Grant, P., Morales, G.J., and Pincus, P. *J. Chem. Phys.*, 1984, **80**, p. 5776.

[14] Bernstein, I.B., and Rabinovitz, I.V. *Phys. Fluids*, 1959, **2**, p. 112.

[15] Laframboise, J.G., and Parker, L.W. *Phys. Fluids*, 1973, **15**, p. 629.

[16] Dougherty, J.E., Porteous, R.K., Kilgore, M.D., and Graves, D.B. *J. Appl. Phys.*, 1992, **72**, p. 3934.

[17] Tsytovich, V.N., Khodataev, Ya.K., and Bingham, R. *Comments Plasma Phys. and Control. Fusion*, 1996, **17**, p. 249.

[18] Bystrenko, T., and Zagorodny, A. *Ukr. J. Phys.*, 2002, **47**, No. 4, p. 341.

[19] Goree, J. *Phys. Rev. Lett*, 1992, **69**, p. 277.

[20] Lampe, M., Joyce, G., Ganguli, G., and Gavrishchaka, V. *Phys. Plasmas*, 2000, **7**, p. 3851.

[21] Lampe, M., Gavrishchaka, V., Ganguli, G., and Joyce, G. *Phys. Rev. Lett.*, 2001, **86**, p. 5378.

[22] Lampe, M. et al. *Proceedings of the Second Capri Workshop on Dusty Plasmas*, (ed. de Angelis, U., and Nappi, C.), p. 13, 2001.

[23] Zobnin, A.V., Nefedov, A.P., Sinel'shchikov, V.A., and Fortov, V.E. *JETP*, 2000, **91**, p. 483.

[24] Pal', A.F., Starostin, A.N., and Filippov, A.V. *Plasma Physics Reports*, 2001, **27**, p. 143; *Fizika Plasmy*, (in Russian), 2001, **27**, p. 155.

[25] Pal', A.F., Serov, A.O., Starostin, A.N., et al. *JETP*, 2001, **92**, p. 235.

[26] Pal', A.F., Sivokhin, D.V., Starostin, A.N., Filippov, A.V., and Fortov, V.E. *Plasma Physics Reports*, 2002, **28**, p. 28; *Fizika Plazmy*, (in Russian), 2002, **28**, p. 32.

[27] Bystrenko, O., and Zagorodny, A. *Phys. Lett. A*, 1999, **255**, p. 325; *Cond. Matt. Phys.*, 1998, **1**, p. 169.

[28] Bystrenko, O., and Zagorodny, A. *Phys. Lett. A*, 1999, **262**, p. 72.

[29] Bystrenko, O., and Zagorodny, A. *Phys. Lett. A*, 2000, **274**, p. 47.

[30] Bystrenko, T., and Zagorodny, A. *Phys. Lett. A*, 2002, **299**, p. 383.

[31] Bystrenko, O., and Zagorodny, A., (to be published).

[32] See for instance: Roberts, S.M., and Shipman, J.S. (1972). *Two Point Value Problems: Shooting Methods*. New York: Elsevier.

[33] See for instance: (1979). *Monte-Carlo methods is statistical physics*, (ed. Binder, K.). Springer.

[34] Schram, P.P.J.M., and Trigger, S.A. *Contrib. Plasma Phys.*, 1997, **37**, p. 251.

[35] Löwen, H., Madden, P.A., Hansen, J.P. *Phys. Rev. Lett.*, 1992, **68**, p. 1081.

[36] Ewald, P. *Ann. d. Phys.*, 1921, **64**, p. 253.

[37] Tsytovich, V.N., and Havnes, O. *Comm. Plasma Phys. Contr. Fusion*, 1993, **14**, p. 267.

[38] Konopka, U., Morfill, G.E., and Ratke, L. *Phys. Rev. Lett.*, 2000, **84**, p. 891.

[39] See for instance: Hockney, R.W., and Eastwood, J.W. (1981). *Computer Simulations Using Particles*. McGrow-Hill.

[40] Chandrasekar, S. *Rev. Mod. Phys.*, 1943, **15**, p. 1.

[41] Hacken, H. (1978). *Synergetics*. Berlin: Springer-Verlag.

DIPOLAR FLUID INCLUSIONS IN CHARGED MATRICES

A replica Ornstein-Zernike integral equation approach

M.J. Fernaud , E. Lomba
Instituto de Química Física ROCASOLANO, CSIC
Serrano 110, E–28006 Madrid, Spain

Abstract   The replica Ornstein-Zernike formalism is applied to analyze the quenching of a simple model of non-primitive electrolyte. This model, which might as well represent the inclusion of a dipolar solvent in a random matrix of charged particles has proven to be one of the first cases in which the effects of partial quenching are vividly exposed by changes in the equilibrium properties. In this particular example we will consider a mixture of dipolar hard spheres and equal sized charged hard spheres. It will be shown how the quenching induces a completely different electrostatics as a consequence of the changes in the screening behavior due to the lack of ionic mobility.

Keywords: Replica Ornstein-Zernike equation, partially quenched systems, dipolar fluids, charged random matrices

1.   INTRODUCTION

Partly quenched systems have been a challenge for quite a long time. The advent of the replica method [1] cleared the path for a systematic treatment of this class of systems. Particularly, the pioneering work of Given and Stell [2] incorporated the Replica method into the well established set of methods available in liquid state theory, giving rise, among others, to the replica Ornstein-Zernike (ROZ) approach. A problem that can be cast into the form of a partly quenched mixture is that of fluid inclusions in random porous matrices. This is a topic of interest both from the scientific and technological standpoints, due to its implications in purification processes, catalysis, and nanotechnology. A wide variety of simple models for these systems have been developed during the last decade with different degrees of complexity, ranging from the simple hard sphere fluid in a hard sphere matrix system [3, 4], to models incorporating dispersion forces [5], molecular fluids [6, 7], associating fluids [8, 9], mixtures [10], etc. These studies are all based on the partial quenching of the positions

D. Henderson et al. (eds.), Ionic Soft Matter: Modern Trends in Theory and Applications, 315–332.
© 2005 *Springer. Printed in the Netherlands.*

of one of the components in a multicomponent mixture. Additionally, more realistic matrix structures can be generated using templated materials [11], or from polydisperse mixtures [12].

A natural step forward has been the addition of charges in these models [13–15]. Charges play a key role in adsorption processes involved in liquid chromatography and consequently the modelling of a charged random matrix goes beyond the mere academic interest. As found in [14, 15] the partial quenching of the matrix particles modifies the screening behavior of the fluid-fluid correlations when the adsorbed particles are also charged. An essentially new behavior is found under some conditions, observing an attractive component in the interaction between ions of the same sign and repulsion between ions of different sign. Large effects should also be expected for dipolar fluid inclusions in charged matrices. Actually, these mixtures of charged particles are one of the first systems in which the effect of quenching in the presence of disorder is strongly evidenced in the equilibrium properties. Until now it was found that in many instances the properties of the partially quenched system departed very little from those of the corresponding fully equilibrated mixture [16]. For the charged matrix systems, however, the freezing of the matrix positions hinders the ionic reorganization required to attain the screening that one observes in equilibrated mixtures of electrically active particles. These has been analyzed in some detail by Fernaud and coworkers [17]. Additionally, the simple case of a dipolar fluid adsorbed in neutral random matrices has also been studied in depth, in particular the behavior of the dielectric constant [18, 19]. It is our intention in this work to put in perspective and summarize the essentials of the dielectric behavior of dipolar fluids – and its relation with the microscopic structure – constrained inside random matrices, both charged and uncharged, in terms of the ROZ approach.

To that aim we will first consider a model consisting in a matrix constituted by quenched charged hard spheres, whose particles interact via a potential given by

$$
\beta_0 u_{00}^{\alpha\gamma}(r) = \begin{cases} \infty, & \text{if } r < \sigma_{00}^{\alpha\gamma}, \\ \dfrac{\beta_0 Z_\alpha Z_\gamma e^2}{r}, & \text{if } r \geqslant \sigma_{00}^{\alpha\gamma}, \end{cases} \tag{1}
$$

where e is the electron charge, Z_ν is the charge of the ions of type ν, $\beta_0 = 1/k_B T_0$ is the inverse temperature at which the charged hard spheres have been quenched, and $\sigma_{00}^{\alpha\gamma}$ is the overlap diameter between two matrix particles of types α and γ. We have restricted ourselves to a symmetric case, where $|Z_\alpha| = |Z_\beta| = Z = 1$, and $\sigma_{00}^{++} = \sigma_{00}^{--} = \sigma_{00}^{+-} = \sigma_{00}$. As to the matrix-fluid (ion-dipole) interaction, it will be given by

$$\beta u_{\pm d}(r, \omega_2) = \begin{cases} \infty, & \text{if } r < \sigma_{0d}, \\ -\frac{\beta Z_{\pm}\mu e}{r^2} \cos\theta_{r1}, & \text{if } r \geqslant \sigma_{0d}, \end{cases} \tag{2}$$

where $\sigma_{0d} = (\sigma_{00} + \sigma_{dd})/2$, σ_{dd} is the hard sphere diameter of the dipolar particles, $\beta = 1/k_{\mathrm{B}}T$ the inverse temperature of the dipoles in the matrix, μ is the dipole moment, and θ_{r1} is the angle formed by the dipole moment and the vector $\mathbf{r}$ joining the centers of particles 0 and 1.

Finally, the dipole-dipole interaction is given as usually by

$$\beta u_{dd}(r, \omega_1, \omega_2) = \begin{cases} \infty, & \text{if } r < \sigma_{dd}, \\ \frac{-\beta\mu^2}{r^3}\left(3(\hat{s}_1 \cdot \hat{r})(\hat{s}_2 \cdot \hat{r}) - \hat{s}_1 \cdot \hat{s}_2\right), & \text{if } r \geqslant \sigma_{dd}, \end{cases} \tag{3}$$

where $\hat{s}_i$ and $\hat{r}$ are unit vectors describing the orientation of the dipole moment on particle i and the orientation of the interparticle axis respectively. Notice that not only the spatial distribution of the matrix particles will be important, but the charge distribution as well.

We will see that, as expected, the presence of charges in the matrix strongly modifies the long range (i.e. screening) behavior of the fluid-fluid correlation functions. In order to illustrate the effects of confinement on the dipolar fluid properties, calculations for the corresponding fully equilibrated system – i.e. an electrolyte with explicit solvent – have also been carried out in the hypernetted chain (HNC) approximation. This is known to be accurate enough for the thermodynamic states under consideration.

2. Replica Ornstein-Zernike integral equation approach

The extension of the ROZ formalism to confined molecular fluids has recently been carried out for adsorbed diatomic molecules [6] and dipolar fluids confined in hard sphere matrices [18, 19]. In the case of ionic matrix, new features of the system have to be taken into account. On one hand, we have now a two component matrix (with positive and negative ions). This case was already considered in [14, 15] for the primitive model electrolyte adsorbed in an electroneutral charged matrix. On the other hand, we have to deal with two different temperatures: the matrix temperature, β_0 (at which the ionic fluid is equilibrated before quenched) and the fluid temperature β_1, at which the fluid is adsorbed in the solid matrix. As usual when dealing with molecular fluids one starts with an expansion of the correlation functions in terms of spherical harmonics as follows,

$$f_{ij}^{00}(\mathbf{r}_{12}) = f_{ij}^{00}(r_{12}),$$

$$f^{01}(\mathbf{r}_{12}, \omega_2) = \sqrt{4\pi} \sum_{l_2 m} f_{0\,l_2\,m}^{01}(r_{12}) Y_{l_2 m}(\omega_2),$$

$$f^{10}(\mathbf{r}_{12}, \omega_1) = \sqrt{4\pi} \sum_{l_1 m} f_{l_1\,0\,m}^{10}(r_{12}) Y_{l_1 m}(\omega_1),$$

$$f^{11}(\mathbf{r}_{12}, \omega_1, \omega_2) = 4\pi \sum_{l_1 l_2 m} f_{l_1\,l_2\,m}^{11}(r_{12}) Y_{l_1 m}(\omega_1) \, Y_{l_2 \bar{m}}(\omega_2),$$

$$f^{12}(\mathbf{r}_{12}, \omega_1, \omega_2) = 4\pi \sum_{l_1 l_2 m} f_{l_1\,l_2\,m}^{12}(r_{12}) Y_{l_1 m}(\omega_1), Y_{l_2 \bar{m}}(\omega_2),$$

where $i, j = +, -$ for the positive and negative ions and the superscript 12 denotes a replica-replica correlation. This expansions are carried out in the usual axial reference frame (where the z-axis is placed alongside the line joining the molecular centers). With this, the ROZ equations decouple in Fourier space and can be cast into matrix form as,

$$\mathbf{\Gamma}^{d-} = -\mathbf{C}^{d-} + \mathbf{G}_m \left[S_- \mathbf{C}^{d-} + \rho_+ \tilde{h}_{+-} \mathbf{C}^{d+} \right] \tag{4}$$

$$\mathbf{\Gamma}^{d+} = -\mathbf{C}^{d+} + \mathbf{G}_m \left[S_+ \mathbf{C}^{d+} + \rho_- \tilde{h}_{-+} \mathbf{C}^{d-} \right] \tag{5}$$

$$\mathbf{\Gamma}_m^{dd} = -\mathbf{C}_m^{dd} + \mathbf{G}_m \Big[\mathbf{C}_m^{dd} + (-1)^m \Big(\rho_- S_- \mathbf{C}^{d-} \mathbf{C}^{-d} \mathbf{G}_m$$

$$+ 2\rho_- \rho_+ \tilde{h}_{-+} \mathbf{C}^{d-} \mathbf{C}^{+d} \mathbf{G}_m$$

$$+ \rho_+ S_+ \mathbf{C}^{d+} \mathbf{C}^{+d} \mathbf{G}_m + \rho_d \mathbf{C}_m^{dd'} \mathbf{G}_m \mathbf{C}_m^{c} \Big) \Big] , \tag{6}$$

$$\mathbf{\Gamma}_m^{c} = (-1)^m \rho_d \mathbf{G}_m \mathbf{C}_m^{c\,2} \tag{7}$$

with

$$\mathbf{G}_m = \left[\mathbf{I} - (-1)^m \rho_d \mathbf{C}_m^{c} \right]^{-1} ,$$

$$S_- = 1 + \rho_- \tilde{h}_{--} ,$$

$$S_+ = 1 + \rho_+ \tilde{h}_{++} .$$

Here, $\mathbf{I}$ is the identity matrix, $\tilde{h}_{\alpha\beta}$ are the Fourier transforms of the matrix-matrix total correlation function, and the elements of the remaining matrices are the following: $[\mathbf{\Gamma}_m^{dd}]_{kl} = \tilde{h}_{klm}^{dd} - \tilde{c}_{klm}^{dd}$, $[\mathbf{C}_m^{dd}]_{kl} = \tilde{c}_{klm}^{dd}$, $[\mathbf{\Gamma}^{d\pm}]_k = \tilde{h}_{k00}^{d\pm} - \tilde{c}_{k00}^{d\pm}$ and likewise for $\mathbf{C}^{d\pm}$, being $\tilde{c}^{dd}$, $\tilde{c}^{d\pm}$ the Fourier transform of the dipole-dipole and dipole-charge direct correlation functions respectively, ρ_d is the dipole number density and $\rho_\pm$ the matrix anion or cation densities, and $\tilde{h}^{dd}$, $\tilde{h}^{d\pm}$ are the dipole-dipole and dipole-charge (i.e. fluid-matrix) total correlation functions. The superscripts (klm) identify the corresponding coefficients of the expansion of the correlation functions in spherical harmonics [6] in the axial reference frame. Notice that in order to transform from r to k-space and back one must first rotate the reference frame from the axial system to a space fixed frame, by which a new set of coefficients $f_{\alpha\beta}^{kln}$ are obtained [20]. Finally, $[\mathbf{C}_m^{dd'}]_{kl} = \tilde{c}_{klm}^{dd'} = \tilde{c}_{klm}^{12}$ is the Fourier transform of the replica-replica direct correlation function (blocking function), and the connected function is defined as usual by $c^c = c^{dd} - c^{dd'}$, and similarly for h^c. Let us recall that the replicated particles are the dipolar hard spheres, i.e. the annealed fluid in the partly quenched mixture.

These equations must be complemented with a closure relation in r-space, for which in this work we have chosen the hypernetted chain (HNC) approximation. This equation is known to give reasonable results for ionic fluids, and consequently we can expect a similar accuracy here. In the present instance the HNC can be written as,

$$h_{l00}^{d\pm}(r) = \left\langle \exp\left[-\beta u^{d\pm}(r,\omega_1) + h^{d\pm}(r,\omega_1) - c^{d\pm}(r,\omega_1) \right] |l00 \right\rangle - \delta_{l0},$$

$$h_{l_1 l_2 m}^{dd}(r) = \left\langle \exp\left[-\beta u^{dd}(r,\omega_1,\omega_2) + h^{dd}(r,\omega_1,\omega_2) \right.\right.$$

$$\left.\left. - c^{dd}(r,\omega_1,\omega_2) \right] |l_1 l_2 m \right\rangle - \delta_{l_1 l_2 m,000}, \qquad (8)$$

$$h_{l_1 l_2 m}^{dd'}(r) = \left\langle \exp\left[h^{dd'}(r,\omega_1,\omega_2) - c^{dd'}(r,\omega_1,\omega_2) \right] |l_1 l_2 m \right\rangle - \delta_{l_1 l_2 m,000},$$

where $\langle \ldots |l_1 l_2 m \rangle$ denotes the projection onto the spherical harmonic basis function $Y_{l_1 m}(\omega_1)Y_{l_2 \bar{m}}(\omega_2)$. Additionally, the HNC equations for the matrix components decouple completely and can be solved by the standard procedures devised for primitive model electrolytes [21].

3. Long-range behavior of the correlation functions and dielectric constant

The long range behavior of the correlation functions of the system, can be studied by analyzing Eqs. (4)–(7) in the asymptotic limit ($k \to 0$). It has been observed [22] that for the general case of a ion-dipole fluid adsorbed in a ion-dipole matrix, the long-range behavior of some of the correlation functions (screening potentials) is essentially modified with respect to the equivalent equilibrated mixture. As our system is a particular case of this general one, we also expect to find these modifications in relation to the equilibrated ion-dipole mixture as well as to the pure dipolar fluid.

Beginning with the dipole-dipole angular correlation projection (in the space fixed frame) $h_{dd}^{112} = h_{110}^{dd} + h_{111}^{dd}$, it can be seen that for the bulk dipolar fluid, the long-wave length limit of this function is given by [23],

$$\lim_{k \longrightarrow 0} \tilde{h}_{dd}^{112}(k) = -\frac{4\pi}{3\epsilon} \beta \mu^{\text{eff}\,2} \tag{9}$$

with

$$\mu^{\text{eff}} = \frac{\epsilon - 1}{3y} \mu \,,$$

being ϵ the static dielectric constant and $y = 4\pi\beta\rho_d\mu^2/9$. In the r-space, this corresponds to,

$$\lim_{r \to \infty} h^{112}(r) = \frac{\beta\mu^{\text{eff}\,2}}{\epsilon r^3} \,. \tag{10}$$

When one includes the ionic component (ion-dipole equilibrated mixture), due to screening effects, this asymptotic limit changes [23], being now

$$\lim_{k \longrightarrow 0} \tilde{h}_{dd}^{112}(k) = 0 \,.$$

Finally, if the ionic component is quenched, it can be seen from Eqs. (4)–(7) that the behaviour shown in Eqs. (9) and (10) is recovered, but now for the connected part of the fluid-fluid correlation function,

$$\lim_{r \to \infty} h_c^{112}(r) = \frac{\beta\mu^{\text{eff}\,2}}{\epsilon r^3} \,. \tag{11}$$

This connected part is the one playing the role of the dipole-dipole correlation function in the quenched-annealed system.

The other angular coefficient to be analyzed is the $h_{dd}^{110} = h_{110}^{dd} - 2h_{111}^{dd}$. We find for this function a particularly relevant result: the long-range behavior of one of the terms of this coefficient, the $h_{110}^{dd}(r)$, gets Coulombic due to quenching, namely,

$$\lim_{k \longrightarrow 0} \tilde{h}_{110}^{dd}(k) = \frac{4\pi\beta\mu^2 e^2}{3k^2(1 - \tilde{c}_{110}^c(0))} \left(Z_-^2 \rho_-(1 + \rho_-\tilde{h}_{--}(0)) \right.$$
$$\left. + Z_+^2\rho_+(1 + \rho_+\tilde{h}_{++}(0)) + 2\rho_-\rho_+ Z_- Z_+\tilde{h}_{-+}(0) \right).$$

This result is consistent with the limiting behavior found by Holovko and Polishchuk [22] for an ion-dipole system adsorbed in ion-dipolar matrices. The Coulombic tail has to be properly dealt with in any numerical procedure to solve the ROZ equations in Fourier space. In this respect, an adequate treatment of the long range behavior of the correlation functions can be constructed according to the prescriptions of Chen and Forstmann [24] for the ion-dipole mixture. Following these authors, the long range contribution of the functions can be treated analytically, and the result is added to the numerical calculations carried out on the short ranged part.

In the bulk dipolar fluid, the long wave length behavior of h_{dd}^{110} is related to the static dielectric constant ϵ by the expression,

$$\frac{(\epsilon - 1)(2\epsilon + 1)}{3\epsilon} = \frac{4\pi}{3} \rho_d \mu^{*2} \left[1 + \frac{\rho_d}{3} \tilde{h}_{dd}^{110}(0) \right], \qquad (12)$$

being the reduced dipole moment defined by $\mu^{*2} = \beta \mu^2 / \sigma_{dd}^3$. However, in the equilibrated ion-dipole mixture one finds [23],

$$\frac{\epsilon - 1}{3y} = 1 + \frac{\rho_d}{3} \tilde{h}_{dd}^{110}(0). \qquad (13)$$

When quenching the ions to form the matrix, the expression(12) is formally recovered, but once again is the connected part of the fluid-fluid correlation function the one reflecting the response of the fluid to an external field,

$$\frac{(\epsilon - 1)(2\epsilon + 1)}{3\epsilon} = \frac{4\pi}{3} \rho_d \mu^{*2} \left[1 + \frac{\rho_d}{3} \tilde{h}_c^{110}(0) \right]. \qquad (14)$$

This expression was derived combining the response theory results and the replica method in the nonpolar matrix case [18]. For the ionic matrix the expression is formally the same, as the positionaly frozen charges are no able respond to any external electric field, acting in this respect as if they were neutral. Nevertheless the situation is not identical for both systems. In the neutral matrix, $\tilde{h}_c^{110} = \tilde{h}_{dd}^{110}$, being the blocking contribution identically null. This is not the case when the matrix particles are ions, where the convolution of the non-vanishing angular components of the matrix-fluid interaction leads to a non-vanishing blocking function $h_{dd'}^{110}$.

The values of the dielectric constant obtained from the relations (11) and (14) should be consistent, and this is indeed the case in our calculations. To end this section, we should mention that Eq. (14) is similar to the expression derived by Klapp and Patey [25] for positionally frozen dipolar fluids once the local freezing order parameters are set to zero.

4.   Thermodynamic properties

The replica method can be also applied to express some of the thermodynamic properties of a confined fluid in terms of the correlation functions of the system. The details of the procedure can be found elsewhere [6, 26]. In our system, we need to rearrange some expressions so as to avoid the difficulties in the numerical calculations due to the long range behavior of some correlation functions.

- **Excess internal energy.** The expression for the net fluid-fluid and fluid-matrix contribution to the internal energy is the following,

$$\frac{\beta U_1^{\text{ex}}}{V} = \rho_d \rho_+ \int u_{100}^{d+}(r) h_{100}^{d+}(r) \mathrm{d}\mathbf{r} + \rho_d \rho_- \int u_{100}^{d-}(r) h_{100}^{d-}(r) \mathrm{d}\mathbf{r}$$

$$+ \frac{\rho_d^2}{2} \int \left(u_{110}^{\mathrm{dd}}(r) h_{110}^{\mathrm{dd}}(r) + 2 \int u_{111}^{\mathrm{dd}}(r) h_{111}^{\mathrm{dd}}(r) \right) \mathrm{d}\mathbf{r}, \tag{15}$$

where the coefficients of the dipole-dipole and ion-dipole interaction are

$$u_{110}^{\mathrm{dd}}(r) = -\frac{2}{3} \frac{\mu^2}{r^3}, \tag{16}$$

$$u_{111}^{\mathrm{dd}}(r) = -\frac{1}{3} \frac{\mu^2}{r^3}, \tag{17}$$

$$u_{100}^{\mathrm{d}\pm} = \frac{Z_\pm \mu}{\sqrt{3} r^2}. \tag{18}$$

The first two terms in Eq. (15) give the dipole-matrix (and matrix-dipole) contribution, $\beta U_{10}^{\text{ex}}/V$, and the last term the dipole-dipole contribution, $\beta U_{11}^{\text{ex}}/V$

- **Dipolar fluid excess chemical potential.** A closed expression for the evaluation of this quantity can be easily derived in the HNC approximation using Lee's star function technique [27] and the replica trick [18], leading to

$$\beta\mu_1^{ex} = -\rho_+ \tilde{c}_{000}^{d+}(0) - \rho_- \tilde{c}_{000}^{d-}(0) - \rho_d \tilde{c}_{000}^{dd}(0) + \rho_d \tilde{c}_{000}^{dd'}(0)$$

$$+ \rho_+ \left[\sum_{l_1 \neq 1} \int h_{0l_10}^{+d}(r)\gamma_{0l_10}^{+d}(r)\mathrm{d}\mathbf{r} + \int h_{010}^{+d}(r)\gamma^{+d*}_{010}(r)\mathrm{d}\mathbf{r} \right]$$

$$+ \rho_- \left[\sum_{l_1 \neq 1} \int h_{0l_10}^{-d}(r)\gamma_{0l_10}^{-d}(r)\mathrm{d}\mathbf{r} + \int h_{010}^{-d}(r)\gamma^{-d*}_{010}(r)\mathrm{d}\mathbf{r} \right]$$

$$- \frac{\rho_- Z_+ e\mu}{\sqrt{3}} \int h_{010}^{+d}(r)\mathrm{d}\mathbf{r} - \frac{\rho_+ Z_- e\mu}{\sqrt{3}} \int h_{010}^{-d}(r)\mathrm{d}\mathbf{r} \qquad (19)$$

$$+ \frac{\rho_d}{2} \sum_{l_1 l_2 m} \int \left(h_{l_1 l_2 m}^{dd}(r)\gamma_{l_1 l_2 m}^{dd}(r) - h_{l_1 l_2 m}^{dd'}(r)\gamma_{l_1 l_2 m}^{dd'}(r) \right) \mathrm{d}\mathbf{r},$$

where

$$\gamma_{l_1 l_2 m}^{\alpha\beta} = h_{l_1 l_2 m}^{\alpha\beta} - c_{l_1 l_2 m}^{\alpha\beta}, \text{ and } \gamma^{d\pm *} = \gamma^{d\pm} + \beta u^{d\pm}.$$

In this way it is possible to deal with short range functions and the long range behavior of the ion-dipole potential can be treated explicitly.

- **Isothermal compressibility.** As seen in [18], this quantity is a response function, and consequently it will solely depend on the connected part of the fluid-fluid correlation function. Here, as in [18] one gets,

$$\beta \left(\frac{\beta P}{\rho_d} \right)_T = 1 - 4\pi\rho_d \int \mathrm{d}r r^2 c_{000}^c(r). \qquad (20)$$

5. Results

We have solved the ROZ equations in the HNC approximation in a rather simple case, where matrix and fluid particles have equal sizes, and, as additivity is assumed, $\sigma_{00} = \sigma_{0d} = \sigma_{dd} = \sigma$.

The solution of the ROZ equations was carried out on a discretized mesh of 8192 points with a grid size 0.01σ. The same conditions were used to solve the HNC equation for the corresponding equilibrated mixtures.

The reduced dipole-dipole and charge-dipole couplings have been set to $\mu^{*2} = 2.75$ and $\beta\mu e/\sigma^2 = 1.658$. To study the influence of the charge distribution in the matrix on the behavior of the dipolar fluid, we have considered two different matrix configurations: one corresponding to an ionic fluid quenched

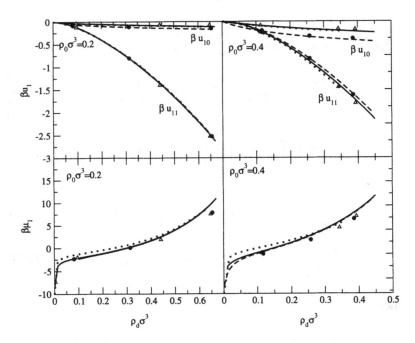

Figure 1. Excess internal energy (split in ion-dipole ($\beta u_{10} = \beta U_{10}^{ex}/V$) and dipole-dipole ($\beta u_{11} = \beta U_{11}^{ex}/V$) contributions) and chemical potential for a dipolar fluid inclusion in ionic matrices quenched at low temperature $\beta_0 e^2/\sigma = 1$ (solid line and filled triangles) and high temperature $\beta_0 e^2/\sigma = 0.005$ (dashed line and filled circles). The dotted lines correspond to the ion-dipole equilibrated mixture at $\beta e^2/\sigma = 1$ for the corresponding charge densities. Curves denote ROZ results and symbols GCMC data[18].

at a reduced temperature of $\beta_0 e/\sigma = 1$ and the other at $\beta_0 e/\sigma = 0.005$. In the latter case the charge distribution in the matrix is practically random and the spatial distribution is identical to that of the hard sphere fluid. While the first system can be compared to an equilibrated mixture of ions and dipoles, the second one is specific of a quenched-annealed model, with no equilibrated equivalent. In both cases we have studied two different matrix densities $\rho_0 \sigma^3 = (\rho_+ + \rho_-)\sigma^3 = 0.2$ and 0.4, and various fluid densities.

In Fig. 1 we present the results for the internal energy and chemical potential for the two model matrices under consideration compared with Grand Canonical Monte Carlo calculations [17]. Some differences are observed in the internal energy values when comparing the dipolar fluid within the matrix quenched at the same temperature and inside the high temperature (i.e. with randomly distributed charges) matrix. In this latter case the charge-dipole contribution seems substantially larger than that of the low temperature quench. This is probably due to the fact that in this latter instance, both the matrix charge and particle position distributions correspond to a primitive electrolyte, and consequently exhibit a certain degree of pairing and charge screening. As a result, the dipoles will interact with positive and negative charges in many cases distributed in pairs, by which a significant portion of the interaction energy will cancel. This partial cancellation will certainly not take place when the charges are randomly distributed. This has important consequences in the convergence properties of the integral equation for low matrix and fluid densities for the high temperature matrix case, to the point that the integral equation breaks down as fluid density approaches zero. We will analyze these states more in detail in the last part of this section.

Aside from this, a comparison with the equilibrated mixture results shows that we are now dealing with a situation in which the partial quenching alters the behavior of the chemical potential, and to a much lesser extent the internal energy. As the dipolar density is increased, the HNC equation of the electrolyte breaks down. This is very likely due to a demixing transition, as we could conclude from the stability analysis carried out following the prescriptions of Chen and Forstmann [24, 28]. According to them, it is possible to analyze the stability of the grand potential functional for the case of a ion-dipole mixture with equal size particles. The fluctuations in this quantity for the present case can be cast in the form [24],

$$\delta\Omega = \frac{1}{2\beta V}(\delta\tilde{\rho}(0)\ \delta\tilde{c}(0)) \begin{pmatrix} M_{\rho\rho} & M_{\rho c} \\ M_{c\rho} & M_{cc} \end{pmatrix} \begin{pmatrix} \delta\tilde{\rho}(0) \\ \delta\tilde{c}(0) \end{pmatrix}, \qquad (21)$$

where the density and concentration fluctuations are,

$$\delta\tilde{\rho}(0) = \rho^{-\frac{1}{2}}(\delta\tilde{\rho}_0(0) + \delta\tilde{\rho}_d(0)), \qquad (22)$$

$$\delta c(0) = \rho^{-\frac{3}{2}}(c_0 c_d)^{-\frac{1}{2}}(\rho_d\delta\tilde{\rho}_0(0) - \rho_0\delta\tilde{\rho}_d(0)), \qquad (23)$$

with $\rho = \rho_0 + \rho_d$, $c_i = \rho_i/\rho$ and $\delta\tilde{\rho}_d(0) = \sqrt{4\pi}\delta\tilde{\rho}_d^{00}(0)$. This latter quantity is the Fourier transform of the radial average of the one particle dipole density fluctuation. The coefficients of the symmetric M matrix are [24],

$$M_{\rho\rho} = 1 - \rho\left[c_0^2\tilde{c}_{00}(0) + c_d^2\tilde{c}_{000}^{dd}(0) + 2c_0c_d\tilde{c}_{000}^{+d}(0)\right],$$

$$M_{cc} = 1 - \rho c_0 c_d\left[\tilde{c}_{00}(0) + \tilde{c}_{000}^{dd}(0) - \tilde{c}_{000}^{+d}(0)\right],$$

$$M_{\rho c} = \rho\sqrt{c_0 c_d}\left[c_d\tilde{c}_{000}^{dd}(0) - c_0\tilde{c}_{00}(0) - (c_d - c_0)\tilde{c}_{000}^{+d}(0)\right],$$

with $\tilde{c}_{00} = (\tilde{c}_{++} + \tilde{c}_{+-})/2$. One can then determine the eigenvalues of the matrix M, which are given by

$$\lambda_{1,2} = \frac{M_{\rho\rho} + M_{cc} \mp \sqrt{(M_{\rho\rho} - M_{cc})^2 + 4M_{\rho c}^2}}{2}. \tag{24}$$

The minimum eigenvalue $\lambda_{\min} = \min(\lambda_1, \lambda_2)$ decides the stability of the phase. If $\lambda_{\min} \to 0$ the phase will be unstable. Depending on the components of the eigenvector we will have a demixing transition or a gas-liquid instability. Note that a gas-liquid instability will also be signaled by a divergence in the isothermal compressibility. These two quantities, obtained in the HNC approximation, are plotted in Fig. 2 for the equilibrated ion-dipole mixtures of interest in this paper. We see that the minimum eigenvalue shows an evident decrease as the dipole density increases whereas the isothermal compressibility is monotonously decreasing. This is a clear indication of the tendency to demix. The oscillations observed in the $\lambda_{\min}$ curve at high dipole densities

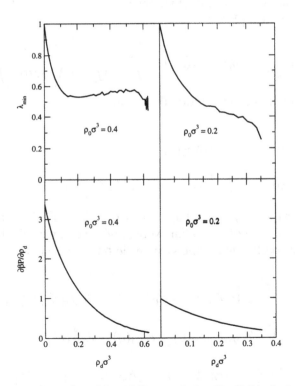

Figure 2. Minimum eigenvalue of the stability matrix (see Eq. (24) in the text) and isothermal compressibility for equilibrated mixtures of charged and dipolar hard spheres.

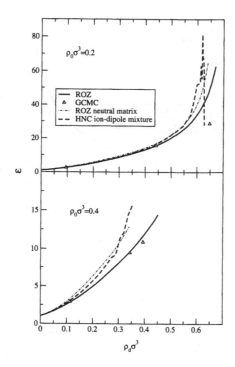

Figure 3. Dielectric constant of the dipolar fluid embedded in a charged matrix quenched at low temperature $\beta_0 e^2/\sigma = 1$. ROZ vs GCMC results. HNC results for the corresponding equilibrated mixture and ROZ results for an equivalent system with a neutral (hard sphere) matrix are included for comparison.

are due to numerical instabilities in the solution of the HNC equation as the correlation functions become more and more long ranged.

The dielectric constant is plotted in Figs. 3 and 4 for the two types of system. Large discrepancies of the HNC results are observed when compared to GCMC in the high fluid density regions. Despite the lack of accuracy of the HNC for these high density states, one must also bear in mind that the GCMC results for a quantity like the dielectric constant at these states are plagued with appreciable uncertainties. Again, the anomalous behavior of the dielectric constant for the equilibrated mixture can be ascribed to the vicinity of a demixing transition.

We also compare in both figures the dielectric constant values of the dipolar fluid when adsorbed in the ionic and in the hard sphere matrix. A clear influence of the presence of charges in the matrix is then observed. The ionic matrix lowers the response of the dipoles to an external field, i.e. lowers the value of the dielectric constant of the fluid in the given state. This can easily be understood, since the local electric field that the matrix charges generate

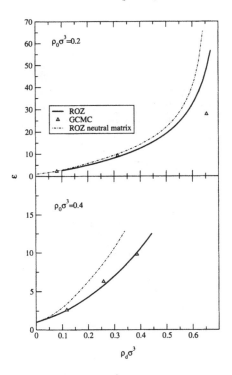

Figure 4. Dielectric constant of the dipolar fluid embedded in a charged matrix quenched at high temperature $\beta_0 e^2/\sigma = 0.005$. ROZ vs GCMC results. ROZ results for an equivalent system with a neutral (hard sphere) matrix are included for comparison.

will compete with the external one, impeding somehow the reorientation of the dipoles as compared with the neutral matrix. This tendency seems to be similar no matter how the charges are distributed, as it is seen when comparing the low and high quenching temperature cases.

Now we observe that the charge distribution does not seem to have a major influence on the macroscopic properties of the fluid (either thermodynamic or dielectric) but if we look into the microscopic structure things are somewhat different. In particular, the low fluid density states evidence a clear difference between the two matrix topologies: the integral equation in the high temperature quench and low matrix densities breaks down as the fluid density is lowered, which does not occur in the low temperature quench. Most of the relevant correlation functions behave in the same way as in the bulk and mixture systems independently of the charge distribution of the matrix. Nevertheless, the fluid-matrix leading angular coefficient, $h_{100}^{d\pm}$, offers some insight on what might be going on. This function is plotted for low fluid and matrix densities in Fig. 5, illustrating the difference between the matrix-fluid correlations for the two matrix topologies. We observe that for the high temperature matrix the angular correlations become appreciably long ranged (up to five diameters). Actually, in this case we could not reach the simulation density due to convergence problems in the ROZ equations. The contact values are similar for both matrix configurations but the correlations die out more rapidly for the low temperature quench, and this time there was no problem to lower the density in the ROZ equations. We can conclude from these results that the breakdown of the integral equations is connected to diverging ion-dipole correlations.

Finally, we have plotted in Fig. 6 the dipole-dipole angular coefficient h_{dd}^{110} for the same low fluid density states. The shape of the functions show a clear indication that in both cases the dipoles exhibit a dominant head-to-tail alignment, extended in the high temperature matrix up to the second neighbors. It seems rather clear that, at low ρ_0 the charged matrix particles induce a head-to-tail alignment of the dipoles around them. This alignment is somewhat impeded in the low temperature matrix, since in this case ions of different signs are clustered together screening each other – see Fig. 7 for a comparison of the unlike pair distribution functions – and distort the ordering of the dipoles. From all these results we can conclude that the local density of ions is higher in the case of the low temperature matrix. Obviously the same effect can be achieved by a simple increase of the matrix density, and no convergence difficulties appear in the ROZ equations at moderate matrix densities. On the other

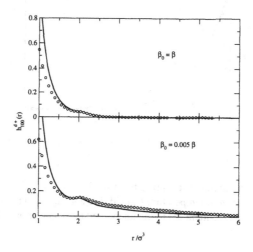

Figure 5. Leading angular coefficient of the ion-dipole correlation function for the two matrix topologies for low density states ($\rho_0\sigma^3 = 0.2$, $\rho_d\sigma^3 = 0.095$ – upper graph, and $\rho_d\sigma^3 = 0.081$ – lower graph) by means of the ROZ integral equation (lines) and from GCMC simulation (symbols). In the case of the high temperature matrix quench (lower graph) the ROZ results correspond to the lowest attainable dipole density, $\rho_d\sigma^3 = 0.089$.

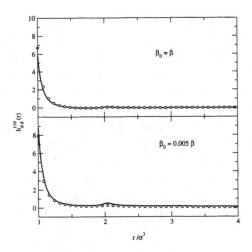

Figure 6.   Dipole-dipole h_{dd}^{110} component of the total correlation function for the two matrix topologies for low density states ($\rho_0\sigma^3 = 0.2$, $\rho_d\sigma^3 = 0.095$ – upper graph, and $\rho_d\sigma^3 = 0.081$ – lower graph) by means of the ROZ integral equation (lines) and from GCMC simulation (symbols). In the case of the high temperature matrix quench (lower graph) the ROZ results correspond to the lowest attainable dipole density, $\rho_d\sigma^3 = 0.089$.

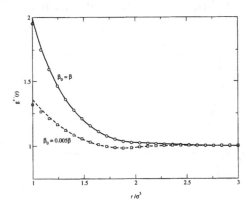

Figure 7.   Matrix-matrix unlike distribution function for $\rho_0\sigma^3 = 0.2$ for the low temperature (solid line and circles) and high temperature (dotted line and squares) matrix. In the latter case like and unlike correlations are identical to the uncharged hard sphere pair distribution function.

hand, the break down of the HNC equation for bulk dipolar systems in the low density regime is a well known feature, and has been attributed to the inability of the equation to account for clustering effects beyond pairing. It is clear that the lack of screening in the high temperature matrix enhances dipole association, and one can easily understand why the ROZ equations break down here in the low density regime as well.

6. Summary

In this article we have presented a study of a dipolar fluid inclusion in charged matrices. In order to analyze the influence of the matrix charge distribution on the behavior of the adsorbed dipolar fluid, we have considered two matrix topologies: one obtained by quenching the charged hard sphere positions at the same temperature than the dipolar fluid and another quenched at a temperature two hundred times higher. This latter case implies a random distribution of charges, and hence the absence of screening. As expected, the two systems present substantial differences with respect to the fully equilibrated mixture of ions and dipoles, specially as far as their dielectric behavior is concerned. However, the charge distribution of the matrix does not seem to have a relevant influence on the thermodynamic and dielectric properties of the fluid even if significant differences are found in the low density microscopic structure. In this case the lack of screening in the high temperature matrix enhances dipole-dipole association to the point that the ROZ equations with the HNC closure break down for rather low dipole moments. In the domain where solutions can be found, the results provided by the ROZ equations are generally in fair agreement with the simulations.

Acknowledgements

M.J.F. wishes to thank the organization of the NATO Advanced Research Workshop on *Ionic soft matter* in Lviv, Ukraine, for supporting her stay in Lviv. E.L. acknowledges support from the Dirección General de Investigación Científica under grant No. BFM2001–1017–C03–01.

References

[1] Edwards, S.F., and Jones, C. *J. Phys. A*, 1976, **9**, p. 1595.

[2] Given, J., and Stell, G. *J. Chem. Phys.*, 1992, **97**, p. 4573.

[3] Madden, W.G., and Glandt, E.D. *J.Stat.Phys.*, 1988, **51**, p. 537.

[4] Lomba, E., Given, J., Stell, G., and Weis, J.J. *Phys. Rev. E*, 1993, **48**, p. 233.

[5] Alvarez, M., Levesque, D., and Weis, J.J. *Phys.Rev. E*, 1999, **60**, p. 5495.

[6] Fernaud, M.J., Lomba, E., and Weis, J.J. *Phys. Rev. E*, 2001, **64**, p. 051501.

[7] Kovalenko, A., and Hirata, F. *J. Chem. Phys.*, 2001, **115**, p. 8620.

[8] Trokhymchuk, A., Pizio, O., Holovko, M., and Sokolowski, S. *J. Phys.Chem*, 1996, **100**, p. 17004.

[9] Padilla, P., Pizio, O., Trokhymchuk, A., and Vega, C. *J. Phys.Chem*, 1998, **102**, p. 3012.

[10] Schöll-Paschinger, E., Levesque, D., Weis, J.J., and Kahl, G. *Phys.Rev. E*, 2001, **64**, p. 011502.

[11] van Tassel, P.R. *Phys. Rev. E*, 1999, **60**, p. R25.

[12] Rzysko, W., Sokolowski, S., and Pizio, P. *J. Chem. Phys.*, 2002, **116**, p. 4286.

[13] Bratko, D., and Chakraborty, A.K. *J. Chem. Phys.*, 1996, **104**, p. 7700.

[14] Hribar, B., Pizio, O., Trokhymchuk, A., and Vlachy, V. *J. Chem. Phys.*, 1997, **107**, p. 6335; *ibid.*, 1998, **109**, p. 2480.

[15] Hribar, B., Vlachy, V., and Pizio, O. *Mol. Phys.*, 2002, **100**, p. 3093.

[16] Ford, D.M., Thompson, A.P., and Glandt, E.D. *J. Chem. Phys.*, 1995, **103**, p. 1099.

[17] Fernaud, M.J., Lomba, E., Martín, C., Levesque, D., and Weis, J.J. *J. Chem. Phys.*, 2003, **119**, p. 364.

[18] Fernaud, M.J., Lomba, E., Weis, J.J., and Levesque, D. *Mol. Phys.*, 2003, **101**, p. 1721.

[19] Spöler, C., and Klapp, S.H.L. *J. Chem. Phys.*, 2003, **118**, p. 3628.

[20] Fries, P.H., and Patey, G.N. *J. Chem. Phys.*, 1985, **88**, p. 429.

[21] Høye, J.S., Lomba, E., and Stell, G. *Mol. Phys.*, 1992, **75**, p. 1217.

[22] Holovko, M.F., and Polishchuk, Z.V. *Condens. Matter Phys.*, 1999, **2**, p. 267.

[23] Levesque, D., Weis, J.J., and Patey, G.N. *J. Chem. Phys.*, 1980, **72**, p. 1887.

[24] Chen, X.S., and Forstmann, F. *Condens. Matter Phys.*, 2001, **4**, p. 679.

[25] Klapp, S.H.L. and Patey, G.N. *J. Chem. Phys.*, 2001, **115**, p. 4718.

[26] Kierlik, E., Rosinberg, M.L., and Tarjus, G. *J. Chem. Phys.*, 1997, **106**, p. 264.

[27] Lee, L.L. *J. Chem. Phys.*, 1992, **87**, p. 8606.

[28] Chen, X.S., and Forstmann, F. *Mol. Phys.*, 1992, **76**, p. 1203.

SOLUTE IONS AT ICE/WATER INTERFACE

A molecular dynamics study of the ion solvation and excess stress

A.D.J. Haymet[1,2], T. Bryk[1,3], E.J. Smith[1]

[1]*Department of Chemistry, University of Houston,*
Houston, Texas 77204–5003,USA

[2]*CSIRO Marine Research,*
GPO Box 1538, Hobart Tasmania 7001, Australia

[3]*Institute for Condensed Matter Physics*
of the National Academy of Sciences of Ukraine,
1 Svientsitskii Str., 79011 Lviv, Ukraine

Abstract   A molecular dynamics study of the I_h ice/water interface and the behavior of solute ions at the basal ice/water interface is presented. The excess stress at the interface of pure ice and pure water is discussed. We compare the solvation of Na^+ and Cl^- solute ions in bulk water and ice, and discuss their behavior at the ice/vacuum interface. The free energy profiles for Na^+ and Cl^- ions across the ice/water interface is estimated from calculations of potential of mean force.

Keywords: Ice/water interface, interfacial excess stress, hydration structure, free energy profile

1.   Introduction

Over the last two decades the exploration of microscopic processes at interfaces has advanced at a rapid pace. With the active use of computer simulations and density functional theory the theory of liquid/vapor, liquid/liquid and vacuum/crystal interfaces has progressed from a simple phenomenological treatment to sophisticated *ab initio* calculations of their electronic, structural and dynamic properties [1]. However, for the case of liquid/crystal interfaces progress has been achieved only in understanding the simplest density profiles, while the mechanism of formation of solid/liquid interfaces, emergence of interfacial excess stress and the anisotropy of interfacial free energy are not yet completely established [2].

D. Henderson et al. (eds.), Ionic Soft Matter: Modern Trends in Theory and Applications, 333–359.
© 2005 *Springer. Printed in the Netherlands.*

The aim in this study is to investigate the behaviour of Na^+ and Cl^- ions at the equilibrium ice/water interface. It is well-known that salt is excluded from freezing water. The most interesting remaining issue is whether a charge separation can be induced by the ice/water interface and, if so, to characterize the reason for it. The remainder of this paper is organized as follows. The next section summarizes the current base of knowledge on condensed phase interfaces. In Sec. 3 we present the details of the molecular dynamics (MD) computer simulations and free energy calculations. In Sec. 4 we discuss specific structural characteristics of ice/water interfaces such as the 10–90 width. Sec. 5 contains a full treatment of the interfacial excess stress at the ice/water interface and discussion about its relevance to the solid/liquid interfacial free energy. Further to this, we present results of a solvation shell analysis in bulk ice, at the ice surface, and different regions of the two-phase ice/water system. The results on the free energy profile estimation for solute ions at the ice/water interface are given in Sec. 6. The conclusions of this study are collected in Sec. 7.

2.   Interfaces between two condensed phases

Theoretical treatment of solid/liquid interfaces is extremely difficult as they possess a variety of properties for which sophisticated theoretical models are required. Furthermore, large computational resources may be required to equilibrate and simulate such interfaces. The interfaces are lengthy objects with the width of order 10–15 Å, therefore the liquid/crystal interfaces require very large MD simulation boxes in order to reflect correctly the properties of bulk regions and the interfaces. Moreover the liquid/crystal interfaces are not stable in time formations with well-defined boundaries, but show large fluctuations of the interfacial structure at the temperatures of two-phase coexistence, which are usually close to the melting point (solid and undercooled liquid phases can coexist also below the melting point). That is why realistic computer simulations of equilibrium liquid/crystal interfaces up to date are limited to only classical molecular dynamics simulations. On the other hand, not all empirical two- and three-body interactions can correctly reflect the real melting temperatures in simulations. For example, the Tersoff [3] and Stillinger-Weber [4] potentials for silicon yield a difference of approximately 900 K for the estimated melting point [5]. In order to improve melting point calculations from simulations of liquid/crystal coexistence, several studies were focused on *ab initio* corrections to the classical simulations [6, 7].

Another important issue for liquid/crystal interfaces is their roughness. It was shown in [8] that treatment of fluctuations in the interfacial width as a function of in-plane coordinates permits calculation of the stiffness for various interfaces, which can be used for calculations of the anisotropy of interfacial free energy. This method was successfully used for interfacial free energy calculations in metallic Al [9], Au and Ag [10] and Lennard-Jones [11] liquid/crystal systems. Liquid/crystal interfacial free energy calculations [11] for Lennard-Jones systems were in remarkable agreement with results of calculations using artificial "cleaving" approach by Davidchack and Laird [12], which built upon the original method of Broughton and Gilmer [13].

Among the different two-phase liquid/crystal systems, ice/water interfaces are of great interest because of their fundamental presence in nature and importance in chemical, biological, environmental and atmospheric processes [14]. Systematic studies of ice/water interfaces by molecular dynamics simulations began in 1987, when Karim and Haymet [15] simulated for the first time the two-phase coexistence using the SPC model of water molecules. Since 1987, ice/water interfaces were studied with TIP4P [16], CF1 [17], SPC/E [18, 19] and six-site [20] models of water molecules.

Even more challenging are the problems focused on the behaviour of solutes at the interfaces. More than fifty years ago, Workman and Reynolds [21] claimed to observe a charge separation occurring during the freezing of dilute aqueous solutions and postulated its relevance to thunderstorm electricity. Recently Shibkov *et al.* [22] observed electromagnetic emission accompanying freezing of dilute aqueous NaCl solutions and discussed the relationship between this phenomenon and the Workman-Reynolds effect. Although it was claimed that the generation of the freezing potential explained in part by a simple kinetic model, using the different probabilities of positive/negative ion trapping into ice during the freezing front propagation [23], to the best of our knowledge, there are no computer simulation studies with realistic models attempting to provide more microscopic insight into processes occurring with ions close to the ice/water interface. Also of interest in the study of solute particles at the ice/water interface is the fact that HCl and HBr acids can dissolve at the ice surface, [24–26] possibly due to the presence of a liquid-like layer at the ice surface. This would imply that the ice/water interface can also possess catalytic properties, but no computer simulation evidence exists. Recently Toubin *et al.* [27] reported free energy profile calculations for non-ionizable HF and HCl molecules at the ice surface, and as they penetrate into the ice crystal. Strong melting of ice structure was observed at 235 K when the molecules were constrained to the penetration path, while at the lower temperature of 190 K the ice structure was stable. For the latter case the free energy profiles showed an adsorption minimum on the ice surface and sharp increase towards the bulk ice region. In both cases, for HF and HCl molecules, the free energy difference between the bulk region and ice surface was estimated to be approximately 50 kcal/mol.

Considerably more studies consider solute behaviour at the liquid/liquid and liquid/vapor interfaces. Molecular mechanism and dynamics of ion transfer across a liquid/liquid interface are the focus of the molecular dynamics study reported by Benjamin [28]. It was shown that surface roughness and capillary distortions play important roles in the transfer process, and the dynamics of ion transfer is governed by a barrier in the free energy profile across the interface. In the case of the water/dichloroethane interface, it was observed that thermal fluctuations superimpose capillaries as long as 8 Å on the sharp interface and generate quite a rough surface. These capillaries are constantly moving "fingers" of water that protrude into the opposing liquid on the time scale of tens of picoseconds, even though the time average over the long observation time results in a relatively smooth density profile that gives rise to a surface thickness of about 10 Å. It was stressed that the interaction between the water and the ion is still quite appreciable even when the ion reaches the "bulk" dichloroethane phase. In this case the simulations showed that the ion carries at least part of the hydration shell with it.

The molecular picture of ion behaviour at liquid/liquid interfaces obtained from MD simulations is supplemented with phenomenological treatments of ion transfer across the liquid/liquid interface. The actual mechanism is not specified within a phenomenological approach, although a small but finite diffusion is assumed for crossing a flat interface. In [29] it was proposed to improve the phenomenological treatment by taking into account the effect of protrusions of one solvent extending into the other. The overall process of ion transfer across the liquid/liquid interface was formulated as consisting of several steps:

(i)   an ion initially in solvent A first attaches itself to a protrusion of solvent B,

(ii) once attached ion diffuses across the interface toward solvent B,

(iii) and at some point, when the change of solvation is nearly complete, the ion begins to detach itself from the interface, which may now be the tip of a protrusion of solvent A extending into solvent B [29].

The important role of interfaces in atmospheric chemistry is discussed in [30]. The authors showed that the combination of experimental observations, computer kinetic modelling with the use of only gas and bulk aqueous-phase chemistry, molecular dynamics simulations, and quantum chemical calculations leads to the suggestion that the chemistry at the interface is responsible for the observed production of Cl_2. Dang and Chang [31] explored the role of polarizability of water molecules in the molecular mechanism of iodide binding to the water/vapor interface. A free energy minimum was observed near the Gibbs dividing surface for simulations that employed polarizable models in contrast to simulations with nonpolarizable models, when no surface state was obtained as iodide crossed the water/vapor interface. More detailed study of molecular transport of I^-, Br^-, Cl^- and Na^+ ions across the water/vapor interface was undertaken in [32]. The larger I^- and Br^- anions were found to bind more strongly to the water/vapor interface than the smaller Cl^- ion. Also, in agreement with the proposed molecular mechanism of ion transport across the liquid/liquid interface [28, 29], it was shown in MD simulations [32] that the I^- ion carried a water molecule with it as it crossed the water/dichloromethane interface.

3. Method and molecular dynamics simulations

3.1 *Model potential energies*

In this study of ice and water bulk phases and ice/water interfaces, we use the rigid SPC/E model [33] of water which is constrained with intramolecular OH distances of 1 Å and H–O–H angle of 109.47°. Hydrogens and oxygens bear effective point charges with the values $+0.4238$ and -0.8476, respectively. Short-range interaction between water molecules is represented by a $12 - 6$ Lennard-Jones (LJ) two-body potential between the oxygen sites of different molecules. The LJ parameters for the oxygen-oxygen interaction are: $\epsilon = 0.1554$ kcal/mol and $\sigma = 3.16557$ Å.

The solute-water interactions are modelled by the sum of the Coulomb and LJ contributions

$$U_{ij}(r) = 4\epsilon_{ij}\left[\left(\frac{\sigma_{ij}}{r}\right)^{12} - \left(\frac{\sigma_{ij}}{r}\right)^{6}\right] + \frac{q_i q_j}{r},$$

where the charges of sodium and chlorine ions q_i, and parameters of their LJ interaction with water molecules were taken from Smith and Dang [34]. All the short-range interactions were cut-off in simulations at $r_c = 10$ Å.

We simulated the SPC/E basal ice/water interface with Na^+ and Cl^- ions at a temperature of 225 K, which was estimated as the melting point for the SPC/E model of water in our previous study of two-phase coexistence [19]. We have used a collection of 2304 rigid water molecules plus a single solute ion in the NVT ensemble. The time step was chosen to be 1.5 fs.

The long-range Coulomb interactions were treated using the 3D Ewald summation method with Ewald convergence parameter $\alpha = 0.284$ Å^{-1} and an Ewald sum precision of $1 \cdot 10^{-5}$ (from the standard in the DL_POLY package [35]).

We stress that special care must be paid during the preparation of solid/liquid interfaces in order to avoid improper stress in the solid phase. The size of the MD box must be carefully adjusted to obtain the same reference pressure in solid and liquid phases. With this in mind we created the solid/liquid coexistence in following steps:

(i) the solid phase was equilibrated at the reference temperature T and pressure P using the isotropic NPT ensemble;

(ii) using the NP_nT ensemble (when the only fluctuating side of MD box is the longest one, L_z, normal to a chosen face of crystal, and P_n is the normal component of pressure tensor) we adjusted L_x and L_y during a series of small runs with the aim of obtaining average values such that $\langle P_{xx} \rangle = P$ and $\langle P_{yy} \rangle = P$ at the reference pressure P;

(iii) after the values L_x and L_y were set we cut the middle part of the crystal, and melted it again using the NP_nT ensemble at a temperature much higher than the reference temperature T. After several picoseconds, and the analysis of pair correlation functions to make sure we had arrived at the liquid state, we slowly reduced the temperature of the liquid to the reference temperature T. As the liquid does not oppose the external stress, from this step we obtained the liquid at the reference temperature T and pressure P and the same area dimensions L_x and L_y as for solid phase;

(iv) and the last step consisted in adjoining solid and liquid phases into one MD box and simulating them in the NP_nT ensemble to make sure that the interface was properly relaxed after the first contact of the liquid and solid phases. Thus, using these four steps one is able to prepare a solid/liquid interface with a reference pressure in bulk regions and an equilibrium relaxed interface. For such a system the deviation from the reference pressure tensor should be only in the interfacial region. Note that, for any other reference temperature one cannot use this geometry of MD box – one must prepare the two-phase system from scratch for each temperature needed. Otherwise, the solid phase will have some residual stress and a pure interfacial excess stress would not be obtained.

A specific feature of simulations of ice/water interfaces with solute ions is a need to generate several initial positions for the ion in the two-phase system. Because the solute diffusivity at 225 K is very small, it is not possible to observe in a single MD run the transfer of an ion from one phase to the other. Hence, we were forced to generate the solute ion in ice and water regions at different fixed distances from the interface. In order to generate an ion in the system, we slowly increased the Lennard-Jones radius first, and in the second step the neutral full size particle was slowly charged with the step of $\pm 0.1\,e$ per 500 timesteps. The stability of the ice/water interface during the process of ion generation was monitored by the inspection of density profiles.

3.2   *Free energy calculations*

Two approaches exist for estimation of the free energy of solute ion transfer across the interface. The first one is calculation of the potential of mean force, when the reaction coordinate for ion transfer is considered to be the z_s position of the ion. With respect to a reference state where the ion is at z_0, the change in Helmholtz free energy (for simulations in the NVT ensemble) when the ion is located at z_s is given by the following [32],

$$\Delta A(z_s) = A(z_s) - A(z_0) = -\int_{z_0}^{z_s} \langle f_z(z_s') \rangle \, dz_s' , \qquad (1)$$

where $\langle f_z(z_s) \rangle$ is the ensemble averaged z component of the total force exerted on the ion at a given reaction coordinate. Here $A(z_0)$ is a reference free energy (which is generally not known) and which represents the free energy of the system with the ion located initially in the bulk water (or ice) region. During the simulations the z coordinate of the ion was constrained to the same value, and the average force acting on the ion was then evaluated. To avoid a shift of ice/water interface in a system with a constrained ion, the simulation cell is constrained by removing the z-component of the velocity at every step during the molecular dynamics simulations.

Calculation of the free energy of ion transfer across the interface via the thermodynamic integration method [36] has been introduced by us earlier [37]. For the ice/water interface, the simulation cell contains an ion initially in the ice phase. The ion is "transferred" to the liquid using the thermodynamic integration method. The free energy difference between the ion in the different positions in the system is then measured. The calculation begins when the ion is in the ice phase of the system and is fully interacting with the water molecules (system A) and ends when the ion is in the liquid phase of the system and is fully interacting with the water molecules (system B). As λ changes from 1 to 0 the ion will effectively be transferred from the ice region into the liquid region of the ice/water interface system.

4.   Structure of the ice/water interface

The density-temperature phase diagram for SPC/E water at the pressure of 1 bar has been reported by us earlier [40, 41]. The famous density maximum of water, located experimentally at 277 K, is found for this model at the temperature 240 K. Our simulations of two-phase ice/water coexistence for SPC/E model resulted in the melting temperature, which is approximately 50 K below the estimated value of 279 K from the free energy study of melting point for SPC/E water by Arbuckle and Clancy [42]. A similar tendency was pointed out by Morris [43], when the melting point from two-phase coexistence simulations for aluminium was lower than the value from free-energy calculations. To date, the origin of the discrepancy in melting point estimation between direct two-phase simulations and free-energy methods is not known. On the other hand, the melting temperature of 225 ± 5 K obtained in [19, 40] with the SPC/E model, is in closer agreement with the results of free-energy study of melting temperature using TIP4P model [44], where the value of 238 ± 7 K was obtained. Also, an *ab initio* Car-Parriello dynamics study of ice surface disordering [45] indicated, that the melting temperature of I_h ice is between 190 K and 230 K, *i.e.*, in quite good agreement with results obtained in [19]. Very recent calculations using thermodynamic integration for SPC/E model [46] located the melting point of I_h ice at 215 ± 7 K.

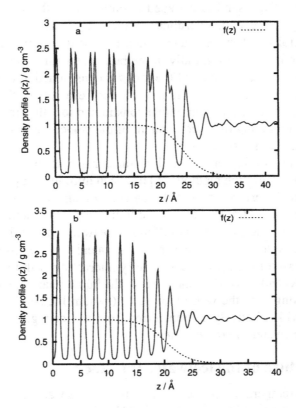

Figure 1. Mass-density profiles for ice/water interfaces (solid lines): basal interface (a) and $(2\bar{1}\bar{1}0)$ interface (b). Translational order parameter $f(z)$ changing from 1 in crystal bulk region to 0 in liquid phase is shown by short-dashed lines.

Mass-density profiles for stable ice/water coexistence at 225 K for basal and $(2\bar{1}\bar{1}0)$ ice planes are shown in Fig. 1. The large difference in the atomic mass between oxygen and hydrogen yields, in the ice region, a sharp double-layer structure from the oxygens and rather small mass-density between the double molecular layers due to hydrogens chemically bonded to those oxygens. Clearly in the water region, $\sim 32 - 42$ Å, the mass-density profiles show a flat density characteristic of bulk water. In the region $\sim 12 - 30$ Å one can observe the gradual change in amplitude and width of the density maxima, which reflects the frustration of crystal structure across the ice/water interface. The density profile of the ice crystal, $(2\bar{1}\bar{1}0)$, shows just single-peak density maxima. The main characteristic of crystal/liquid interfaces is a translational order parameter which reflects the change in crystal structure across the interface. The translational order parameter is equal to unity in the bulk crystal region and zero in the bulk liquid phase.

Table 1. 10–90 width of the basal, prizm and $(2\bar{1}\bar{1}0)$ ice/water interfaces at T = 225 K.

interface	10–90 width (Å)
basal	10.54
prizm	8.30
$(2\bar{1}\bar{1}0)$	6.64

The translational order parameter permits an estimate of the width of the interfaces. The 10–90 width is defined to be the length over which a specific interfacial order parameter changes from 10% to 90% of the bulk solid value. We have estimated the 10–90 widths of the interfaces using a fit by a simple hyperbolic tangent function, used frequently in earlier studies [17]. In the case of the mass-density profile, the translational order parameter may be extracted from a fitting procedure,

$$\rho(z) \overset{\text{fit}}{=} \rho_l\left[1 - f(z)\right] + \rho_s(z)f(z) , \tag{2}$$

where $\rho_s(z)$ is a complicated function, reflecting the double-peak structure of the mass-density profile in the solid region, and is modelled by a sequence of overlapping Gaussians [17] which, in their turn, are connected with the Debye-Waller factor of the ice crystals. In Eq. (2), ρ_l is the liquid density and,

$$f(z) = \frac{1}{2}\left[1 - \tanh\left(\frac{z - z_0}{w}\right)\right] , \tag{3}$$

is the envelope function. The parameters used in the fit reflect the width of interface w, and the location of the mid-point of the profile, z_0. The 10–90 width of the profile is equal [47] to $2.1971 \times w$. The estimated 10–90 width for three different ice/water interfaces is given in Table 1. It follows, that the basal ice/water interface is the most wide among those studied.

Other order parameters can be used to characterize ice/water interfaces via change in average density [17, 19] and diffusivity [17] across the interface. Local order parameters can also be defined which depend only on the position of a reference molecule. It was shown in [40] that the local tetrahedral order parameter, which reflects the change in tetrahedral environment, leads to a 10–90 width of 11 Å for the basal ice-water interface which is in good agreement with the estimated value from the translational order parameter.

5. Excess stress at the ice/water interface

The origin of the excess stress on liquid/vapor interfaces follows from the tendency of the liquid surface to contract. As a molecule inside a mass of liquid is under the effect of the forces of the surrounding molecules, while a molecule on the surface is only partly surrounded by other molecules, some work is necessary to bring molecules from the inside to the surface. This indicates that the force must be applied along the surface in order to increase the area of the surface. This force on the surface appears as excess stress (a difference between normal and transverse components of pressure tensor in the region of the interface) and defines the surface tension of the liquid. Excess stress σ, for liquid/vapor interfaces, is always a positive quantity and is equal to the interfacial free energy.

The processes on solid/vapor interfaces (or solid surfaces) and solid/liquid interfaces differ sufficiently from the liquid/vapor systems. Due to huge relaxation times in the solid phase, the atoms or molecules in the interior are not capable of moving to the surface to accommodate the new area created, as in the case of liquid surfaces. It was noted in [1, 48] that the excess stress at solid surfaces and solid/liquid interfaces can have opposite sign. However, there was no clear explanation of that fact. The relation between the surface stress σ, and solid surface free energy γ_{SV}, was first pointed out by Shuttleworth [49],

$$\sigma_{ij} = \gamma_{SV}\delta_{ij} + \frac{\partial\gamma_{SV}}{\partial\varepsilon_{ij}}, \qquad i,j = x,y, \tag{4}$$

where the second term in the right hand side of the above equation reflects the change in surface free energy per unit charge in the elastic strain of the surface. For most solids $\partial\gamma_{SV}/\partial\varepsilon_{ij} \neq 0$. In [50] this term was called the *driving force* of the reconstruction of surfaces and it was shown that reconstruction permits the surface to evolve towards a "liquid-like" ground state in which the stress is isotropic, with a vanishing shear component (as it is known that the liquids do not support shear stress). Thus, in systems with high surface atomic mobility, one would expect very small values of $\partial\gamma_{SV}/\partial\varepsilon_{ij}$, making these interfaces more similar to liquid surfaces. It is supposed [1, 48] that the relation shown in Eq. (4) is valid for solid/liquid interfaces as well. Excess interfacial stress itself is an interesting quantity for study. This is because there are no indications as to the sign the excess interfacial stress *a priori* will have in solid/liquid systems. Recently, Sibug-Aga and Laird [51] reported that the simulations of interfaces between a one-component fcc crystal and a binary fluid mixture using hard spheres, led to an excess interface stress σ of negative sign, *i.e.*, in contrast to liquid/vapor interfaces, the hard sphere solid/liquid interface had the transverse component of pressure tensor larger than the normal to the interface one. In MD simulations of Lennard-Jones solid/liquid interfaces, Broughton and Gilmer [13] observed negative values of surface stress σ for (111) and (110) interfaces with fcc crystal phase, while for the (100) interface the surface stress was found to be zero.

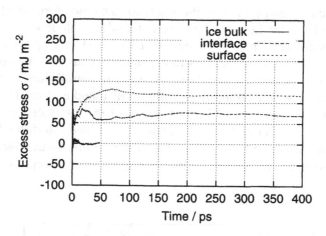

Figure 2. Running averages of the excess stress σ for basal ice surface (short-dashed line) and basal ice/water interface (long-dashed line). For comparison it is shown that in bulk ice (solid line) the excess stress is absent.

It is possible to estimate the interfacial excess stress σ, directly in MD simulations via fluctuations of normal and transverse components of the virial [52–54],

$$\sigma = \frac{1}{2L_x L_y} \left\langle V_{zz} - \frac{1}{2}\left(V_{xx} + V_{yy}\right) \right\rangle , \qquad (5)$$

where $V_{ii}, i = x, y, z$ are the components of virial and $\langle \ldots \rangle$ are the statistical averages. This method of direct evaluation of interfacial excess stress is equivalent to the integration of pressure profiles,

$$\sigma = \frac{1}{2} \int_{-\frac{L_z}{2}}^{\frac{L_z}{2}} \left[p_{zz} - \frac{1}{2}\left(p_{xx} + p_{yy}\right) \right] dz , \qquad (6)$$

where the diagonal components of pressure tensor $p_{ii}(z)$, $i = x, y, z$ are functions of z. The approach of Eq. (5) avoids the artificial dividing of the MD box into small bins and requires only long MD runs for estimation of reliable statistical average values. In contrast to Lennard-Jones crystal/liquid interfaces, a positive sign for the excess stress is observed in the case of the ice surface and ice/water interface. In Fig. 2, the surface excess stress is more positive than the interfacial excess stress. There is an obvious effect of reducing the magnitude of the surface excess stress when the bulk water is in equilibrium contact with the ice.

We stress again that for the liquid/vapor interfaces the excess interfacial stress is always a positive quantity equal to the surface tension γ (interfacial free energy), and is defined by the tendency of the liquid surface to contract. We have seen a similar picture in this study for the case of the basal ice/water interface, while for the Lennard-Jones solid/liquid two-phase system we found the negative sign of σ, which in our opinion is defined by the opposite tendency of the Lennard-Jones solid surfaces to expanding. Thus we suppose, that the sign of the solid/liquid interfacial excess stress in equilibrium is defined by the relation between the equilibrium volumes of liquid and solid phases, *i.e.*, $\text{sign}(\sigma) = \text{sign}(v_{sol} - v_{liq})$. Different signs of the interfacial excess stress in the case of ice/water and Lennard-Jones solid/liquid interfaces imply different contribution from the second term in the Shuttleworth relation shown in Eq. (4), because the interfacial free energy γ must always be a positive

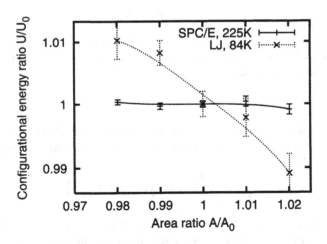

Figure 3. Dependence of the configurational energy of solid/liquid two-phase systems on the small changes in area of the interface: basal ice/water interface at 225 K ("plus" symbols and solid interpolated line) and Lennard-Jones solid/liquid interface at 84 K ("cross" symbols and dotted interpolated line).

quantity. The derivative $\partial\gamma/\partial\varepsilon_{ij}$ is very difficult to calculate. However a tendency for different systems can already be seen from the dependence of the configurational energy of two-phase systems on small change in the interfacial area. In Fig. 3 we show that for the case of the ice/water interface the configurational energy is an extremely weak function of the area, while for Lennard-Jones two-phase systems it decays against the area stretch. If such a tendency was the same for the dependence of the interfacial free energy on infinitesimal strain, then one may conclude that the interfacial excess stress is almost equal to the interfacial free energy for ice/water two-phase systems, while in the case of the Lennard-Jones solid/liquid interfaces the derivative $\partial\gamma/\partial\varepsilon_{ij}$ would be a negative number, leading to possible opposite sign of the excess stress σ according to Eq. (4). In such a case the excess stress of the two-phase ice/water system (shown in Fig. 2 for two interfaces in MD box) gives the value of $\sigma = 39\pm4$ mJ·m^{-2} for a single ice/water interface, is in good agreement with the experimental value of the ice/water interfacial free energy of $\gamma_{\mathrm{exp}} = 29.1\pm 0.8$ mJ·m^{-2} [55].

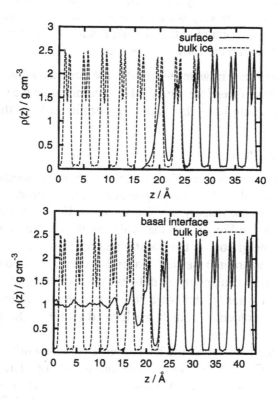

Figure 4. Mass-density profiles $\rho(z)$ for bulk I_h ice (dashed line), ice surface cut at the basal face and basal ice/water interface (solid lines).

Another interesting observation leads to some conclusions about the spatial distribution of the excess stress at the solid/liquid interfaces. Let us consider density profiles of the ice surface and ice/water interface compared with bulk ice density profile. In Fig. 4 we show that the ice cut forms a specific layer on the ice surface, which shows on the density profile the tendency of a small shift towards the bulk ice. When the water is in contact with the ice cut another density maximum is formed at the contact, reducing, at the same time, the initial surface stress and the shift of the top density peak of the ice cut. This implies that the first strongly smeared solid-like layer at approximately $z = 20$ Å is responsible for the main contribution to the interfacial excess stress of the ice/water interface.

6.   Hydration shell around ions in water and I_h ice

6.1   *Ions in bulk water and in bulk I_h ice*

In order to understand the behavior of solute ions in the two-phase ice/water system we first consider the hydration of Na^+ and Cl^- ions in the bulk ice and water. Ice structure is very sensitive to any impurity, which causes strong defects in the crystal structure at the melting point. In Fig. 5 two snapshots of Na^+ ion in I_h ice crystal are shown for two cases:

(i)   when the oxygens in water molecules are constrained and only orienta-
      tional degrees of freedom remained to accommodate the crystal structure
      around the solute ion - this approach permits to keep the ice structure
      rigid;

(ii) and regular SPC/E ice.

In the constrained ice lattice the Na^+ ion takes symmetric location in the middle of a hexagonal hole, which corresponds to a minimum density position in the crystal region in the density profiles shown in Fig. 1. Running coordination numbers shown in Fig. 6a provide evidence that the Na^+ ion has six water molecule nearest neighbors in the constrained ice structure. When the water molecules are unconstrained the Na^+ ion tends to penetrate into the region with larger oxygen density causing many defects in the ice structure (see Fig. 5b).

As a result the hydration shell of a Na^+ ion in the ice structure consists of five SPC/E water molecules, which are located closer to the ion than in the rigid ice lattice (see Fig. 6a).

The effect of the Cl^- ion on the ice structure is sufficiently different due to size effects, as the Cl^- ion has a larger core than a Na^+ ion. The rigid ice lattice is too dense for a Cl^- ion, which in unconstrained ice forms a hydration shell of six nearest neighbours at a distance of $\sim 3.4-3.5$ Å (Fig. 6b). The tendency to push out the nearest molecules causes more defects in the ice crystal structure essentially making it disordered on the scale of several lattice periods.

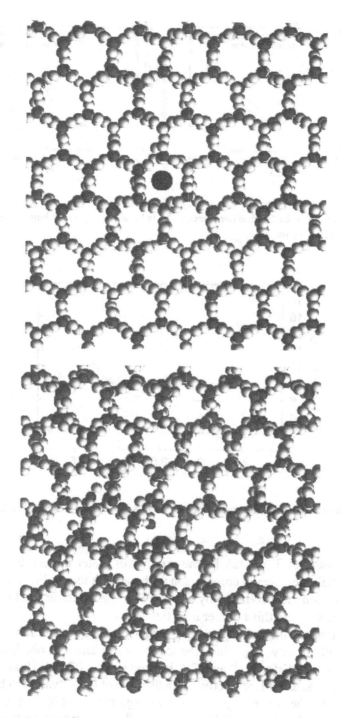

Figure 5. Snapshot of Na$^+$ ion (black sphere) in bulk (rigid) ice with constrained oxygens (a) and in unconstrained ice (b). Oxygens and hydrogens are shown by dark and light grey spheres, respectively.

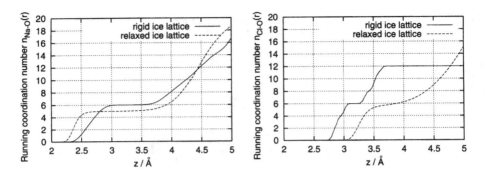

Figure 6.   Running coordination numbers $n_{Na-O}(r)$ and $n_{Cl-O}(r)$ for bulk ice and rigid ice lattice with constrained oxygens.

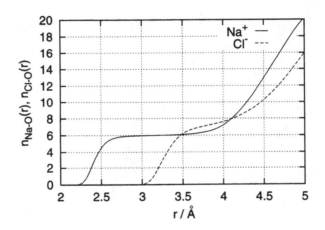

Figure 7.   Running coordination numbers $n_{Na-O}(r)$ and $n_{Cl-O}(r)$ for SPC/E water at 225 K.

In bulk water at T $=$ 225 K, the number of water molecules in the first hydration shell increases in comparison with the bulk ice. In Fig. 7 one can see that the Na$^+$ ion is surrounded by six nearest neighbours in the liquid phase, while for the Cl$^-$ ion this number is $\sim 6.9 - 7.3$.

Solvation properties of Na$^+$ and Cl$^-$ ions in bulk ice and bulk water were the focus of another study [37], where the solvation free energies obtained in liquid water at T $=$ 230 K for Na$^+$ and Cl$^-$ ions were -90.07 ± 3.18 kcal/mol and -91.08 ± 3.16 kcal/mol, respectively. In the case of ion solvation in bulk I_h ice at T $=$ 220 K, the free energies were less negative: -82.68 ± 1.14 kcal/mol for Na$^+$ and $-80.02 \pm$ 3.05 kcal/mol for Cl$^-$, meaning a more favourable solvation of the ions in the bulk water phase.

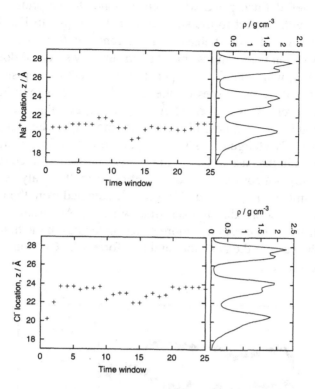

Figure 8.   Location of Na$^+$ and Cl$^-$ ions on the ice basal surface over the simulation time of 1.5 ns (symbols). Each symbol '+' corresponds to the position of a maxima in a single-ion density profile accumulated over each time window of 60 ps. The solid line in the left frame corresponds to the mass-density profile accumulated over the first 60 ps.

6.2   *Ions at the basal plane of the I_h ice/vacuum interface*

Ice surfaces at temperatures close to melting are often said to have a "liquid-like" layer at the contact with a vacuum [14]. (Fig. 4 above shows that for the SPC/E model at least, this is an over-simplification: thee are several different non-ice-like layers at the vacuum.) Therefore it is interesting to study how the Na$^+$ and Cl$^-$ ions behave at the ice/vacuum interface. The ions were initially placed at $z = 15$ Å in the MD box, *i.e.*, approximately at the distance of 2 Å from the surface. We then allowed the ions to approach the interface over a period of 20 ps, and only after that started the production runs. In order to characterize the location of solute ions with respect to the surface, we introduced a single-ion density profile accumulated over a certain time window. We have found that such single-ion density profiles for cases such as our system, being accumulated over the time of 60 ps, have the width of ≈ 2 Å

and contain a well-defined peak, which corresponds to the preferential location of the ion with respect to the surface over the 60 ps. In Fig. 8 one can
see that the Na^+ and Cl^- ions behave in a different way from being on the ice
surface. The Na^+ ion is moving within the top strongly smeared double layer,
20 Å$\leqslant z \leqslant 22$ Å, while the Cl^- ion penetrates through the top surface layer
during the first 120 ps and settles on the top side, in the second layer of the
surface double layer, 22 Å$\leqslant z \leqslant 24$ Å. Fig. 8 suggests that the ice surface
simulated with the SPC/E model of rigid water molecules can produce a separation of oppositely charged ions with respect to the surface. The origin of
charge separation in the case of Na^+ and Cl^- ions on the basal ice surface was
interpreted in [56] as a consequence of size effect from the analysis of running
coordination numbers $n_{Na-O}(r)$ and $n_{Cl-O}(r)$, obtained from the relevant radial distribution functions during each time window. According to [56], the
larger size of the Cl^- ions requires more water molecules in the first hydration
shell than in the case of the Na^+ ion, and this forces the Cl^- ion to penetrate
deeper into ice surface.

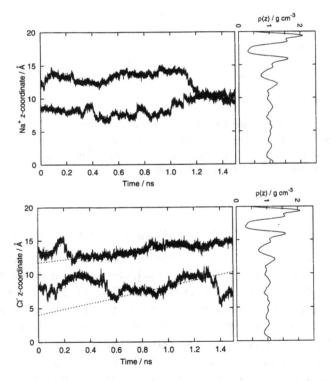

Figure 9.   Time-dependent location of Na^+ and Cl^- ions in the two-phase basal ice/water
system. Dashed lines show the approximate asymptotes in ion movement. The left frames show
the mass-density profiles of basal ice/water interface.

6.3   *Movement of ions at the basal ice/water interface*

Motivated by simulations of ions at the water/dichloroethane interface reported by Benjamin [28], we measured the behaviour of Na^+ and Cl^- ions at the ice/water interface by observing in MD simulations the time-dependent z-coordinate of ions over a production time of 1500 ps. Initially the Na^+ ions were constrained, in different runs, to the positions $z = 8$ Å and $z = 12$ Å in an MD box over 105 ps. No melting of the interface structure was observed in the density profiles. Then the ion was unconstrained and moved freely over 150 ps. We then started the production runs saving the z-coordinate of the ion every 0.15 ps. In the top frame of Fig. 9, the motion of the Na^+ ion along the direction normal to the ice/water interface is shown. For convenience the mass-density profile of the interface is shown on the right. In general, two tendencies in the behaviour of the Na^+ ion at the ice/water interface can be seen. The ion placed at the liquid side of the ice/water interface slowly moves towards the bulk region of water, although some strong fluctuations affect the asymptotic migration. More interesting is the behaviour of the Na^+ ion that was initially grown deep in the interface. We observed motion, for approximately 400 ps, towards the bulk water while there existed periods lasting 100–200 ps during which the ion penetrated deeper into the interface. However, after approximately 1200 ps the Na^+ ion, initially placed at two different distances with respect to the interface, in both cases ended up at $z \approx 10$ Å, implying the existence of a free energy minimum for a Na^+ ion on the liquid side of basal ice/water interface.

In the case of the Cl^- ion initially constrained at $z = 12$ Å (see Fig. 9 lower frame) a long lasting (~ 500 ps) diffusive motion deeper into the interface was observed. A specific feature of simulations of solid/liquid interfaces at the melting temperature with unconstrained ions, is the possibility of local interface melting. Namely this effect was observed in the case of Cl^-. In Fig. 10 we show two snapshots of the ice/water interface containing a Cl^- ion at the beginning of the production run and after 660 ps. One can see that strong frustration of hexagonal structure advanced into the bulk ice region close to the ion location, which can be treated as interface melting by the ion. In the case of the Na^+ ion, we did not observe such an obvious melting of ice/water interface structure. Evidently the different sizes of Na^+ and Cl^- ions play a role in the different effects observed: the Na^+ ion is smaller and causes less defects in interfacial structure.

6.4   *Free energy profiles of solutes at vacuum/ice and water/ice interfaces*

Free energy profiles are an important characteristic of ion behaviour in heterogeneous systems such as crystal/liquid interfaces. It is obvious that the

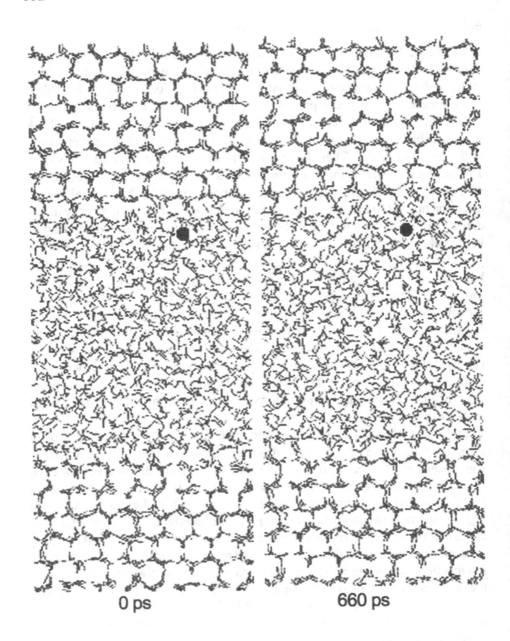

0 ps   660 ps

Figure 10.   Snapshots of changes in structure at the basal ice/water interface with Cl⁻ ion corresponding to the configurations at 0 ps (left) and 660 ps (right) from previous figure.

free energy profile depends strongly on the structure of crystal/liquid or crystal/vacuum interfaces, and the path of profile sampling. That is why an important requirement for the profile calculations is to avoid the local melting of crystal structure along the ion path. In the liquid phase it is not so important how to move a particle along a certain path, because processes in the liquid phase have rather small relaxation time. In the crystal medium, where the relaxation time is extremely large, the behaviour is quite different. For this purpose we applied in our calculations the following procedure for free energy profile estimation. As the solute diffusion in crystal is an activated process it is impossible to observe, the ion transfer across the interface in the time window of computer simulations. Therefore, an important role in the correct sampling of the ion path across the interface to cause the minimum possible changes in the structure of the interface is played by "adiabatically" generating the next position of the ion. This was done by removing the particle at the previous position z and slowly growing the ion into the system, firstly the ionic core at the new z position and then slowly charging it. This allows for the particle to

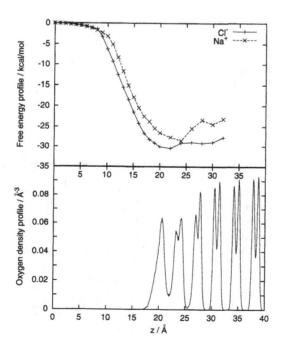

Figure 11. Free energy profile $\Delta A(z)$ for Na^+ and Cl^- ions at the vacuum/ice basal plane interface at T = 225 K. In lower frame the oxygen density profile is shown.

find an energy minimum at the given z without causing drastic changes in the crystal structure.

In Fig. 11 we show the free energy profile for Na^+ and Cl^- ions at the vacuum/ice interface. The profiles were set with the reference zero energy $A(z_0)$ (1) in the middle point of the vacuum region in the simulation box. Both curves have a so-called adsorption minimum at the liquid-like layer on the top of the ice crystal: $z_{min} = 23$ Å for Cl^- and $z_{min} = 24$ Å for Na^+ ion. The ice structure close to the surface at 225 K was very unstable when the ion is present, and easily melted. Similar impurity melting of the ice surface was observed in [27, 56]. Therefore, the minima of the free energy profiles were not located in the top liquid-like layer but in the region of the second strongly smeared density peak with strongly frustrated (and partially molten) crystal structure. Sampling the ion positions in the solid-like region caused stronger disorder in the ice layers in the case of Cl^-, while due to smaller size of the Na^+ ion, it was more easily incorporated into the crystal structure. We stress that for the case of the ice surface at 225 K, the rather low free energy difference for the Cl^- ion between the adsorption minimum and z_{max} sampled

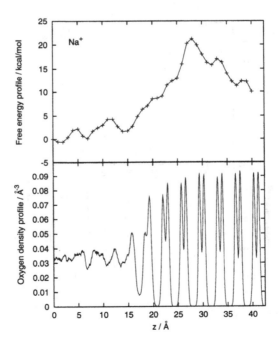

Figure 12.   Free energy profile $\Delta A(z)$ for Na^+ ion at the basal ice/water interface at T = 225 K. In lower frame the oxygen density profile is shown.

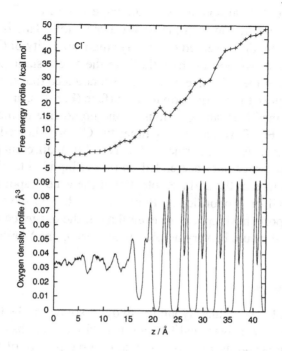

Figure 13. Free energy profile $\Delta A(z)$ for Cl$^-$ ion at the basal ice/water interface at T = 225 K. In lower frame the oxygen density profile is shown.

in simulations, is a consequence of surface structure instability when the Cl$^-$ ion is present.

We observed much better stability of ice/water interfaces during the free energy profile calculations. In Figs. 12 and 13, the free energy profiles across the basal ice/water interface are shown for Na$^+$ and Cl$^-$, respectively. In general, the free energy profiles are in agreement with the profiles obtained for the stable vacuum/ice interface at 190 K in [27], where large free energy differences of ~ 50 kcal/mol were obtained between the ice bulk region and the liquid-like layer on the ice surface for rigid HF and HCl molecules. However, due to size effects, the smaller Na$^+$ ion has several well-pronounced local minima especially on the liquid side of the ice/water interface and in bulk ice, while for the larger Cl$^-$ ion, only one well-pronounced minimum deep in the ice/water interface was obtained. Another interesting difference of both estimated free energy profiles is in the region of bulk ice where the Na$^+$ tended to the middle point of the bulk ice region, while for the Cl$^-$ ion the maximum of free energy profile exists in this region. These different tendencies imply that the interface may have an electrostatic potential that is attractive for Cl$^-$

and repulsive for Na^+, although due to the large size of the Cl^- ion, the total minimum of this solute ion is in the bulk water region. This is in agreement with the experimentally observed extremely small solubility of Cl^- in ice.

Two more results support the hypothesis of the electrostatic potential at the ice/water interface: the existence of the well-pronounced free energy minimum for the Cl^- ion close to the Gibbs dividing surface (Fig. 13), and the behaviour of the unconstrained Na^+ and Cl^- ions on the liquid side of the ice/water interface shown in Fig. 9, when for long times the Cl^- ion showed a tendency to be attracted to the ice/water interface, while Na^+ ion at different positions with respect to the ice/water interface ended up on the liquid side of ice/water interface. Calculations of the surface potential of the water/vapor interface arise from work by Wilson, Pohorille and Pratt [57, 58], but for the ice/water interface no such reports have appeared, even though the existence of such an interfacial potential would be very helpful in addressing the Workman-Reynolds effect [21, 23].

7.   Summary

We performed a molecular dynamics study of solutes at the basal ice/water interface. Our study of Na^+ and Cl^- ion behaviour at the basal ice/water interface aimed at shedding light on the microscopic picture of the Workman-Reynolds effect, which is claimed to be a massive charge separation and emergence of the so-called freezing potential during freezing of aqueous solutions.

Simulations of solute ions at the liquid side of the ice/water interface revealed that for long times, in SPC/E water, the Cl^- ion shows a tendency to be attracted toward the ice/water interface, while the Na^+ ion, at different positions with respect to the ice/water interface, tends to move towards the liquid side of the ice/water interface.

Our free energy calculations, using the thermodynamic integration technique [37], show different results for solvation in the bulk phases and for the ion transfer across the interface in the two-phase system, which may be understood by hypothesizing the existence of an electrostatic potential at the ice/water interface. Calculations of the free energy profiles across the ice/water interface show opposite tendencies of the ions in the bulk crystal phase, which may also can be explained by that same hypothesis, namely the existence of an interface potential attractive to Cl^- ions and repulsive to Na^+ ions.

To study the interfacial excess stress, we compared the density profiles for interfaces and surfaces of relevant crystal faces. We have found a correlation between the sign of the interfacial excess stress and the tendency of the crystal surface either to contract or expand in comparison with the bulk crystal phase. We have found that the interfacial excess stress in the solid/liquid system is smaller than the corresponding surface excess stress, *i.e.*, the liquid phase softens the excess stress induced by a crystal surface.

This study is consistent with the idea that crystal surfaces at temperatures close to melting have some kind of disordered layer or layers , often called "liquid-like". Due to the different equilibrium volumes of the liquid and solid phases, this region makes the surface either contract (as in the case of the ice surface) or expand (as it is for Lennard-Jones systems). The positive interfacial excess stress of the ice/water interface therefore makes it similar to liquid/vapor interfaces, and the water/vapor interface in particular, for which the excess stress is equal to the interfacial free energy (surface tension).

Acknowledgements

This research was supported by the Welch Foundation grant E–1429 for which grateful acknowledgment is made. TB was partially supported by the STCU Project No. 1930. The calculations were performed at the Texas Advanced Computing Center under NPACI award and at Gallaxy cluster at University of Houston. DL_POLY is a package of molecular simulation routines written by W. Smith and T.R. Forester, copyright The Council for the Central Laboratory of the research Councils, Daresbury Laboratory at Daresbury, Nr. Warrington (1996).

References

[1] Howe, J.M. (1997) *Interfaces in Materials*. New York: Wiley.

[2] Laird, B.B., Haymet A.D.J. *Chem. Rev.*, 1992, **92**, p. 1819.

[3] Tersoff, J. *Phys. Rev. B*, 1988, **38**, p. 9902.

[4] Stillinger, F.H. and Weber, T.A. *Phys. Rev. B*, 1985, **31**, p. 5262.

[5] Yoo, S., Zeng, X.C and Morris, J.R. *J. Chem. Phys.*, 2004, **120**, p. 1654.

[6] Alfè, D., Gillan, M.J. and Price, G.D. *J. Chem. Phys.*, 2002, **116**, p. 6170.

[7] Belonoshko, A.B., Ahuja, R. and Johansson, B. *Phys. Rev. Lett.*, 2000, **84**, p. 3638.

[8] Hoyt, J.J., Asta, M. and Karma, A. *Phys. Rev. Lett.*, 2001, **86**, p. 5530.

[9] Morris, J.R. *Phys. Rev. B*, 2002, **66**, p. 144104.

[10] Hoyt, J.J. and Asta, M. *Phys. Rev. B*, 2002, **65**, p. 214106.

[11] Morris, J.R. and Song, X. *J. Chem. Phys.*, 2003, **119**, p. 3920.

[12] Davidchack, R.L. and Laird, B.B. *J. Chem. Phys.*, 2003, **118**, p. 7651.

[13] Broughton, J.Q. and Gilmer, G.H., *J. Chem. Phys.*, 1986, **84**, p. 5759.

[14] Petrenko, V.F. and Whitworth, R.W. (1999) *Physics of Ice*. Oxford: University Press.

[15] Karim, O.A. and Haymet, A.D.J. *Chem. Phys. Lett.*, 1987, **138**, p. 531.

[16] Nada, H. and Furukawa, Y. *Jpn. J. Appl. Phys.*, 1995, **34**, p. 583.

[17] Hayward, J.A. and Haymet, A.D.J. *J. Chem. Phys.*, 2001, **114**, p. 3713.

[18] Bàez, L. and Clancy, P. *J. Chem. Phys.*, 1995, **103**, p. 9744.

[19] Bryk, T., Haymet, A.D.J. *J. Chem. Phys.*, 2002, **117**, p. 10258.

[20] Nada, H. and van der Eerden, J.P.J.M. *J. Chem. Phys.*, 2003, **118**, p. 7401.

[21] Workman, E.J. and Reynolds, S.E. *Phys. Rev.*, 1950, **78**, p. 254.

[22] Shibkov, A.A., Golovin, Yu.I., Zheltov, M.A., Korolev, A.A. and Leonov, A.A. *J. Cryst. Growth*, 2002, **236**, p. 434.

[23] Bronshteyn, V.L. and Chernov, A.A. *J. Cryst. Growth*, 1991, **112**, p. 129.

[24] Gertner, B.J. and Hynes, J.T. *Science*, 1996, **271**, p. 1563.

[25] Clary, D.C. and Wang, L. *Faraday Trans.*, 1997, **93**, p. 2763.

[26] Svanberg, M., Pettersson, J.B. and Bolton, K. *J. Phys. Chem. A*, 2000, **104**, p. 5787.

[27] Toubin, C., Picaud, S., Hoang, P.N.M., Girardet, C., Lynden-Bell, R.M. and Hynes, J.T. *J. Chem. Phys.*, 2003, **118**, p. 9814.

[28] Benjamin, I. *Science*, 1993, **261**, p. 1558.

[29] Marcus, R.A. *J. Chem. Phys.*, 2000, **113**, p. 1618.

[30] Knipping, E.M., Lakin, M.J., Foster, K.L., Jungwirth, P., Tobias, D.J., Gerber, R.B., Dabdub, D. and Finlayson-Pitts, B.J. *Science*, 2000, **288**, p. 301.

[31] Dang, L.X. and Chang, T.-M. *J. Phys. Chem. B*, 2002, **106**, p. 235.

[32] Dang, L.X. *J. Phys. Chem. B*, 2002, **106**, p. 10388.

[33] Berendsen, H.J.C., Grigera, J.R. and Straatsma, T.P. *J. Phys. Chem.*, 1987, **91**, p. 6269.

[34] Smith, D.E. and Dang, L.X. *J. Chem. Phys.*, 1994, **100**, p. 3757.

[35] http : //www.dl.ac.uk/TCS/Software/DL_POLY/

[36] Frenkel, D. and Smith, B. (2002) *Understanding Molecular Simulation.* San Diego: Academic Press.

[37] Smith, E.J., Bryk, T. and Haymet, A.D.J. *J. Chem. Phys.*, 2004, (submitted)

[38] Arthur, J.W. and Haymet, A.D.J. *J. Chem. Phys.*, 1998, **109**, p. 7991.

[39] Heyes, D.M. *Phys. Rev. B*, 1994, **49**, p. 755.

[40] Bryk, T. and Haymet, A.D.J. *Mol. Simul.*, 2004, **30**, p. 131.

[41] Gay, S.C., Smith, E.J. and Haymet, A.D.J. *J. Chem. Phys.*, 2002, **116**, p. 8876.

[42] Arbuckle, B.W. and Clancy, P. *J. Chem. Phys.*, 2002, **116**, p. 5090.

[43] Morris, J.R., Wang, C.Z., Ho, K.M. and Chan, C.T. *Phys. Rev. B*, 1994, **49**, p. 3109.

[44] Gao, G.T., Zeng, X.C. and Tanaka, H. *J. Chem. Phys.*, 2000, **112**, p. 8534.

[45] Mantz, Y.A., Geiger, F.M., Molina, L.T., Molina, M.J. and Trout, B.C. *J. Chem. Phys.*, 2000, **113**, p. 10733.

[46] Sanz, E., Vega, C., Abascal, J.L.F. and MacDowell, L.G. *Phys. Rev. Lett.*, 2004, **92**, p. 255701.

[47] Karim, O.A. and Haymet, A.D.J. *J. Chem. Phys.*, 1988, **89**, p. 6889.

[48] Broughton, J.Q. and Gilmer, G.H. *Acta Metall.*, 1983, **31**, p. 845.

[49] Shuttleworth, R. *Proc. Phys. Soc. London Sect. A*, 1950, **63**, p. 444.

[50] Wolf, D. *Phys. Rev. Lett.*, 1993, **70**, p. 627.

[51] Sibug-Aga, R. and Laird, B.B. *J. Chem. Phys.*, 2002, **116**, p. 3410.

[52] Alejandre, J., Tildesley, D.J. and Chapela, G.A. *J. Chem. Phys.*, 1995, **102**, p. 4574.

[53] Harris, J.G. *J. Phys. Chem.*, 1992, **96**, p. 5077.

[54] Lindahl, E. and Edholm, O. *J. Chem. Phys.*, 2000, **113**, p. 3882.

[55] Hardy, J.H. *Phil. Mag.*, 1977, **35**, p. 471.

[56] Bryk, T., Haymet, A.D.J. *J. Molec. Liq.*, 2004, **112**, p. 47.

[57] Wilson, M.A., Pohorille, A. and Pratt, L.R. *J. Chem. Phys.*, 1988, **88**, p. 3281.

[58] Wilson, M.A., Pohorille, A. and Pratt, L.R. *J. Chem. Phys.*, 1989, **90**, p. 5211.

PROTON TRANSPORT IN POLYMER ELECTROLYTE FUEL CELL MEMBRANES

An overview over the recent experimental, theoretical and simulation studies

E. Spohr
Institut für Werkstoffe und Verfahren
der Energietechnik (IWV–3)
Forschungszentrum Jülich
D–52425 Jülich, Germany

Abstract   At the heart of every fuel cell there is an electrolyte separating anode and cathode compartments. In low and intermediate temperature fuel cells its principal purpose is *efficient* and *selective* transport of anodically generated protons to the cathode where they can combine with oxygen ions to form water. The electrolyte in many current fuel cell designs such as the hydrogen-fuelled polymer electrolyte membrane fuel cell and the methanol-fuelled direct methanol fuel cell consists of a thin membrane of a polymer electrolyte in which protons are the only mobile charge carriers. Current membrane materials are *efficient* proton conductors but due to their high permeability for water and methanol also quite *unselective*. Understanding the mechanisms and bottlenecks of proton transport in such materials is thus key for the design of improved materials which are needed for introduction of fuel cells as a power supply for electrical appliances. It is the purpose of this chapter to give an overview over the recent literature of experimental, theoretical and simulation studies on proton transport in polymer electrolyte membranes.

Keywords: Proton transport, polymer electrolyte membranes, fuel cells, molecular dynamics simulations, atomistic modelling

1. Introduction

As a consequence of their high thermodynamic efficiency, low temperature fuel cells such as the hydrogen-powered polymer electrolyte fuel cell (PEFC) or the methanol-powered direct methanol fuel cell (DMFC) are considered as a most promising technology to compete and eventually replace batteries or the combustion engine in mobile and portable applications. Fuel cells transform the chemical energy of their fuel to electrical energy by separating the anodic

D. Henderson et al. (eds.), Ionic Soft Matter: Modern Trends in Theory and Applications, 361–379.
© 2005 *Springer. Printed in the Netherlands.*

oxidation reactions ($H_2 \rightarrow 2H^+ + 2e^-$ in the PEFC and $CH_3OH + H2O \rightarrow CO_2 + 6H^+ + 6e^-$ in the DMFC, respectively) from the cathodic reductive recombination reaction ($O_2 + 4e^- + 4H^+ \rightarrow 2H_2O$) with a proton conducting membrane.

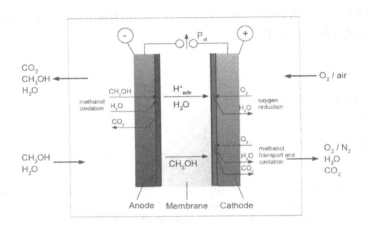

Figure 1.   Sketch of a low temperature fuel cell.

Figure 1 shows a sketch of a PEFC or DMFC, which consists of a central polymer electrolyte membrane (PEM) which is sandwiched by the cathode and the anode. Each electrode consists of a thin chemical reaction zone, the catalyst layer, and a thicker diffusion layer, whose purpose is fast transport of fuel and oxygen to the reaction zone and of reaction products away from the reaction zone. This membrane-electrode assembly (MEA) is pressed between two electronically conducting plates made out of carbon or metal, into which flowfields (channel-like structures of varying form) are worked. These are responsible for delivering fuel and oxygen homogeneously across the entire cell and for removing the cathodically produced water and the anodically produced CO_2 (in the DMFC) from the reaction zone. Reviews on fuel cell science and technology can be found, e. g., in [1–4]

The high theoretical efficiency of a fuel cell is substantially reduced by the finite rate of dynamic processes at various locations in the cell. Substantial efficiency losses at typical operating temperatures occur already in the anodic and cathodic catalyst layers due to the low intrinsic reaction rates of the oxygen reduction and, in the case of the DMFC, of the methanol oxidation reaction. (The catalytic oxidation of hydrogen with platinum catalysts is very fast and thus does not limit PEFC performance.) In addition, at low temperatures, turnover may be limited by noble metal catalyst poisoning due to sulfur impurities (in hydrogen produced from a reforming reaction) or carbon monoxide which binds rather tightly particularly to the platinum surface. Currently, the standard solution to this problem consists in rather high, and thus expensive, catalyst loadings. Various efforts are under way to develop more active and more CO tolerant catalysts increase its specific surface, to increase the turnover by making the catalyst more accessible to the reactants, to deposit the catalyst electrochemically directly in the carbon/Nafion interphase and to decrease aging (see, e. g., [2, 3, 5–22] and references therein).

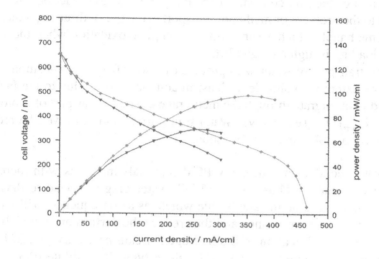

Figure 2. Current vs. Voltage curves of DMFC fuel cells with Nafion membranes and different amounts of black (pure metal) catalysts [96].

These efficiency losses due to the finite velocity of the chemical reactions manifest themselves as reaction overpotentials and lead already to a steep decrease of the cell voltage (and thus of efficiency) when only small currents are drawn from the cell. Figure 2 shows typical current-voltage curves from measurements of a direct methanol fuel cell. The slope of the intermediate regime with an almost linear decrease of voltage as a function of current, the Ohmic regime, is largely a consequence of the ohmic resistance of the cell. In addition to contact resistances between membrane, catalyst, carbon (in the diffusion layer) and contact in the flow field, this resistance is determined substantially by the proton conductance of the PEM material. Contact resistances are reduced by hotpressing (thermal tempering) of the MEA prior to its use, and the membrane resistance can be reduced by making the membrane as thin as possible [23, 24]. It cannot, however be made arbitrarily thin (typical thickness is between 50–200 μm) because otherwise it may fail mechanically when pressed between current collectors and flow fields.

Even more important is the fact that most PEM materials (see below) are also quite permeable for water and methanol. Thus, thin membranes lead to substantial transport of these molecules from the anode side to the cathode (e. g., [25–29]). The permeation of methanol in the DMFC is undesirable for the obvious reason that it reduces the cell power ("mixed potential formation"), because no electrical work is generated in a cathodic oxidation reaction. Furthermore methanol on the cathode is unfavorable because it can block adsorption sites needed for the oxygen reduction reaction. The presence of methanol may even alter the rate constant of the oxygen reduction reaction. A typical solution to the problem of methanol transport is the use of dilute aqueous solutions of methanol, which assures almost complete oxidation when the anodic catalyst loading is high enough [30].

The co-transport of water with protons (the so-called electro-osmotic drag) is to some extent unavoidable, at least in aqueous media, due to the fact that protons do not migrate through the membrane as such but as part of protonated clusters $H_{2n+1}O_n^+$ (see below). Water transport should, however, be reduced as much as possible for several reasons:

(i) proton conductance in many PEM materials increases with increasing water content. Thus, in the PEFC, water drag will lead to drying of the membrane on the anode side which, as a consequence, will increase the membrane resistance and thus reduce the performance, which often makes humidification of the hydrogen stream necessary [31, 32]. This is not a problem for the DMFC, since typically a dilute (0.5 to 2 m) solution of methanol in water is used as fuel;

(ii) due to the small solubility of oxygen in water, excess water in the cathode compartment in addition to the oxygen water will block oxygen access to the catalyst ("flooding"), thereby reducing further the efficiency of the cell [33–36];

(iii) a rising water transport from the anode to the cathode disturbs the water and heat balance of the cell.

Often it is necessary to transport excess water from the cathode side to the anode side, where it is used either for humidification of the hydrogen stream in the PEFC or to dilute the methanol fuel. (In order to increase energy and power density, methanol, while being used in solution, will usually be stored as the pure liquid.) Water management needs additional aggregates or devices. This adds to the cost of the fuel cell system and further reduces its efficiency. All these transport limitations give rise to "diffusion overpotentials" which lead to the rapid breakdown of the cell at high current densities in Fig. 2.

This introduction illustrates that the membrane is one key material component of a fuel cell whose properties limit the achievable performance. The ideal fuel cell membrane should

(i) possess a high proton conductance,

(ii) possess low conductivity for anions and low permeability for water, methanol, hydrogen gas and air,

(iii) be mechanically stable and chemically stable against acidic, oxidative and reductive environments,

(iv) and, last but not least, inexpensive.

Substantial efforts are being invested world-wide into the design and development of such a "dream membrane". Understanding the mechanisms and bottlenecks of proton transport in existing materials is one necessary prerequisite for the design of improved materials. Computer simulation and theory contribute significantly to this endeavor.

In the following sections, the key properties of PEM materials used in fuel cells are briefly reviewed and theoretical and simulation results on proton transport are discussed.

2. Polymer electrolyte membrane materials and their properties

While several homogeneous electrolytes such as phosphoric and sulfuric acid or $CsHSO_4$ have been used in commercial and laboratory fuel cells, most modern low temperature fuel cells employ adducts of polymers with oxo-acids such as poly-benzimidazole/phosphoric acid (PBI/H_3PO_4) or polymers functionalized with sulfonic acid groups. In these materials the electrolyte is confined within the polymer phase. The polymer/oxo-acid adducts suffer from slow degradation due to leaching out of the acid component, a problem which does not exist with the covalently bound sulfonic acid groups, where the electrolyte phase consists of hydrated protons which are formed through dissociation of the strong sulfonic acids in the presence of absorbed water. Reviews of these materials can be found, e. g., in [37–44].

Among the sulfonated membrane materials are perfluorinated polymers with sulfonated perfluorinated side chains such as Nafion (DuPont®), Dow (Dow Chemicals®), Gore(Gore ®) and Acipex (Asahi®) and stochastically sulfonated polymeric phenylene-ether-keto compounds (such as $(CO–C_6H_4–O–C_6H_4)_n$ (S-PEK), $(CO–C_6H_4–O–C_6H_4–O–C_6H_4)_n$ (S-PEEK), $(CO–C_6H_4–CO–C_6H_4–O–C_6H_4–O–C_6H_4)_n$ (S-PEEKK) etc.) [45–47]. These polymers combine non-polar backbones with the high hydrophilicity of the sulfonic acid functional groups, which leads to hydrophilic/hydrophobic separation on the nanometer

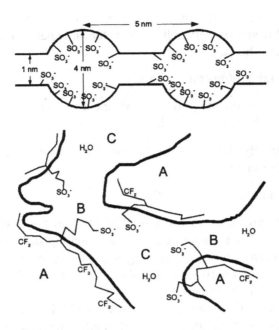

Figure 3. Two models describing the microphases of swollen Nafion membranes. Top: Gierke's [48] suggestion of aqueous inverse spherical micelles connected by water-filled cylindrical channels. Bottom: Yeager and Steck's [49] three-region model of a water/ionomer mixture without regular structure. Regions A, B and C are the hydrophobic polymer, the solvent bridges and the hydrophilic regions, respectively.

scale. All materials absorb water when exposed to liquid water or water vapor, and the sulfonic acid groups become hydrated in the aqueous domains. With increasing water content (Nafion 117, e. g., absorbs at room temperature more than 14 water molecules per sulfonic acid group) the aqueous domains grow and at some point merge, thereby reaching the percolation threshold at which the aqueous phase becomes continuous; protons dissociate from the sulfonic acid groups and the material becomes proton conducting.

Nafion and other fluorinated ionomers are copolymers of tetrafluorethylene (TFE) and perfluorinated vinyl ethers

$$CF = CF_2$$
$$|$$
$$O\text{-}(CF_2\text{-}CF\text{-}O)_n\text{-}(CF_2)_m\text{-}SO_3H$$
$$|$$
$$CF_3.$$

Nafion ($n = 1$, $m = 2$) is most frequently used in fuel cell applications. Small angle X-ray (SAXS) and neutron scattering (SANS) data were compiled into structural models. Figure 3 shows two such structural models, the inverted

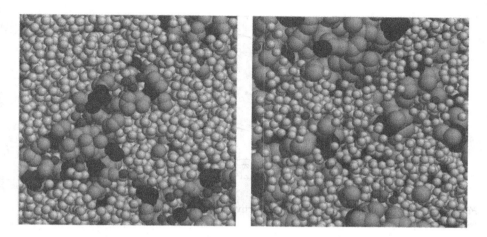

Figure 4. Snapshots from simulations of water/Nafion systems [76]. Left: $\lambda = 5$ (160 H_2O (lightgray), 40 H_3O^+ (gray) and 2 Nafion-20-meres (white, sulfonate groups gray/black). Right: $\lambda = 10$ (360 H_2O, 40 H_3O^+ and 2 Nafion-20-meres).

spherical micelle model by Gierke [48] and the more disordered three-region model by Yeager et al [49]. Simulation snapshots of water/Nafion systems at intermediate water content (Fig. 4) show irregularly shaped aqueous domains which grow together with increasing water content.

By changing the ratio of vinyl ether and TFE, Nafion polymers with varying equivalent weight (molecular weight per sulfonic acid group) can be produced. If the ratio is about 1:6.5, a polymer with an equivalent weight of about 1100 g/mol is obtained. A widely used membrane is Nafion **117** with an equivalent weight of **1100** and a thickness of 7/1000 of an inch (about 175 μm). Similarly, by changing the degree of stochastic sulfonation in the phenylene group, the equivalent weight can be manipulated in the phenylene-ether-keto polymers. Proton conductivity as well as water uptake increases with decreasing equivalent weight ([44] and references therein). At equivalent weights lower than 800 g the two-phase system becomes gel-like and looses its structural integrity, thereby becoming unusable as fuel cell membrane. Figure 5 shows the s-shaped dependence of proton conductivity on the water content $\lambda = n_{H_2O}/n_{SO_3}$. At low water content the conductivity increases slowly due to the fact that the path length for diffusion is rather long and hindered by dead ends. At high water content the conductivity levels off at rather high values.

The thermal activation energy of proton conductivity at intermediate and high water content is similar to that in dilute aqueous acidic solutions (see Fig. 6). From this similarity it has been concluded that proton transfer in PEMs proceeds, like in the bulk, according to the Grotthus structural diffusion mechanism (see below). For decreasing water content a very slight trend towards

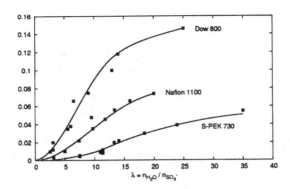

Figure 5. Room temperature proton conductivities of several membranes. Data compiled from [97–99].

higher activation energies is visible in Fig. 6; other groups have observed drastically higher activation energies in dry membranes [23].

3. Theory and computer simulation of proton transport in membranes

Atomistic modelling of proton transport in PEMs has to deal with a wide range of relevant time and length scales. On the one hand, the membrane is not stationary but dynamic. Shape and size of the aqueous proton-conducting pores change on the time scale of nanoseconds to microseconds or even longer times. A fully adequate description of such processes requires coarse-graining simulation methods that allow very long simulation times. On the other hand, both experiments [23, 50] and theory [51–53] indicate that proton transport in wet PEM membranes with a water content that is characteristic for an operating fuel cell membrane occurs to a large extent according to the so-called Grotthus mechanism [54] of structural diffusion. In this non-classical transport mechanism a proton from a hydronium ion H_3O^+ is transferred to a neighboring water molecule via intermediate adducts of composition $H_5O_2^+$ (Zundel ions). This very efficient transport mechanism is active on the time scale of femto- and picoseconds and, in principle, requires a detailed description of the electronic structure and dynamics of proton defects in water. The essential validity of the mechanism in aqueous solutions has been demonstrated recently [55–60] using the Car-Parrinello method for the simultaneous description of electronic and atomic dynamics. While providing a very realistic description of local properties, Car-Parrinello simulations are, due to their large computational requirements, viable presently only for relatively small systems of up to a few hundred atoms and over times of the order of 10 ps.

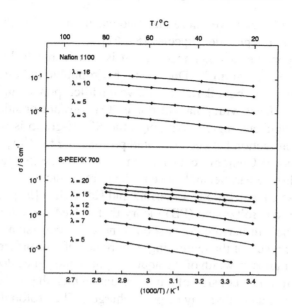

Figure 6. Proton conductivity of Nafion and s-PEEKK as a function of temperature and water content λ (Data courtesy of K.-D. Kreuer [50].)

In light of this problem of scales, to date most efforts to calculate proton transport coefficients have been based on the assumption that proton transport in a *single pore* is characteristic for proton transport in the *entire membrane*, thereby decoupling the scales of motion.

The high proton diffusion coefficient in bulk aqueous solutions of strong acids is considered to be the result of a combination of two transport mechanisms: the classical ion diffusion mechanism (vehicle mechanism) and the so-called Grotthus structural diffusion mechanism. Based on the size and hydration properties of the hydronium ion one would expect its diffusion coefficient to be similar to that of Na^+ and K^+. The significant excess of proton mobility above that of these alkali ions is considered to be due to the decoupling between proton transport and charge transport. In the Grotthus scenario, the smallest hydrogen ion unit is a hydronium ion H_3O^+. If one OH bond in this ion is elongated, the proton can be transferred to the neighboring water molecule (which initially forms a hydrogen bond). The intermediate is a so-called Zundel ion $H_5O_2^+$ in which the charge is delocalized between two water molecules. In the next step another proton from the newly-formed H_3O^+ cation can then be transferred to another hydrogen-bonded water molecule. As a result, individual protons move about 0.4 Å while the center of charge moves about 2.5 Å in one step. Although detailed analysis shows that the positive charge can be delocalized over more than two water molecules, the H_3O^+ and the $H_5O_2^+$ ion are the prevalent species [59, 61].

In the PEM pore yet another mechanism is considered to be active: the so-called "surface mechanims". Due to the hydrophobic nature of the PEM backbone polymer, the boundaries of the water-filled pores can be regarded as hydrophobic surfaces with polar sulfonate SO_3^--groups or side chains protruding into the aqueous region. Although the SO_3H-group is a strong acid, the electrostatic attraction between solvated protons and the SO_3^--groups can lead to an additional Coulomb contribution to the activation energy of those protons which diffuse parallel and close to the pore surface.

Based on these considerations, Eikerling and Kornyshev [53] showed on the basis of the Poisson-Boltzmann (PB) theory that indeed the activation energy for proton transport in narrow pores should increase (consistent with experiment [23]) as the result of the increasing importance of the surface mechanism with decreasing radius or width of the aqueous pores and hence decreasing water content. Their model consists of slit pores onto which the sulfonate ions are attached in a rigid quadratic array of point charges. The proton distribution is calculated via Poisson-Boltzmann theory. The authors proceed to calculate the effective activation energy for protons as a function of distance from the pore surface and obtain an effective temperature dependence of proton transport by convoluting activation energies and proton distribution. While the model is able to predict the pore size dependence of the activation energy, there is no intrinsic dynamic description of proton transport. Hence, numerical estimates of proton diffusion coefficients or conductivities cannot be obtained.

In order to obtain more insight into the dynamics, molecular dynamics (MD) simulations of similar slit systems have been performed [62–64]. Since ab initio MD methods are not applicable, an effort was made to incorporate the Grotthus mechanism into the MD simulations via a simplified empirical valence bond (EVB) model [65].

The EVB formalism [66] allows the making and breaking of bonds in a classical MD force field. The adiabatic ground state is described as a superposition of two or more diabatic states, each of which represents a specific arrangement of chemical bonds. In the case of a proton defect this bond arrangements corresponds, e. g., in a $H_5O_2^+$ complex to two different states, state 1, $H_2O - H \cdots OH_2$, and state 2, $H_2O \cdots H - OH_2$, with different bond topologies($-$ denotes a covalent bond, $\cdots$ denotes a hydrogen bond). These diabatic states are described via classical force fields. Several such EVB models for proton transfer in bulk water have been developed using a larger basis of up to 20 states (corresponding to a delocalization of the proton defect over a large cluster of water molecules) [67–71]. In the simulations of proton transport in PEM pores, a simplified two-state version has been used. In this version the two states are coupled by a local coupling function V_{12} which is, like the diabatic state force field functions, a continuous function of the atomic positions in the $H_5O_2^+$ ion. Together with an empirical charge-switching function, this EVB model makes the simulation of systems with high proton concentrations possible by decoupling the individual protons. In the course of a simulation there is a continuous transition between H_3O^+ and $H_5O_2^+$ ions. Further details can be found in [62, 65].

In a first series of studies [62] proton transport through the aqueous phase has been studied with the goal to identify those important physical and chemical characteristics of the polymer that influence proton transport most strongly. In this study, the polymer material is not described atomistically but as an excluded volume for the aqueous phase, which is modelled as slab pore. The pores have fixed volume and shape. The influence on proton mobility of the charge delocalization inside the SO_3^- groups (related to the acid strength of the SO_3H group), of the headgroup density (related to the equivalent weight of the polymer), and of the motion of headgroups and sidechains has been studied. It was found that the factors that facilitate proton transfer the most are acid strength of the polymer, (i. e., as wide as possible charge delocalization of the

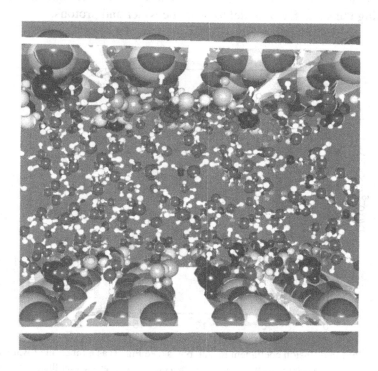

Figure 7.   Snapshot from a slab simulation with SO_3^- groups tethered to a smooth solid wall.

negative charge), fluctuational motions of the headgroup and the side chains, and the water content (thickness) of the pore.

A slab pore model with an atomistic description of the SO_3^- groups only, which are themselves attached via harmonic tethers on the (atomically smooth) pore surface (see Fig. 7) efficiently allows systematic simulation studies of proton transport as a function of temperature, water content, and sulfonate surface density [63]. Proton mobility is characterized as the self-diffusion coefficient of the proton defects calculated from the mean square displacement in the two directions parallel to the slab wall. Figure 8 shows the variation of proton diffusion coefficient with water content for 4 different temperatures. The model correctly describes the experimental observations (see [1] for an overview) that proton mobility is very small at low water content, then increases strongly with water content and finally levels off at high water content. In the simulations, proton diffusion coefficients can hardly be calculated for the water content $\lambda = n_{H_2O}/n_{SO_3^-}$ below about 3, because of increasing statistical uncertainties and alos because the EVB model is not applicable down to such low concentrations where SO_3H states would need to be introduced into the scheme. At large water content ($10 < \lambda < 20$) the proton mobility becomes almost constant at about 50 % of the proton mobility in diluted bulk aqueous solution, when using the same EVB model to describe water and protons.

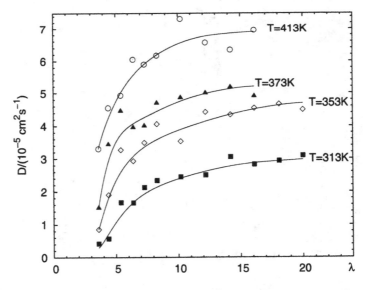

Figure 8.   Proton diffusion coefficient in slab pores (see Fig. 7) as a function of water content λ for 4 different temperatures as indicated. The density of SO_3^- groups on the pore surfaces is $1/58$ Å^2.

From these data is also possible to calculate the activation energy in pores of given water content from an Arrhenius representation of the calculated diffusion coefficients [63]. The results show no indication of a substantial increase in activation energy at water levels as low as $\lambda = 4$, contrary to the results in [23] and more in line with results from Kreuer (see Fig. 6 and [50]). Rather, the activation energy for proton transfer remains low at all studied water contents. Thus, it appears that in an individual pore the Grotthus mechanism remains rate-determining whereas the surface mechanism does not dominate the transport behavior. Similar results as in slab pores were also found in cylinder pores [63].

The apparent discrepancy between the MD simulations and the PB theory can be resolved when employing a more realistic description of the surface charge density in the PB calculations. In [63] we showed that with a more realistic atomic description of the sulfonate group (charge smearing via a form factor) and its motion (via a Debye-Waller factor) *and* by considering the fact that the proton complex has a finite size, PB results come in line with the MD simultaiotns and do not exhibit the increase in activation energy in narrow pores that was found with the simple model [53].

We conclude from our results that the most likely scenario for an increased activation energy is the occurence of fluctuative bridging: at very low water content a situation must arise when there is insufficient water for the aqueous pores to form a percolating network structure. In this case, fluctuations in polymer structure will lead to temporary bridges between isolated water regions which will give rise to macroscopic proton conductance. The rate determining step in such a scenario is, however, not the local Grotthus transport but the polymer motion with larger activation energies.

Several attempts have been made to simulate transport in realistic fully atomistic MD simulations of water/Nafion mixtures. Vishnyakov and Neimark [72–74] investigated alkali transport in aqueous and methanolic solution (and in mixed solvents) in the presence of Nafion. They found indications for the existence of the fluctuative bridging mechanism. The group by Kokhlov and Khalatur has also performed extensive yet unpublished studies of simple ion transport in Nafion. Goddard and coworkers [75] compared structural and dynamical properties of two different copolymerisation patterns, in order to estimate the effect of statistical vs. regular copolymerisation of TFE with the sulfonated vinyl ether.

Our group has combined the EVB model with a complete atomistic description of the Nafion subphase [76]. The pore structure obtained (see Fig. 4) supports the disordered Nafion model by Yeager and Steck [49] more than the symmetrical one by Gierke [48]. Activation energies obtained for proton transport at $\lambda = 5$ and $\lambda = 10$ are similar to the ones discussed above for the simpler pore model. This is not unexpected since the total simulation time of only a few nanoseconds does not allow to observe the possible effect of fluctuative bridging. We subsequently rigidified the Nafion polymer and also employed rigid water and hydronium models [77, 78] thus being able to extend the simulation times to several tens of nanoseconds. At $\lambda = 5$ we indeed observe polymer motions that lead to the opening and closing of aqueous channels. However, these events are too rare to allow an accurate determination of their rate over the simulation time; obtaining the activation energy from the temperature dependence of these rates is impossible from our simulation data.

Going beyond an atomistic description of the aqueous phase and the membrane, Paddison and coworkers [79–88] employed statistical mechanical models, incorporating solvent friction and spatially dependent dielectric properties, to the calculation of the proton diffusion coefficient in Nafion and PEEKK membrane pores. They concluded from their studies that, in accordance with NMR based evidence [50], the mechanism of proton transport is more vehicular (classical ion transport) in the vicinity of the pore surface and more Grotthus-like in the center.

4. Proton generation

Paddison, Eikerling and coworkers performed quantum chemical calculations and ab initio MD simulations of crystal hydrates [89–91]. The studies reveal the strong influence of the acid functional group on proton mobility at low humidity; they show how chemical modifications of side chain entities can make the dissociation of the acid more complete.

As has been mentioned above, contact resistances between catalyst layer and membrane and the intrinsic speed of the electrocatalytical reactions also influence fuel cell performance significantly. Thus, it is of importance to study mechanism and reaction rates for the electrooxidation of methanol on Pt and Pt/Ru catalysts which are commonly used in low temperature fuel cells. Recently we employed the ab initio MD method on the basis of gradient-corrected quantum-mechanical density functional theory to study the mechanism of the initial steps of methanol oxidation at the Pt/water interface [92]. We observed a mechanism substantially different from the proposed gas phase mechanism, which requires, after initial dissociation of a C–H bond, the molecular reorientation before the OH bond can be cleaved [93, 94]. On the positively charged Pt surface in the presence of water the reaction cascade appears to be initiated by the formation of a hydrogen bond between a water molecule and the OH group of a methanol molecule which is absorbed on the surface via the methyl group. After formation of the hydrogen bond, one of the C–H bonds is cleaved and very rapidly the hydrogen atom from the OH group becomes shared between the adsorbed CH_2OH and the water molecule in an analogous way as in the Zundel complex $H_5O_2^+$. This proton is then rapidly carried away into the liquid via a sequence of Grotthus steps, without any contact with the metal surface. An adsorbed formaldehyde molecule remains on the surface and does not react any further during the simulation time, in agreement with experimental evidence pointing towards a stepwise oxidation mechanism with desorption and readsorption of intermediates formaldehyde and formic acid [95].

5.   Summary and outlook

Theory and atomistic MD simulations have been extensively used to study the transport of simple ions and protons in polymer electrolyte membrane models. The continuous making and breaking of OH bonds characteristic of the Grotthus mechanism has been described successfully using empirical valence bond models. Due to the large separation of time and length scales between proton and polymer dynamics most models are rather generic in character and describe general features of the polymer rather than specific chemical compounds. Nevertheless, an understanding of the relative importance and the interplay between the essential microscopic mechanism, simple ion transport, non-classical Grotthus transport and surface transport is evolving. Theory and simulation will become particularly useful for fuel cell technology if it also becomes possible, among others, to account for the differences between polymers with different chemical structures, to understand the molecular origin and predict the amount of electro-osmosis (water co-transport) and methanol crossover, to describe proton conductance in inorganic/organic compound membranes, to rationalize the influence of surface treatment of the membrane, or to give insight into the electrocatalytic processes in the complex environment of an operating low temperature fuel cell. Efforts along several of these directions are already under way, and progress in the molecular understanding of operating fuel cell membranes and membrane-electrode assemblies can be expected in the near future.

Acknowledgements

The author is thankful to K.-D. Kreuer and H. Schmitz for providing the data in Fig. 6 and Fig. 2. The John von Neumann Institute for Computing in Jülich supported the MD simulations by generous grants of computer time over the last years.

References

[1] Gottesfeld, S., and Zawodzinski T.A. (1997) *Advances in Electrochemical Science and Engineering*, (eds. Alkire R.C., Gerischer, H., Kolb D.M., and Tobias, C.W.), vol. 5, p. 195–301. Weinheim: Wiley-VCH.

[2] Carrette, L., Friedrich, K.A., and Stimming, U. *ChemPhysChem*, 2000, **1**, p. 162.

[3] Carrette, L., Friedrich, K.A., and Stimming, U. *Fuel Cells*, 2001, **1**, p. 5.

[4] Litster, S., and McLean, G. *J. Power Sources*, 2004, **130**, p. 61.

[5] Thomas, S.C., Ren, X., and Gottesfeld, S. *J. Electrochem. Soc.*, 1999, **146**, p. 4354.

[6] Dinh, H.N., Ren, X., Garzon, F.H., Zelenay, P., and Gottesfeld, S. *J. Electroanal. Chem.*, 2000, **491**, p. 222.

[7] Tong, Y.Y., Rice, C., Wieckowski, A., and Oldfield, E. *J. Am. Chem. Soc.*, 2000, **122**, p. 1123.

[8] Passalacqua, E., Lufrano, F., Squadrito, G., Patto, A., and Giorgi, L. *Electrochim. Acta*, 2001, **46**, p. 799.

[9] Waszczuk, P., Wieckowski, A., Zelenay, P., Gottesfeld, S., Coutanceau, C., Léger, J.-M., and Lamy, C. *J. Electroanal. Chem.*, 2001, **511**, p. 55.

[10] Lasch, K., Hayn, G., Jörissen, L., Garche, J., and Besenhardt, O. *J. Power Sources*, 2002, **105**, p. 305.

[11] Roth, C., Martz, N., Hahn, F., Léger, J.-M., Lamy, C., and Fuess, H. *J. Electrochem. Soc.*, 2002, **149**, p. E433.

[12] Friedrich, K., Geyzers, K., Dickinson, A., and Stimming, U. *J. Electroanal. Chem.*, 2002, **524-525**, p. 261.

[13] Neergat, M., Leveratto, D., and Stimming, U. *Fuel Cells*, 2002, **2**, p. 25.

[14] Lu, C., Rice, C., Masel, R.I., Babu, P.K., Waszczuk, P., Kim, H.S., Oldfield, E., and Wieckowski, A. *J. Phys. Chem. B*, 2002, **106**, p. 9581.

[15] Khazova, O.A., Mikhailova, A.A., Skundin, A.M., Tuseeva, E.K., Hvránek, A., and Wippermann, K. *Fuel Cells*, 2002, **2**, p. 99.

[16] Ralph, T.R., and Hogarth, M.P. *Platinum Metals Rev.*, 2002, **46**, p. 3.

[17] Ralph, T.R., and Hogarth, M.P. *Platinum Metals Rev.*, 2002, **46**, p. 117.

[18] Ralph, T.R., and Hogarth, M.P. *Platinum Metals Rev.*, 2002, **46**, p. 146.

[19] Diemant, T., Hager, T., Hoster, H., Rauscher, H., and Behm, R. *Surf. Sci.*, 2003, **541**, p. 137.

[20] Löffler, M.-S., Natter, H., Hempelmann, R., and Wippermann, K. *Electrochim. Acta*, 2003, **48**, p. 3047.

[21] Gasteiger, H., Panels, J., and Yan, S. *J. Power Sources*, 2004, **127**, p. 162.

[22] Taniguchi, A., Akita, T., Yasuda, K., and Miyazaki, Y. *J. Power Sources*, 2004, **130**, p. 42.

[23] Cappadonia, M., Erning, J.W., Niaki, S.M.S., and Stimming, U. *Solid State Ionics*, 1995, **77**, p. 65.

[24] Kreuer, K.D. *J. Membrane Sci.*, 2001, **185**, p. 29.

[25] Zawodzinski, T.A., Davey, J., Valerio, J., and Gottesfeld, S. *Electrochim. Acta*, 1995, **40**, p. 297.

[26] Munichandraiah, N., McGrath, K., Prakash, G.K.S., Aniszfeld, R., and Olah, G.A. *J. Power Sources*, 2003, **117**, p. 98.

[27] Ramya, K., and Dhathathreyan, K. *J. Electroanal. Chem.*, 2003, **542**, p. 109.

[28] Kim, Y.J., Choi, W., Woo, S.I., and Hong, W.H. *J. Membrane Sci.*, 2004, **238**, p. 213.

[29] Gogel, V., Frey, T., Yongsheng, Z., Friedrich, K.A., Jörissen, L. and Garche, J. *J. Power Sources*, 2004, **127**, p. 172.

[30] Dohle, H., Schmitz, H., Mergel, J., and Stolten, D. *J. Power Sources*, 2002, **106**, p. 313.

[31] Liu, F., Yi, B., Xing, D., Yu, J., Hou, Z., and Fu, Y. *J. Power Sources*, 2003, **124**, p. 81.

[32] Chen, J., Matsuura, T., and Hori, M. *J. Power Sources*, 2004, **131**, p. 155.

[33] Zawodzinski Jr, T.A., Springer, T.E., Uribe, F., and Gottesfeld, S. *Solid State Ionics*, 1993, **60**, p. 199.

[34] Geiger, A.B., Tsukuda, A., Lehmann, E., Vontobel, P., Wokaun, A., and Scherer, G.G. *Fuel Cells*, 2002, **2**, p. 92.

[35] Tüber, K., Pócza, D., and Hebling, C. *J. Power Sources*, 2003, **124**, p. 403.

[36] Satija, R., Jacobson, D., Arif, M., and Werner, S. *J. Power Sources*, 2004, **129**, p. 238.

[37] Kreuer, K.-D. *Chem. Mater.*, 1996, **8**, p. 610.

[38] Savadogo, O. *J. New Mater. Eelctrochem. Syst.*, 1998, **1**, p. 47.

[39] Rikukawa, M., and Sanui, K. *Prog. Polym. Sci.*, 2000, **25**, p. 1463.

[40] Vielstich, W., Lamm, A., and Gasteiger, H.A. (2003). *Handbook of Fuel Cells – Fundamentals, Technology and Applications*. U.K., Chichester: Wiley.

[41] *Annu. Rev. Mater. Res.*, 2003, **33**, (Special issue, materials for fuel cells).

[42] Li, Q., Jenson, J.O., and Bjerrum, N.J. *J. Chem. Mater.*, 2003, **15**, p. 4896.

[43] Rozière, J., and Jones, D. *Annu. Rev. Mater. Res.*, 2003, **33**, p. 503.

[44] Kreuer, K.D., Paddison, S.J., Spohr, E., and Schuster, M. *Chem. Rev.*, 2004, dOI: 10.1021/cr020715f.

[45] Kreuer, K.D. *Solid State Ionics*, 1997, **94**, p. 55.

[46] Drioli, E., Regina, A., Casciola, M., Oliveti, A., Trotta, F., and Massari, T. *J. Membrane Sci.*, 2004, **228**, p. 139.

[47] Mikhailenko, S.D., Wang, K., Kaliaguine, S., Xing, P., Robertson, G.P., and Guiver, M.D. *J. Membrane Sci.*, 2004, **233**, p. 93.

[48] Gierke, T.D., Munn, G.E., and Wilson, F.C. *Journal of Polymer Science Part B-Polymer Physics*, 1981, **19**, p. 1687.

[49] Yeager, H.L., and Steck, A. *J. Electrochem. Soc.*, 1981, **128**, p. 1880.

[50] Kreuer, K.D. *Solid State Ionics*, 1997, **97**, p. 1.

[51] Eikerling, M., Kornyshev, A.A., and Stimming, U. *J. Phys. Chem. B*, 1997, **101**, p. 10807.

[52] Eikerling, M., Kornyshev, A.A., Kuznetsov, A.M., Ulstrup, J., and Walbran, S. *J. Phys. Chem.*, 2001, **105**, p. 3646.

[53] Eikerling, M., and Kornyshev, A.A. *J. Electroanal. Chem.*, 2001, **502**, p. 1.

[54] von Grotthus, C.J.D. *Ann. Chim.*, 1806, **LVIII**, p. 54.

[55] Tuckerman, M., Laasonen, K., Sprik, M., and Parrinello, M. *J. Chem. Phys.*, 1995, **103**, p. 150.

[56] Tuckerman, M., Laasonen, K., Sprik, M., and Parrinello, M. *J. Phys. Chem.*, 1995, **99**, p. 5749.

[57] Lobaugh, J., and Voth, G.A. *J. Chem. Phys.*, 1996, **104**, p. 2056.

[58] Tuckerman, M., Marx, D., Klein, M., and Parrinello, M. *Science*, 1997, **275**, p. 817.

[59] Marx, D., Tuckerman, M.E., Hutter, J., and Parrinello, M. *Nature*, 1999, **397**, p. 601.

[60] Marx, D., Tuckerman, M.E., and Parrinello, M. *J. Phys.: Cond. Matter.*, 2000, **12**, p. A153.

[61] Agmon, N. *Chem. Phys. Lett.*, 1995, **244**, p. 456.

[62] Spohr, E., Commer, P., and Kornyshev, A.A. *J. Phys. Chem. B*, 2002, **106**, p. 10560.

[63] Commer, P., Cherstvy, A.G., Spohr, E., and Kornyshev, A.A. *Fuel Cells*, 2002, **2**, p. 127.

[64] Spohr, E. *Mol. Simulation*, **30**, p. 107.

[65] Walbran, S., and Kornyshev, A.A. *J. Chem. Phys.*, 2001, **114**, p. 10039.

[66] Warshel, A., and Weiss, R.M. *J. Am. Chem. Soc.*, 1980, **102**, p. 6218.

[67] Sagnella, D.E., and Tuckerman, M.E. *J. Chem. Phys.*, 1997, **108**, p. 2073.

[68] Schmitt, U.W., and Voth, G.A. *J. Phys. Chem. B*, 1998, **102**, p. 5547.

[69] Vuilleumier, R., and Borgis, D. *Chem. Phys. Lett.*, 1998, **284**, p. 71.

[70] Cuma, M., Schmitt, U.W., and Voth, G.A. *Chem. Phys.*, 2000, **258**, p. 187.

[71] Day, T.J.F., Soudackov, A.V., Čuma, M., Schmitt, U.W., and Voth, G.A. *J. Chem. Phys.*, 2002, **117**, p. 5839.

[72] Vishnyakov, A., and Neimark, A.V. *J. Phys. Chem. B*, 2000, **104**, p. 4471.

[73] Vishynakov, A., and Neimark, A.V. *J. Phys. Chem. B*, 2001, **105**, p. 9586.

[74] Vishynakov, A., and Neimark, A.V. *J. Phys. Chem. B*, 2001, **105**, p. 7830.

[75] Jang, S.S., Molinero, V., Cağin, T., and Goddard III, W.A. *J. Phys. Chem. B*, 2004, **108**, p. 3149.

[76] Seeliger, D., Hartnig, C., and Spohr, E. 2004, (in preparation).

[77] Berendsen, H.J.C., Grigera, J.R., and Straatsma, T.P. *J. Phys. Chem.*, 1987, **91**, p. 6269.

[78] Fornili, S.L., Migliore, M., and Palazzo, M.A. *Chem. Phys. Lett.*, 1986, **125**, p. 419.

[79] Paddison, S.J., Paul, R., and Zawodzinski Jr, T.A. (1999). *Proton Conducting Membrane Fuel Cells II*, vol. 98–27 of *The Electrochemical Society Proceedings Series*, (eds. Gottesfeld, S., and Fuller, T.F.), p. 106–120. New Jersey: The Electrochemical Society.

[80] Paddison, S.J., Paul, R., and Zawodzinski Jr, T.A. *J. Electrochem. Soc.*, 2000, **147**, p. 617.

[81] Paddison, S.J., Paul, R., and Zawodzinski Jr, T.A. *J. Chem. Phys.*, 2001, **115**, p. 7753.

[82] Paddison, S.J., Paul, R., and Pivovar, B.S. (2001). *Direct Methanol Fuel Cells*, vol. 01–04 of *The Electrochemical Society Proceedings Series*, (eds. Narayanan, S., Gottesfeld, S., and Zawodzinski, T.A.), p. 8–13. New Jersey: The Electrochemical Society.

[83] Paddison, S.J., Paul, R., Kreuer, K.-D., and Zawodzinski Jr, T.A. *Direct Methanol Fuel Cells*, vol. 01–04 of *The Electrochemical Society Proceedings Series*, (eds. Narayanan, S., Gottesfeld, S., and Zawodzinski, T. A.), p. 29–33. New Jersey: The Electrochemical Society.

[84] Paddison, S.J., Paul, R., and Kreuer, K.-D. *Phys. Chem. Chem. Phys.*, 2002, **4**, p. 1151.

[85] Paddison, S.J., and Paul, R. *Phys. Chem. Chem. Phys.*, 2002, **4**, p. 1158.

[86] Paul, R., and Paddison, S.J. *J. Chem. Phys.*, 2001, **115**, p. 7762.

[87] Paul, R., and Paddison, S.J. *Phys. Rev. E*, 2003, **67**, p. 016108.

[88] PAul, R., and Paddison, S.J. *Solid State Ionics*, 2004, **168**, p. 245.

[89] Eikerling, M., Paddison, S.J., and Zawodzinski Jr., T.A. *J. New. Mat. Electr. Sys.*, 2002, **5**, p. 15.

[90] Eikerling, M., Paddison, S.J., Pratt, L.R., and Zawodzinski Jr, T.A. *Chem. Phys. Lett.*, 2003, **368**, p. 108.

[91] Eikerling, M., Paddison, S.J., Pratt, L.R., and Zawodzinski Jr, T.A. *Chem. Phys. Lett.*, 2003, **368**, p. 108.

[92] Hartnig, C., and Spohr, E., in preparation.

[93] Greeley, J., and Mavrikakis, M. *J. Am. Chem. Soc.*, 2002, **124**, p. 7193.

[94] Ishikawa, Y., Liao, M.-S., and Cabrera, C.R. *Surf. Sci.*, 2000, **463**, p. 66.

[95] Jusys, Z., Kaiser, J., and Behm, R.J. *Langmuir*, 2003, **19**, p. 6759.

[96] Schmitz, H., unpublished results.

[97] Zawodzinski Jr., T.A., Springer, T.E., Davey, J., Jestel, R., Lopez, C., Valerio, J., and Gottesfeld, S. *J. Electrochem. Soc.*, 1993, **140**, p. 1981.

[98] Edmondson, C., Stallworth, P., Chapman, M., Fontanella, J., Wintersgill, M., Chung, S., and Greenbaum, S. *Solid State Ionics*, 2000, **135**, p. 419.

[99] Edmondson, C.A., and Fontanella, J.J. *Solid State Ionics*, 2002, **152–153**, p. 355.

DNA SALINE SOLUTIONS NEAR SURFACES

A few ideas towards design parameters of DNA arrays

B. M. Pettitt, A. Vainrub, K.-Y. Wong

Department of Chemistry University of Houston,
Houston, TX 77204–5003, USA

Abstract   DNA hybridization occurs with different equilibrium constants and kinetics when near surfaces as opposed to in homogeneous saline solution. Thus melting temperatures are shifted by the presence of a surface. Much effort has gone into understanding DNA in homogeneous solution. The goal of this article is to review the current molecular picture of DNA in salt solution near surfaces which is consistent with known experimental thermodynamic and kinetic data. Calculations with simple models have been used in an attempt to make the most direct possible comparisons with recent structural and biological/biochemical experiments. In particular this research is of relevance to microarray technologies where nucleic acid probes are immobilized on surfaces and duplexes are formed with target strands in solution. We present an overview of the experiments, recent theories and computer simulations relevant to DNA on surfaces.

Keywords: DNA arrays, Poisson-Boltzmann equation, molecular dynamics simulations,

1.   Introduction

Combinatorial analysis has had an irrefutable effect on the very way we think of interrogating DNA sequences in the last 10 years [1]. The ability to perform 104 to 106 separate analysis on the same sample simultaneously would have already changed the very nature of even routine medical analysis if were not for the fact that difficulties in interpretation and even reproducibility. Some of these problems are related to preparation issues associated with microarray DNA production; others are related to the fundamental experimental design for the devices.

DNA as a polyelectrolyte in saline solution presents a unique chemistry and physics in terms of the complementary self pairing, known biologically as Watson-Crick hybridization. The free energies of association in solution are well studied and form the basis for the normal molecular recognition process. This process is fundamentally altered by the presence of a solid interface.

D. Henderson et al. (eds.), Ionic Soft Matter: Modern Trends in Theory and Applications, 381–393.
© 2005 *Springer. Printed in the Netherlands.*

In practice, one must consider the physical characteristics of the template of genomic information needed to answer a given problem. Those physical characteristics include not only the sequence and polymer length (and therefore charge) but also its melting temperature or equivalently its free energy of hybridization. However, the thermodynamic properties of DNA for devices including a metallic or dielectric interface are altered from that of homogeneous solution.

Experimental evidence for anomalous properties of nucleic acids near surfaces abounds but a cogent theoretical analysis has only recently begun to emerge [2–5]. It is well known that the kinetics and binding constant for hybridization in homogeneous saline solution is strongly affected by the presence of a surface [6–8]. Kinetics of hybridization can be drastically altered near surfaces [9]. Also, in the case of oligonucleotides covalently tethered to surfaces, Southern's experiments have shown a striking distance dependence of the hybridization melting temperature [10]. Experiments from the Hogan laboratory [11] have indicated that, near surfaces, it is possible under certain salt/solution conditions to obtain both increased affinity and increased selectivity near surfaces for DNA hybridization. This striking result is contrary to normal solution based intuition where affinity and selectivity are usually counter balanced.

The fundamental biophysics of nucleic acids near surfaces in salt water is of relevance to problems in both biology and in biotechnology [12]. In viruses recent electron microscopy images have demonstrated that nucleic acids can have specific structures near the interior capsid protein surface [13]. DNA hybridization near lipid bilayers has been studied experimentally [14, 15]. The results indicated that the hybridization melting temperature shifted dramatically when the duplex was near the charged bilayer. DNA in the presence of charged colloids in salt water has been seen to exhibit dramatic charge dependent shifts in binding constants as well [16]. Since salt is known to form gradients near surfaces [17] we expect the bulk salt dependence on conformations of DNA near surfaces to be nontrivial. We must consider the homogeneous solution control in hand before we can understand surface effects.

In this article, a few ideas about how the variables of surface composition, sample constitution and electric fields interplay with the thermodynamic variables to determine the outcome of various experiments on DNA binding near surfaces will be explored. Here, we will consider fundamental theoretical principles and computational methods needed in order to control and understand these effects. Theory and simulation will be involved to complete the current picture given by experiment. First, a brief and rudimentary over view of the experiments is given to provide context. We next review the progress in the theory and simulations of DNA near surfaces.

2. Experiment

The experiments done under the guise of DNA chips or DNA microarrays form actually a broad set. The interested reader is referred to the literature for a view of this breadth [18]. Here only a few commonly occurring themes will be presented.

Fundamentally, much of the nature of the experiments derives from the area of combinatorial analysis. Many experiments must be done with very limited sample and should ultimately take advantage of a single experimental preparation. These constraints lead naturally to miniaturization and thus to microarrays. Typically, each individual experiment takes place on a spot on a surface from 10–100 μm in diameter. Up to 10^5 experiments can then be conveniently carried out in a few 10's of square centimeters. The collection is thus a microarray.

The surface of each spot is identified by its own probe sequence and has therefore its own kind of single stranded nucleic acid molecules. The probes may be either simply placed on a surface with unlike charge or covalently attached. Given then a target sequence in saline solution, hybridization occurs at the probe near the surface which most nearly forms a Watson-Crick paired double strand. Typically the target DNA (or RNA) is partially fragmented into smaller oligomers which have better transport properties. Each fragment is often fluorescently labelled for optical detection.

A typical experiment then consists of preparing the surface and placing the probes on the surface (often not in the experimenter's lab) followed by placing the sample saline solution containing the target on the spots. The kinetics of hybridization may then proceed. Once sufficient time has elapsed, saline solution is used to wash the remaining unhybridized nucleic acid away. At this point the fluorophores can be excited and the area normalized intensity of fluorescence at each spot is then related to the affinity and the concentration of duplex nucleic acid.

The experiments can be performed under a variety of thermodynamic and composition conditions. By varying the temperature one can estimate the melting temperature or the temperature at which hybridization is 50%. The melting point for this discussion is defined as the temperature at which the free energy of association of two (usually) complementary strands is zero, $\Delta G = 0$, or alternatively when the enthalpic and entropic contributions for that process cancel, $\Delta H = T\Delta S$. The melting curve is thus an affinity measure. Changes in affinities give rise to specificities, for instance the difference in free energies between a perfect match and strands having one or two mismatches. Throughout this paper we will refer most often to melting curves as they are the defacto standard in experiment but equilibrium constants as a function of temperature are equivalent.

In biochip assays, both affinities and specificities of association are different from those observed in homogeneous solution [8, 10]. Often times the probes are covalently linked to the surface with tether molecules of various, specific lengths [10]. There is some restriction in position as well as orientation that must be accounted for. Thus, the rest of the difference in binding comes from changes in the solvent (and salt) activity and electrostatic fields induced by the surface presence.

Experiments on DNA arrays, especially microarrays, have revealed substantial differences in hybridization thermodynamics of DNA free in solution and surface tethered DNA. The main observations include a considerable decrease in the thermodynamic stability of the DNA duplex on an uncharged nonmetallic surface with a concomitant suppression of the thermal denaturation temperature of the duplex into single strands and a dramatic broadening of the thermal denaturation (duplex melting) curve [10, 19, 20]. Large effects with different trends occur for metal surfaces. Recently, more detailed experiments demonstrated that these effects grow as the surface density of probes increases [8, 21]. Although for common experimental conditions these phenomena can adversely affect the DNA array performance by suppression of the sensitivity and ability to detect mutations (SNP's), they are not well understood. In contrast to a large experimental effort, the theoretical analysis of DNA arrays [12] has gotten much less attention.

We need testable theoretical pictures of these physical contributions to the surface binding thermodynamics and dynamics. To that end both theory and simulation serve to provide a framework for understanding the nature of DNA near surfaces.

3. Theory

We recently considered the effect of the nucleic acid-surface electrostatic interaction on the thermodynamics of the surface hybridization [2–5, 22]. This theory used an analytical solution of the linearized Poisson-Boltzmann boundary value problem for a charged sphere-surface interaction in electrolyte solution and corresponds to the system characterized by a low surface density of immobilized probes. To understand the motivation for that work and extensions, we need to consider the physical effects of a surface in solution and the theoretical tools available for their study.

Surfaces order the solvent and salts in solution, sometimes strongly [17]. Understanding the relationship of a nucleotide's structure and thermodynamics with its solution environment is important for probing the fundamental relation of conformation with biological or chemical activity [23]. Surfaces further complicate the picture. Understanding the homogeneous solution properties of DNA requires delving into the interplay between the solvent structure, that of the salt and DNA equilibrium structures and thermodynamics (see below). Using computational techniques like all atom computer simulations to obtain atomic level hypotheses which might be difficult to view from experimental data requires confidence in and controls on the parameters used in simulations [24]. The equilibrium between A-form and B-form DNA in homogeneous solution as a function of sequence and salt by computer simulation of an atomic force field is one such required control test [25].

We recently reported the development of a general analytic theory with applications in this area [3–5]. We derived exact thermodynamic equations for the surface effects on the Gibbs free energy difference of reaction, regardless of the specific species-interface and electrolyte-interface interactions. Further, we approximated the interactions in solution using the linearized Poisson-Boltzmann equation theory of electrostatic interactions of an ion-penetrable charged sphere with a charged plate immersed in electrolyte to achieve a closed form solution. Theoretical justification for this model comes from both crystallographic data as well as simulations where ion populations in the grooves translate into ion permeability for hard cylinder or hard sphere models [26, 27]. This theory has an analytical solution for the Gibbs free energy difference of interaction [28]. We use it to derive equations for the surface effect on the enthalpy and entropy contributions to the Gibbs free energy difference of binding and equilibrium reaction constants.

The problem considered by Ohshima and Kondo [28] starts with a charged plate and a charged ion-penetrable sphere with radius a, separated by a distance h. Thus we first approximate short oligonucleotides as an ion penetrable sphere which for oligos 7–8 bases long is reasonable but short by experimental standards. The potential field, φ is assumed to obey a linearized PB equation,

$$\nabla^2 \varphi g = g\kappa^2 \varphi s, \qquad \text{in solution,} \qquad (1)$$

$$\nabla^2 \varphi g = g\kappa^2 \varphi - \left(\frac{\rho}{\varepsilon\varepsilon_0}\right), \qquad \text{inside the DNA sphere,} \qquad (2)$$

where ε is the relative dielectric constant of the solution, ε_0 is the permittivity of vacuum and the inverse Debye length is,

$$\frac{1}{\kappa} = \left(\varepsilon\varepsilon_0 k \frac{T}{2} c_0 e^2\right)^{1/2} \qquad (3)$$

for a concentration c_0. The analytic solution expresses the potential of the system through the surface potentials of a non-interacting plate and sphere in salt solution.

We put the charges on the surface of the sphere and obtain the potential at the sphere,

$$\varphi_{so} = \left(\frac{q}{4\pi\varepsilon\varepsilon_0 a}\right)\left[\exp(-\kappa a)\frac{\sinh(\kappa a)}{\kappa a}\right]. \tag{4}$$

If we view our process as charging the sphere such that we go from a single strand to a double strand then we have a Born charging model of the Free Energy, G, of binding DNA near a surface. This can be adjusted for both dielectric and metallic surfaces. Given the free energy we can then calculate entropy and enthalpy via the familiar derivatives,

$$-\Delta S = \frac{\partial G}{\partial T} = \left(\frac{\partial G}{\partial \kappa}\right)_\varepsilon \frac{\partial \kappa}{\partial T} + \left[\left(\frac{\partial G}{\partial \varepsilon}\right)_\kappa + \left(\frac{\partial G}{\partial \kappa}\right)_\varepsilon \frac{\partial \kappa}{\partial \varepsilon}\right] \cdot \frac{\partial \varepsilon}{\partial T}. \tag{5}$$

Substituting Eq. (3) we have the compact result,

$$-T\Delta S = -\left[\kappa\frac{(1+\alpha)}{2}\right]\left(\frac{\partial V}{\partial \kappa}\right)_\varepsilon + \alpha\varepsilon\left(\frac{\partial V}{\partial \varepsilon}\right)_\kappa, \tag{6}$$

where we defined,

$$\alpha = \frac{T}{\varepsilon}\frac{\partial \varepsilon}{\partial T}. \tag{7}$$

These equations give a simple mean field description of the free energy surface and its entropy/enthalpy decomposition as a function of the important physical parameters including the salt concentration and distance from the surface. Given the experimental interest in linker length [10] we have a direct link with experimental design parameters. Below a typical example is given demonstrating the chemical linker length effect on the free energy or equivalently the melting curve for duplex association. Notice the dependence on the surface charge. Surfaces may be charged easily by chemical preparation.

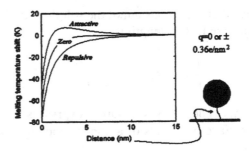

Figure 1. The shift of the melting temperature for an immobilized 8 base pair oligonucleotide duplex at 0.01 M NaCl as a functional of the distance from a dielectric surface charge q.

Further progress can be made considering the system from a surface coverage or Langmuir point of view. Keeping the target concentration, C, constant, this reversible reaction obeys first-order kinetics and thus the hybridization yield θ ($0 < \theta < 1$) at equilibrium is given by the Langmuir isotherm,

$$\theta = \frac{1}{1 + C^{-1}\exp(\Delta G/kT)}, \tag{8}$$

where ΔG is the duplex binding Gibbs free energy. As the binding free energy ΔG is independent of θ, this equation corresponds to the well-known Langmuir adsorption isotherm. To account for the screened electrostatic repulsion of the target from the probe array we use the Gibbs free energies of interaction to calculate the shifts in the binding free energy from homogeneous solution. We found the repulsion of targets depletes targets near the probe array according to a simple Boltzmann average. We thus obtain the free energy of interaction as a function of distance between the sphere surface and plane for any surface concentration of probes and solution concentration of targets [2].

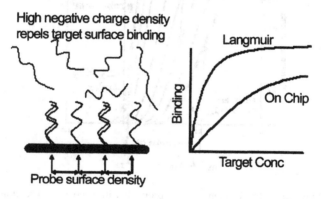

Figure 2.   Coulomb blockage dominates the optimum surface spacing.

We can extend this surface isotherm modelling and find a simple analytical result for optimizing conditions to maximize sensitivity of detection which takes into account the electrostatics and the depletion of the target concentration in solution during binding [5]. Solving for the concentration we obtain,

$$C_0 = \frac{n_D S}{N_A V} + \frac{n_D}{n_P - n_D}\exp\left(\frac{\Delta G_0}{RT}\right)\exp\left[\frac{wZ_P(Z_P n_P + Z_T n_D)}{RT}\right], \tag{9}$$

where ΔG_0 is the reference state free energy for hybridization in isotropic solution, Z_P and Z_T are the length of probe and target oligonucleotides, n_P is the surface density of probes and the mean field parameter w is fixed in comparison with experiments at an appropriate salt and buffer concentration. Given that the DNA concentration C_0 and volume V of the hybridization solution are small for most microarray experiments the target depletes to the concentration, $C_0 - n_D S/N_A V$ given by the first term on the right hand side. The last exponential represents the repulsion from the target-array that causes an electrostatic blocking of hybridization dependant on the surface probe density n_P and therefore the surface probe spacing as shown qualitatively in Fig. 2. This equation gives a good account of recently published experiments for arrays mounted on glass [29–32] and gold [8, 33, 34] surfaces. This formalism makes simple analytic predictions for melting curve widths and sensitivity optima as deviations from homogeneous solution hybridization as demonstrated in Fig. 3.

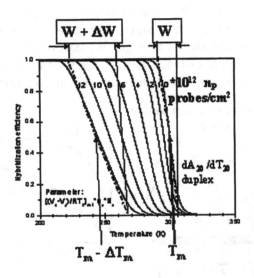

Figure 3. Melting curves for different surface probe densities (coverage) as indicated.
In solution: $T_m = \Delta H_0/(\Delta S_0 - R\ln C)$, $W = 4RT_m^2/\Delta H_0$.
On-array: Isotherms $\Delta T_m = (3wZ^2 n_p)/(2\Delta H_0 + 3wZ^2 n_p)$ $\Delta W = 2/3\Delta T_m$.

While our theoretical analysis has provided a sound basis to consider the thermodynamics, we needed to check the quantitative aspects of the suite of calculations. In particular to consider the effects of fluctuating conformation and geometry, molecular simulations are a particularly useful way to check our assumptions.

4.   Simulation

Using computational techniques like all atom computer simulations to obtain atomic level hypotheses which might be difficult to view from experimental data requires confidence in and controls on the parameters used in simulations [24] as mentioned above. In particular we are interested in the fundamental questions of how the surface might affect the local geometry and conformation of the DNA. One could imagine that the DNA might lie with its helical axis parallel to the surface, or that gross distortions of the normal Watson-Crick geometry could be induced.

To study the range of possibilities the first molecular dynamics simulations of a DNA duplex tethered to a surface was performed [35, 36] The technical aspects of simulations near surfaces are nontrivial, especially as concerns reliable boundary conditions [37]. Molecular dynamics provides a more quantitative picture of the salt gradients and DNA structures near the surface responsible for changes in hybridization affinities and specificities than approximate (PB level mean field) theory and so may be used as a check on the simple analytical picture derived above. In addition simulation provides mechanistic clues which can form additional hypotheses for testing.

Below in Fig. 4 we show a typical snapshot from a molecular dynamics simulation of a duplex in saltwater tethered to a surface confirming that the uncharged organically functionalized surface has little effect on the average conformation of the tethered DNA.

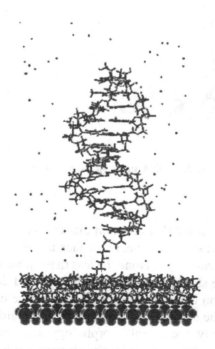

Figure 4.   DNA tethered to a surface in salt solution.

The effects of a surface charge could be expected to have a more severe affect on the available conformations. Experiments have shown that rapid hybridization kinetics can be achieved on positively charged surfaces [38–41]. However, reversibility depends on the systems path or preparation. When a single strand is placed on the surface reversible target binding is found but when duplexes are placed on the surface preformed single strands cannot be removed and only strong washing conditions will remove the duplexes. This indicates that a change in the nature of the single strand probes first placed on the surface with respect to the form expected for solution duplexes exists.

To study that phenomenon we have more recently performed simulations of single strands near a per-aminated surface akin to those used in c-DNA technology [42]. The results show an idea of what the equilibrium conformational manifold of single stranded DNA near surfaces looks like. This is critical to understanding the differences between tethered and untethered system.

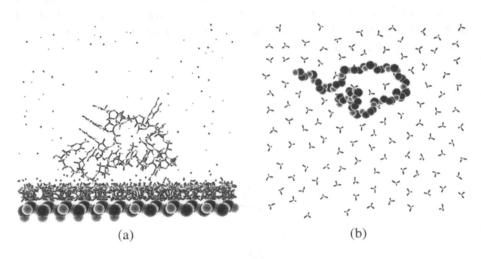

(a)   (b)

Figure 5.   DNA single strand near a surface

In fact the time to reach an equilibrium distribution of ssDNA conformers even for relatively short sequences near a surface is unknown but we speculate it is certainly greater than 0.1 μs. None the less these simulations show that the base stacking in single strands near surfaces is local and does not have longer range order. In addition the association with the surface can be strong, given a positive charge on the surface. So while W–C base pairing is the only way to explain the specificity, the overall morphology may not be a standard W–C double helix but may be an ensemble of structures which are locally base paired but not globally helical.

5. Summary

Design of nucleic acid microarrays not only requires a knowledge of the surface preparations and solution conditions but also the sequences to be assayed to result in maximal information content. We have only considered the affect the surfaces have on DNA hybridization here. The interested reader is referred to other reviews which consider sequence choice [43]. When this field really began to make progress in several different laboratories simultaneously in the early 90's theoretical tools relating to the atomic design of surfaces and conditions for DNA chips were lacking or absent. The application of even the simple tools shown here already gives some understanding of the range of phenomena one is dealing with. Much remains to be done, including the use of theories which go beyond mean field methods used on low resolution models.

Acknowledgements

B.M. P. thanks his collaborators, Mike Hogan and Rosina Georgiadis who have contributed to and stimulated this area. This work was supported in part by the NIH and Robert A. Welch Foundation. Computer time resources are gratefully acknowledged from NPACI at the San Diego Supercomputer Center and Molecular Science Computing Facility (MSCF) in the William R. Wiley Environmental Molecular Sciences Laboratory, a national scientific user facility sponsored by the U.S. Department of Energy's Office of Biological and Environmental Research and located at the Pacific Northwest National Laboratory. Pacific Northwest is operated for the Department of Energy by Battelle.

References

[1] Editorial. *The chipping forecast.* Nature Genetics, 1999, **21** (1).

[2] Vainrub, A. and Pettitt, B.M. Thermodynamics of association to a molecule immobilized in an electric double layer. *Chemical Physics Letters*, 2000, **323** (1–2), p. 160–166.

[3] Vainrub, A. and Pettitt, B.M. Coulomb blockage of hybridization in two-dimensional DNA arrays. *Physical Review E*, 2002, **66** (4), p. 041905.

[4] Vainrub, A. and Pettitt, B.M. Surface electrostatic effects in oligonucleotide microarrays: Control and optimization of binding thermodynamics. *Biopolymers*, 2003, **68**, p. 265–270.

[5] Vainrub, A. and Pettitt, B.M. Sensitive quantitative nucleic acid detection using oligonucleotide microarrays. *Journal of the American Chemical Society*, 2003, **125** (26), p. 7798–7799.

[6] Shoemaker, D.D. and Linsley, P.S. Recent developments in DNA microarrays. *Current Opinion in Microbiology*, 2002, **5** (3), p. 334–337.

[7] Heaton, R.J., Peterson, A.W. and Georgiadis, R.M. Electrostatic surface plasmon resonance: Direct electric field- induced hybridization and denaturation in monolayer nucleic acid films and label-free discrimination of base mismatches. *Proceedings of the National Academy of Sciences of the United States of America*, 2001, **98** (7), p. 3701–3704.

DNA saline solutions near surfaces

[8] Peterson, A.W., Heaton, R.J. and Georgiadis, R.M. The effect of surface probe density on DNA hybridization. *Nucleic Acids Res.*, 2001, **29** (24), p. 5163–5168.

[9] Peterson, A.W., Heaton, R.J. and Georgiadis, R. Kinetic control of hybridization in surface immobilized DNA monolayer films. *Journal of the American Chemical Society*, 2000, **122** (32), p. 7837–7838.

[10] Shchepinov, M.S., CaseGreen, S.C. and Southern, E.M. Steric factors influencing hybridisation of nucleic acids to oligonucleotide arrays. *Nucleic Acids Research*, 1997, **25** (6), p. 1155–1161.

[11] Du, X. Oligonucleotide hybridization on surfaces. *Ph. D. Thesis*, 1995, (Baylor College of Medicine).

[12] Chan, V., Graves, D.J. and McKenzie, S.E. The biophysics of DNA hybridization with immobilized oligonucleotide probes. *Biophysical Journal*, 1995, **69** (6), p. 2243–2255.

[13] Chen, D.H., et al. The pattern of tegument-capsid interaction in the herpes simplex virus type 1 virion is not influenced by the small hexon-associated protein VP26. *Journal of Virology*, 2001, **75** (23), p. 11863–11867.

[14] Fajkus, M. and Hianik, T. Peculiarities of the DNA hybridization on the surface of bilayer lipid membranes. *Talanta*, 2002, **56** (5), p. 895–903.

[15] Tarahovsky, Y.S., et al. High temperature stabilization of DNA in complexes with cationic lipids. *Biophys. J.*, 2002, **82** (1), p. 264–273.

[16] Balladur, V., Theretz, A. and Mandrand, B. Determination of the main forces driving DNA oligonucleotide adsorption onto aminated silica wafers. *Journal of Colloid and Interface Science*, 1997, **194** (2), p. 408–418.

[17] Lozada-Cassou, M. and Saavedra-Barrera, R. The application of the hypernetted chain approximation to the electrical double layer: Comparison with Monte Carlo results for symmetric salts. *Journal of Chemical Physics*, 1982, **77** (10), p. 5150–5156.

[18] Schena M. (1999) *DNA Microarrays: A Practical Approach*. Oxford: Oxford Univ Press, p. 210.

[19] Guo, Z., et al. Direct flourescence analysis of genetic polymorphisms by hybridization with oligonucleotide arrays on glass supports. *Nucleic Acids Res.*, 1994, **22** (24), p. 5456–5465.

[20] Forman, J.E., et al. Thermodynamics of duplex formation and mismatch discrimination on photolithographically synthesized oligonucleotide arrays. *ACS Symposium Series*, 1998, **682**, p. 206–228.

[21] Watterson, J.H., et al. Effects of oligonucleotide immobilization density on selectivity of quantitative transduction of hybridization of immobilized DNA. *Langmuir*, 2000, **16**, p. 4984–4992.

[22] Vainrub, A. and Pettitt, B.M. Theoretical aspects of genomic variation screening using DNA microarrays. *Biopolymers*, 2004, **73** (5), p. 614–620.

[23] Anderson, C.F. and Record, M.T. Salt Nucleic-Acid Interactions. *Annual Review of Physical Chemistry*, 1995, **46**, p. 657–700.

[24] Reddy, S.Y., Leclerc, F. and Karplus, M. DNA polymorphism: A comparison of force fields for nucleic acids. *Biophysical Journal*, 2003, **84** (3), p. 1421–1449.

[25] Feig, M. and Pettitt, B.M. Structural equilibrium of DNA represented with different force fields. *Biophysical Journal*, 1998, **75** (1), p. 134–149.

[26] Feig, M. and Pettitt, B.M. A molecular simulation picture of DNA hydration around A- and B-DNA. *Biopolymers*, 1998, **48** (4), p. 199–209.

[27] Feig, M. and Pettitt, B.M. Sodium and chlorine ions as part of the DNA solvation shell. *Biophysical Journal*, 1999, **77** (4), p. 1769–1781.

[28] Ohshima, H. and Kondo, T. Electrostatic Interaction of an Ion-Penetrable Sphere with a Hard Plate – Contribution of Image Interaction. *Journal of Colloid and Interface Science*, 1993, **157** (2), p. 504–508.

[29] Jin, L., Horgan, A. and Levicky, R. Preparation of end-tethered DNA monolayers on siliceous surfaces using heterobifunctional cross-linkers. *Langmuir*, 2003, **19** (17), p. 6968–6975.

[30] Podyminogin, M.A., Lukhtanov, E.A. and Reed, M.W. Attachment of benzaldehyde-modified oligodeoxynucleotide probes to semicarbazide-coated glass. *Nucleic Acids Research*, 2001, **29** (24), p. 5090–5098.

[31] Charles, P.T., et al. Fabrication and surface characterization of DNA microarrays using amine- and thiol-terminated oligonucleotide probes. *Langmuir*, 2003, **19**, p. 1586–1591.

[32] Walsh, M.K., Wang, X.W. and Weimer, B.C. Optimizing the immobilization of single-stranded DNA onto glass beads. *Journal of Biochemical and Biophysical Methods*, 2001, **47** (3), p. 221–231.

[33] Steel, A.B., Herne, T.M. and Tarlov, M.J. Electrochemical quantitation of DNA immobilized on gold. *Anal. Chem.*, 1998, **70**, p. 4670–4677.

[34] Herne, T.M. and Tarlov, M.J. Characterization of DNA probes immobilized on gold surfaces. *Journal of the American Chemical Society*, 1997, **119** (38), p. 8916–8920.

[35] Wong, K.Y. and Pettitt, B.M. Orientation of DNA on a surface from simulation. *Biopolymers*, 2004, **73** (5), p. 570–578.

[36] Wong, K.Y. and Pettitt, B.M. A study of DNA tethered to surface by an all-atom molecular dynamics simulation. *Theoretical Chemistry Accounts*, 2001, **106** (3), p. 233–235.

[37] Wong, K.Y. and Pettitt, B.M. A new boundary condition for computer simulations of interfacial systems. *Chemical Physics Letters*, 2000, **326** (3–4), p. 193–198.

[38] Belosludtsev, Y., et al. Nearly instantaneous, cation-independent, high selectivity nucleic acid hybridization to DNA microarrays. *Biochemical and Biophysical Research Communications*, 2001, **282** (5), p. 1263–1267.

[39] Belosludtsev, Y., et al. DNA microarrays based on noncovalent oligonucleotide attachment and hybridization in two dimensions. *Analytical Biochemistry*, 2001, **292** (2), p. 250–256.

[40] Lemeshko, S.V., et al. Oligonucleotides form a duplex with non-helical properties on a positively charged surface. *Nucleic Acids Research*, 2001, **29** (14), p. 3051–3058.

[41] Zhang, P.M., et al. Acceleration of nucleic acid hybridization on dna microarrays driven by pH tunable modifications. *Nucleosides Nucleotides & Nucleic Acids*, 2001, **20** (4–7), p. 1251–1254.

[42] Wong, K.Y., et al. A non-Watson-Crick motif of base-pairing on surfaces for untethered oligonucleotides. *Molecular Simulation*, 2004, **30** (2–3), p. 121–129.

[43] Vainrub, A., et al. (2003). *Theoretical Considerations for the Efficient Design of DNA Arrays*. Biomedical Technology and Devices Handbook, eds. James, E., Moore, Jr. and George Zouridakis CRC Press.

CHARGE TRANSPORT IN HIGHLY-RADIOACTIVE SUBSTANCE

Experimental study of silicate alkali-earth glasses from Chornobyl site

O. Zhydkov

Institute for NPP Safety Problems
of the National Academy of Sciences of Ukraine,
Kirova 36A, 07270, Chornobyl, Ukraine

Abstract   The paper is devoted to the experimental study of electron and ion transport in a very special substance, namely, alkali-earth glasses containing the noticeable quantity (up to 10%) of dissolved irradiated uranium nuclear fuel and its fission and daughter products as well. Such a high-radioactive product was formed at the active stage of the well-known heavy nuclear accident which occurred on Chornobyl NPP facility in 1986. The soft matter behavior was established by measuring the temperature dependence of viscosity, where the glassy properties had been identified unambiguously. Static electric conductivity temperature dependence was measured for 80 K–1000 K temperature interval. The transport processes connected with thermal activation of electrons, hopping conductivity in the band tails and variable range hopping (VRH) were identified. The band structure of such glasses manifests the energy gap of 1.8–2.0 eV width, which formed due to long-range order and wide band tails connected with horizontal disorder, which, in turn, may originate from numerous traps and internal radiation damages. The latter makes it possible to identify the investigated matter as devitrified glass. The distinguishing feature of such devitrified glasses is low ionization energy for electrons, providing a high spatial density of electron excitations in α-particle tracks, which leads to such a collective phenomena as the so-called Coulomb explosion.

Keywords: Glass ceramics, radioactive glass, hopping conductivity

1.   Introduction

The paper is devoted to the experimental study of electron and ion transport in a very special complicated substance. Such a matter has been formed as a product of well-known heavy nuclear accident, which was occurred on Chornobyl NPP facility in 1986. Its formation occurred at the active stage

D. Henderson et al. (eds.), Ionic Soft Matter: Modern Trends in Theory and Applications, 395–410.
© 2005 *Springer. Printed in the Netherlands.*

of the accident by dissolving the uranium nuclear fuel and other constructive materials of the reactor core in melted serpentine ($Mg_6(OH)_8 \times [Si_4O_{10}]$); the latter had been used in the reactor construction as a heat-insulating filling. The irradiated fuel heat release served as a heat source for the melting process. The final product of melting process is known as LFCM – lava-like fuel-containing material. There are several sorts of such materials, which vary in color and in uranium content. The most widespread and typical LFCM sort is the so-called brown LFCM.

The averaged element composition for brown LFCM [1] is given below in the Table 1.

Table 1. Element composition (significant quantities) of brown LFCM (mass %).

Element	Al	Mn	Fe	Mg	Si	Ca	Zr	Na	U
Average Content	3.5± 0.7	0.52± 0.2	0.84± 0.2	4± 0.9	30± 3.6	4.7± 0.8	4.8± 1.1	4± 0.4	8.4± 0.2

The direct study of LFCM structure by routine X-ray analysis gives ambiguous results, as far as LFCM themselves are the powerful sources of X-ray radiation, which provides a strong background for detectors. In order to classify at least a condensed matter type (crystalline or amorphous), a temperature dependence of fluidity was measured. Corresponding results are presented in Fig. 1.

From Fig. 1 one can easily identify the absence of a certain melting point for LFCM, but monotonic (and close to exponential) growth of LFCM fluidity with temperature. Such behavior is typical of glasses (or glass ceramics). There are a few theories for viscous flowing of glasses, but all of them are based on the picture of presence of separate moving particles (called kinetic units), which provide flowing [2]. Viscosity itself is a statistical parameter, as far as its value is determined by the whole ensemble behavior of interacting kinetic particles. Proper calculations for such ensemble behavior will lead to the expression describing η as a function of temperature,

$$\eta(T) = \eta_0 \exp\left(U_a/kT\right), \tag{1}$$

where U_a is the activation energy for viscosity. Proper Fig. 1 data processing indicates that for the so-called multicoloured LFCM $U_a \approx 3$ eV; this is because we can find (low temperature) $U_a \approx 1$ eV for brown LFCM. Those initial data enable us to estimate ion diffusion coefficients for LFCM at a room temperature as well as high-temperature LFCM fluidity.

One can identify from the above that LFCM vitrifiability is based mainly on SiO_2 and partially on Al_2O_3 oxides (see Tab. 1). LFCM belongs to a class

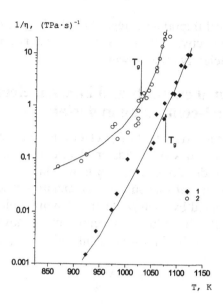

Figure 1.   LFCM dynamic fluidity as a function of temperature: 1 – the so-called multi-coloured LFCM, 2 – brown LFCM. Vitrification temperature T_g corresponds to conventional viscosity value of 1 TPa·s.

of alkali silicate glasses regarding the noticeable Na^+ quantity, which, as it is known [2], enlarges the ion transport in glasses to a great extent. On the other hand, alkali-earth Ca^{++} and Mg^{++} ions are capable of blocking the ion mobility by way of saturating broken valence bounds [2].

Thus, the LFCM are alkali-earth silicate glasses, in which electric transport mechanisms are unclear. The latter, however, is important regarding high radioactivity of LFCM and simultaneous action of high internal electric fields and radiation damages on their behavior.

As one can see, LFCM contain about 8% of uranium irradiated fuel, and, hence, its fission and transuranium daughter products. The internal radioactivity of glasses is provided by the fission products (Cs^{137}, Sr^{90}) and daughter products of irradiated uranium (Pu^{239}, Am^{241}). The specific α-activity of such glasses is 10MBq/cm^3 (the typical value); β-activity can achieve 1GBq/cm^3 value.

The electric transport properties of such a matter present a scientific interest as far as it is a multi-phase heterogeneous non-equilibrium system, where a continuous generation of extra electrons (β-particles) occurs. There is, however, the practical aspect: internal self-sustained currents cause a high inner electric field, especially just beneath the surface of material; those electric fields can indirectly stimulate the action of various factors leading to such a material self-destruction. The items, connected with the matter durable stability are of a crucial practical importance, regarding the large quantity (about 1000 tons) of such high-radioactive material, which now cannot be isolated from the environment in a sufficiently reliable way.

2.   Self-sustained currents and internal electric fields in lava-like fuel-containing material

It is quite clear that a certain part of β-electrons will escape an LFCM volume trough the surface and will provide an electric current density $J = e \cdot \Phi$, where Φ is the surface density of electron flow. One may expect a relatively large current value, taking into account a large quantity of secondary electrons, which are to be originated by each β-electron having typical initial energy of 200–1000 keV. Actually, the volume concentration of moving particles N will provide their flow density (in [3] it was calculated for α-particles),

$$\Phi = N\frac{\lambda}{4} \, , \tag{2}$$

where λ is a whole path of particles in condensed media. The N quantity should exceed the one calculated from specific radioactivity and hence the current density J by two-three orders, as far as J does not depend on the energy of electrons escaping the surface. There was provided a special workbench experiment devoted to direct measurement of electric current provided by electrons escaping LFCM surface. At the same time the possibility was provided to quantitatively identify the low-energy electron spectrum by applying a stopping potential to the anode (the LFCM specimen surface served as a cathode). The corresponding volt-ampere characteristic is indicated in Fig. 2.

One can see in Fig. 2 that there are no expected low-energy electrons input in total electric current as far as the curve indicates a weak $J(U)$ dependence only. Moreover, a thorough qualitative analysis of this curve behavior for $U < 0$ region (stopping external field) indicates that there are no electrons of $E < 50$ eV energy and the input of electrons belonging to 300 eV $< E < 50$ eV energy interval (at $U \approx 300$ V there is a full stopping for low-energy electrons since the curve derivative is close to zero) does not exceed 3–4%. The current density value corresponds to the one estimated from Eq. (2) if N corresponds to the known (mentioned in Introduction) specific β-activity and there is no need to take a large quantity of secondary electrons into account. Observable picture seems contradicting at first glance. The phenomenon under observation, however, has a simple physical explanation based on the consideration that λ parameter in (2) itself depends on E to a great extent. Such a situation was analyzed in detail in [4], where based on the tabulated data [5] it was identified that $\lambda(E)$ function sharply decays with the electron energy decreasing. The latter is the main reason for a negligibly small input of low-energy electrons in the total

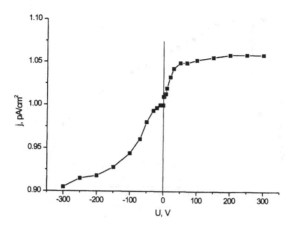

Figure 2. Density of current emitted by LFCM surface in vacuum (0.001 Pa) on the stopping potential dependence.

Φ value, despite their doubtlessly high concentration just beneath the surface. The energy spectrum of the emitted electrons was experimentally estimated in [4], where it was found that it mainly corresponds to a β-spectrum of Cs-137 and Sr-90 [6], which provide a main β-activity in LFCM volume. Thus, one may conclude that, on the one hand, there is a certain surface Coulomb barrier located just beneath the surface, which is non-transparent for low-energy electrons. On the other hand, there is an electron-lacking region, which formed at the depth λ (0.1–0.2 mm for majority of β-electrons) under the surface. The above resulted in the stationary (in a long-time scale) under-surface electron flow directed from LFCM surface within its volume up to the a depth of the order of λ. As it will be clear hereinafter, such a stationary picture can exist only at a room (and lower) temperature, when there are no free electrons, but electric conductivity can be realized through localized states by hopping mechanisms.

It is quite clear from the above that noticeable hopping conductivity can exist when there is a large density of localized states (traps for electrons) only, which can be provided by initial structure defects or by radiation damages. As far as hopping parameters (and possible free electron concentration) do depend on temperature to a large extent, the temperature increase will lead to a redistribution of internal electric fields and currents. The results of some pertinent experiments are presented below.

One of the experiments was devoted to the measuring of spontaneously generated electric field. There was used the plane specimen, the thickness of which (1 mm) was taken larger than λ. The potential difference was measured with

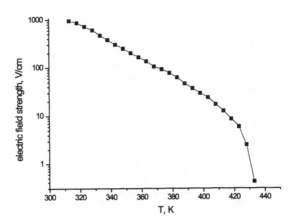

Figure 3. Temperature dependence of a self-sustained electric field generated by brown LFCM specimen.

high-impedance voltmeter. Corresponding temperature dependence is presented in Fig. 3.

One can see in Fig. 3 that spontaneous electric field strength exceeds 1 kV/cm and decays exponentially (with activation energy about 0.43 eV) while temperature increases. The given electric field strength should be understood as an av-

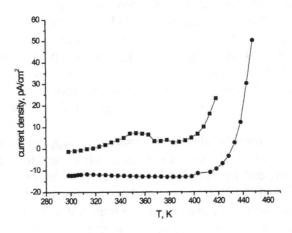

Figure 4. Measured short-circuit regime self-sustained electric current as a function of temperature. Zero level is a conventional quantity. Lower curve corresponds to a very slow heating. Upper curve was measured three days later and corresponds to the heating velocity of 0.025 K/s.

erage integral value for the whole thickness, but the possible local value for the near-surface spatial region is still unclear. Sharp decay may be connected with the following two processes: (a) the release of the localized electron from the existing traps and (b) sharp increase of volume conductivity due to free electron generation by way of their thermal activation on the mobility edge. Both processes will lead to the elimination of the above mentioned surface Coulomb barrier and to the increase of self-sustained current. There was provided a special experiment on self-sustained current in short-circuit regime measurement. Current versus temperature is reflected in Fig. 4.

One can observe in Fig. 4 (the lower curve) the sharp increase of current at $T > 400$ K, though it is clear that current originates from nuclear processes, which cannot depend on temperature. Possible physical explanation for the observed phenomenon is the appearance of thermally activated electrons, which break the surface Coulomb barrier and in this way make possible a large quantity of secondary electrons into the electric transport involvement. Data analysis indicates that thermal activation energy for such a process is about 1 eV, which, as it will be clear hereinafter, corresponds to the energy for electric conductivity in high-temperature region. The LFCM specimen was kept at a room temperature for a few days in order to fill in high-energy traps, if any. Here, a method was used accepted for semiconductors having high concentration of deep impurities, where relatively quick heating leads to liberation of electrons from traps and, hence, to the appearance of thermal-stimulated current [7]. The result of such an experiment is reflected by the upper curve in Fig. 4. It is not difficult to identify the thermal peak at 355 K vicinity. Data processing in a manner described in [7] indicates the presence of the traps of $0.4 \sim$ eV depths. The nature of the trap is unclear yet. The current lowering just after 370 K indicates that the traps are already exhausted at that temperature.

3. Electric transport in LFCM volume

It is clear from the above that to identify the physical mechanisms, responsible for electric transport in LFCM, is a difficult task. Self-sustained currents and spontaneous electric fields will not permit to provide the conductivity measurements in a standard way. The correct way in this situation is to measure volt-ampere characteristic at every temperature point and then to find its derivative, when electric field is close to zero. A heap of such a derivative values will provide a temperature dependence of low-field (ohmic) electric conductivity. There was provided the special investigation for LFCM specific resistance on temperature dependence, where a 80 K $< T <$ 1000 K temperature region has been covered. The temperature dependence of LFCM specific resistance (high temperature region) is indicated in Fig. 5.

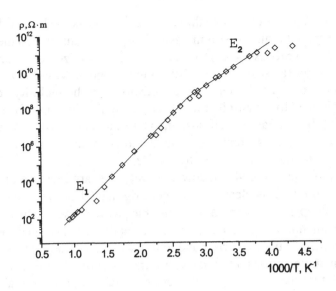

Figure 5.   Dependence of brown LFCM specific resistance on reciprocal temperature. Both E_1 and E_2 activation energies can be found from the slope.

One can see in this figure that (at $T > 350$ K) such a dependence can be satisfactorily described in a form of a simple activation law, namely, as

$$\rho(T) = \rho_0 \exp(E_1/kT),\qquad\qquad (3)$$

where E_1 is the corresponding activation energy and ρ_0 is the pre-exponent coefficient. As it was mentioned earlier, there are two possible mechanisms for electric transport. In principle, the first one may be connected with ionic hopping conductivity, where an ion hopping is thermally activated. A simple evidence in favor of such a mechanism may be an observable additional hopping stimulated by electric field; those can be identified as the rise of hopping conductivity with external electric field increasing even at moderate electric field values. In order to testify the above, the volt-ampere characteristic of LFCM specimen at T = 550 K is performed in Fig. 6.

One can see in this figure the high linearity of the characteristic up to external electric field strength ≈ 1.5 kV/cm, which permits to identify the non-hopping mechanism of electric transport at this temperature range. One cannot exclude, however, a certain input of ionic hopping in high-temperature conductivity; it is known that in homology (composition) raw of complicated silicate glasses where ionic hopping mechanism changes by thermal-activated electron conductivity, nobody can identify for sure the noticeable activation energy changes [2]. Besides this, the high volt-ampere characteristic linearity in principle may be connected with the small hopping range for ions and, therefore,

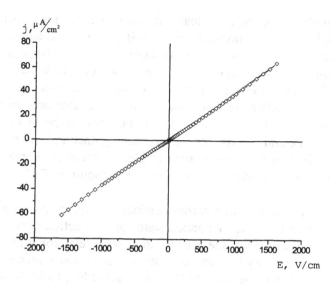

Figure 6. LFCM volt-ampere characteristic at $T \approx 550$ K. Specific conductivity can be found from its slope.

non-linearity can be manifested at extremely high electric fields only. Another possible mechanism is a thermal activation of electrons from a Fermi level into the states lying beyond a mobility edge [2,8]. Such a mechanism is typical of semiconductor glasses, where alkali ions (such as K^+, Na^+ or Li^+) are absent, or ionic hopping mechanism is suppressed by two-valence alkali-earth ions, such as Ca^{++}, Mg^{++} or Ba^{++}. As it was mentioned in Introduction, in our system such a suppression is doubtlessly connected with noticeable Ca and Mg content (see also Tab. 1). The additional argument in favor of non-hopping mechanism of electric transport (in a given temperature range) is the absence of noticeable frequency dispersion of conductivity up to 1 kHz frequency. Determined from (3) activation energy $E_1 \approx 0.74$ eV value corresponds to the energy distance between the Fermi level and the mobility threshold, which, in its turn, corresponds to a percolation level in a conduction band. The conduction band existence, as it is known, is possible due to short-range order and valence bounds between neighboring atoms coordination. As far as usually (in glasses) a Fermi level is fixed in the middle of mobility gap [2,8], one can estimate the (thermal) energy gap width as ≈ 1.5 eV. Real (traditionally determined as an optical) mobility gap width at a room temperature can be estimated taking into account a real LFCM elastic constants [9] and thermal expansion coefficient, as $1.8 \div 2$ eV.

There is an additional question, connected with an origination of noticeable electron conductivity. In accordance with a conventional point of view [2], in oxide glasses having electron conductivity there are conditions for weakened manifestation of the tendencies typical of transient metal oxides. As far as pure transient metal oxides have no tendency to vitrification, the usual way to obtain noticeable electron conductivity is to add them into a glassy composition. The electron conductivity does depend on concentration of cations having uncompleted d-shells. There are well-known oxide glassy compositions, containing Ti^{4+} and V^{4+}; noticeable electron conductivity has in fact been observed for glasses containing Fe, Co and Mn oxides. In our case, at the first glance, one can suggest that Fe and U ions (see Tab. 1), having uncompleted d-shells, can provide a domination of free electrons in high-temperature conductivity regarding simultaneous blocking of ion hopping transport by Ca^{++} and Mg^{++} ions.

Let us estimate an order of electron conductivity value. As far as electron transport is provided through extended states of conduction band, which are not far from a mobility edge, the extreme disorder of electron scattering takes place, where between any two acts of scattering the phase coherence of electron wave turned out to be lost. In this case, according to Mott [8, 10] one may expect that pre-exponent coefficient ρ_0 will correspond to the minimum metallic conductivity quantity,

$$\sigma_{\min} = \frac{Ae^2}{\hbar\delta}, \tag{4}$$

where A is the numerical coefficient being 0.026, if coordination number is 6 and vertical disorder dominates; δ is the average distance between impurity centres being $\approx 0.5539N^{-1/3}$ in a case of their random spatial distribution. Here N is their concentration. One can find from Fig. 5 data that $\rho_0^{-1} = \sigma_0 \approx 20\,\Omega^{-1}m^{-1}$. This value, however, is smaller, than a typical one for both glasses having electron conductivity [2] and amorphous semiconductors [8]. If one compares that value with (4) relation, it will give $N \approx 10^{13}$ cm^{-3}, which is a very small value. Such a small N value can be explained by self-compensation phenomena, which are frequently observable in semiconductors having deep captures (for example, vacancies and interstitial atoms) [14]. Another approach based on the consideration that the typical drift mobility value μ for glasses, when extended states transport occurs (near a mobility edge), is of the order of 1 cm$^2/V \cdot s$. Then, taking into account that $\sigma_0 = Ne\mu$, one can obtain $N \approx 10^{18}$ cm^{-3}, which is more reasonable and provide the density of states ($T = 550$ K) about $2 \cdot 10^{19}$ cm^{-3}eV^{-1}.

As one can identify from Fig. 5, at $T < 350$ K we do have another conductivity mechanism. In the temperature interval 350 K $< T <$ 270 K it can be described by the activation law,

$$\rho(T) = \rho_2 \exp(E_2/kT), \tag{5}$$

where the corresponding activation energy is 0,47 eV. The pre-exponent co-efficient for conductivity is three-orders smaller than σ_0 value, which points indirectly to the conductivity mechanism through localized states. It is a well-known mechanism for numerous glasses and amorphous semiconductors, when activation occurs from the Fermi level to the localized states belonging to a conduction (and valence) band tails. A conventional convincing proof for such conductivity mechanism is a noticeable conductivity increase even in moderate electric fields. There was measured the volt-ampere characteristic at $T = 294$ K, which belongs to the above temperature interval. Results are presented in Fig. 7.

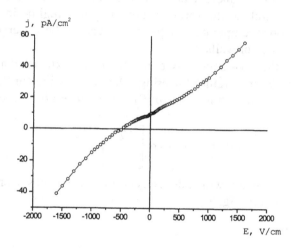

Figure 7. LFCM volt-ampere characteristic at $T = 294$ K. One can identify non-ohmic regime connected with hopping conductivity in the band tails.

One can see in Fig. 7 a noticeable deviation of the presented characteristic from linear law since ≈ 500 V/cm. The deviation corresponds to non-ohmic hopping conductivity regime on localized states belonging to the conduction band tail and become sufficient when $eER \approx kT$, where R is the hopping range. Rough estimation from Fig. 7 data provides $R \approx 250$ nm. In order to identify the physical sense of such R value, one should estimate the density of localized states in the band tail located in the energy interval between 0.47 eV and 0.74 eV. Considering that the band tail width is $\Delta E \approx E_1 - E_2 \approx 0.27$ eV, which is connected with Coulomb interaction between neighboring localized states, one can use the known relation, which is commonly used for impurity band in semiconductors [11], namely,

$$\Delta E \approx \frac{e^2 N^{1/3}}{4\pi\varepsilon_0\varepsilon}, \tag{6}$$

where N is the localized state concentration, ε_0 and ε are the dielectric constant and LFCM dielectric permeability, correspondingly. Taking into account $\varepsilon \approx 10$ for LFCM [12], one can establish from (6) the $N \approx 2 \cdot 10^{22}$ cm^{-3} and the corresponding density of states will be about $8 \cdot 10^{22}$cm^{-3}eV^{-1}. These are the upper-limit estimations as far ΔE may be increased due to compensation processes [11], but no doubt the order of values is correct. The average distance between neighboring localized states will be $\delta \approx 3$ nm at the above concentration. One can conclude hear that the above mentioned hopping range of 250 nm corresponds to the variable range hopping regime in the band tail transport, but proper temperature dependence cannot be observed while E_2 activation mechanism prevails.

At $T < 270$ K the E_2 – conductivity mechanism is exhausted and activation energy decays with temperature lowering. Data processing for 270 K $< T <$ 80 K temperature region shows that $\rho(T)$ dependence obeys the well-known Mott law,

$$\rho(T) = \rho_m \exp\left[\left(\frac{T_0}{T}\right)^{1/4}\right], \tag{7}$$

where T_0 is the parameter, which depends on localized states overlapping. The results are presented in Fig. 8.

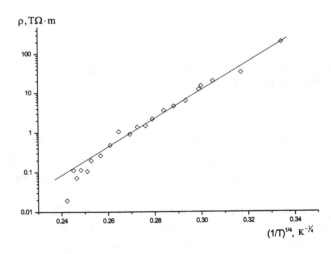

Figure 8.   Dependence of low-temperature LFCM specific resistance on temperature. VRH-regime indicated by the straight line.

One can detect from this figure that (7) dependence covers more than three orders of $\rho(T)$ value. From the straight line slope it is possible to determine $T_0 \approx 4 \cdot 10^7$ K value. No doubt, this is a variable range hopping process, which occurs in a narrow band of the order of kT width in the Fermi level vicinity. There were provided special calculations for T_0 value [8,11,13] based on the percolation approach. Those calculations give the similar result, which is

$$T_0 \approx (16.9 \div 17.5)\frac{\gamma^3}{N(E_F)}, \tag{8}$$

where γ^{-1} is the localization radii for screened Coulomb potential of $\exp(-\gamma r)$ type, $N(E_F)$ is the density of localized states on the Fermi level and numerical coefficient in the round brackets originates from percolation considerations. Then, if we suggest $\gamma^{-1} \approx 0.5$ nm as a reasonable value for numerous amorphous semiconductors and oxide glasses [13], one can establish from (8) relation, taking into account the experimental T_0 value, that $N(E_F) \approx 2 \cdot 10^{19}$ cm^{-3}eV^{-1}. At last, let us give the rough lower-value estimation for possible energy width for the stripe of localized states on the Fermi level. Thorough calculation of proper percolation problem [11], we can estimate an upper temperature edge; below that edge the conductivity mechanism (7) should dominate. That estimation gives,

$$kT_c \approx 0.3 \cdot \Delta E_f, \tag{9}$$

where ΔE_f is the energy width and T_c is the corresponding temperature edge. One can easily identify from Fig. 8 data that $T_c \approx 260$ K, so one can obtain from (9) the $\Delta E_f > 0.07$ eV. In the case of classical gapless glasses, where the mobility edges exist only, one should expect (at $T > T_c$) the hopping process with low activation energy (of the order of 0.1 eV) observation, connected with nearest-neighbor hopping (usual ε_3-conductivity [11]). We can observe, however, the E_2 value only, which is not in coincidence with expected hypothetic ε_3 one. There is one distinct physical explanation for such divergence – the localized states having $E > E_2$ energy and the ones situated just near the Fermi level simply belong to the different bands of localized states. It means, that the density of states between E_2 and the Fermi level (except ΔE_f) is close to zero, i.e. that the real energy gap does exist between them.

Finally, the lowest estimation for localized states on the Fermi level concentration will be (one may consider $N(E_F)$ to be practically constant regarding the good obeying of Fig. 8 experimental data to the (7) law) the $N(E_F) \cdot \Delta E_f \approx 1.5 \cdot 10^{18}$ cm^{-3} value.

4. Summary

The heap of the above described experimental results makes it possible to determine preliminarily the LFCM energy structure and, hence, to classify it as a condensed matter sort. The suggested LFCM energy structure is presented in Fig. 9.

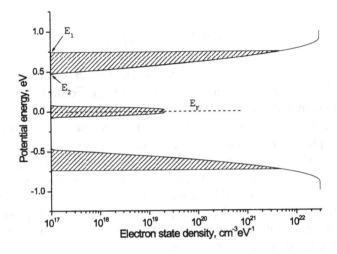

Figure 9. Suggested LFCM energy structure. Localized states are identified by the dashed area.

The main outcomes are as follows:

1 LFCM are the alkali-earth silicate glasses, whose vitrification temperature belongs to 1000 K ÷ 1100 K temperature interval.

2 LFCM is an disordered semiconductor having an energy gap which formed due to a short-range order and a wide band tails connected, as it is common for glasses, with horizontal disordering (see Fig. 9), which, in its turn, may be connected with numerous traps and radiation damages. Estimated thermal energy gap is 1.5 eV, possible optical energy gap at a room temperature is 1.8 ÷ 2.0 eV. There is a noticeable concentration of localized states just on a Fermi level.

3 There are no evidences for ionic transport, which can be explained by its blocking by alkali-earth ions, such as Ca^{++} and Mg^{++}. At 1000 K < T < 350 K the main mechanism responsible for electric transport is a thermal activation of electrons on the mobility edge E_1. Such type of conductivity (pre-exponent coefficient σ_0 value) is anomalously low, which can be explained by self-compensation processes. Another possible factor for it might be a non-equilibrium energy spectrum for free

electrons in a conduction band originated from fast β-electrons and a great amount of secondary electrons, which leads to chaotization of free electrons movement and prevents their directed movement along the external electric field. At 350 K $< T <$ 270 K an electric transport is provided by hopping conductivity in the band tails, whose energy threshold is E_2. The high localized states density can be provided by U^{4+} and Fe^{3+} ions, having changeable valence, whose concentration is about 1% (atomic). At $T <$ 260 K the variable range hopping in a stripe of order kT around a Fermi level was identified as the dominating mechanism.

4 An internal irradiation process leads to the appearance of self-sustained currents; observable external current corresponds to internal radioactivity, as far as a surface Coulomb barrier for secondary electrons exists. Such a barrier can be destroyed by heating due to electric conductivity enlarging, which results in the involvement of secondary electrons in electric transport. The thermal-stimulated current, connected with traps liberation can be also observed at a fast heating. Spontaneous electric fields, connected with internal currents, can also be observed. Their strength achieves 1 kV/cm and more at a room temperature.

5 The distinguishing feature of LFCM is low activation energy for electron ionization, which can provide a high spatial density of electron excitations in α-particles tracks and can lead to the group of phenomena, connected with the so-called Coulomb explosion. The latter item is a subject of a special interest [14,15].

References

[1] Pazukhin, E.M. (1997). *Lava-like Fuel-containing Masses of the Unit 4 of Chornobyl NPP: Physical and Chemical Properties, Formation Scenario, Impact on the Environment,* (in Russian). Thesis for doctor's degree in engineering sciences, ISTC "Shelter" NASU, Chornobyl.

[2] Feltz, A. (1983). *Amorphe und Anorganische Festkörper.* Berlin: Akademie-Verlag.

[3] Baryakhtar, V., Gonchar, V., Zhidkov, A., and Zhydkov, V. *Condens. Matter Phys.,* 2002, **5**, No. 3(31), p. 449–471.

[4] Gonchar, V.V., Zhidkov, A.V., and Maslov, D.M. Electron emission from LFCM surface and qualitative picture for electric field distribution in its near-surface layer. *Problems of Chornobyl,* 2004, **14**, (in Ukrainian).

[5] Bronstein, I.M., and Fraiman, B.S. (1969). *Secondary Electron Emission,* (in Russian). Moskow: Nauka.

[6] Alexankin, V.G., Rodichev, S.V., Rubtsov, P.M. et al. (1989). *Beta- and Antineutrinic Radiation of Radioactive Nuclei,* (in Russian). Reference book. Moscow: Energoatomizdat, p. 800.

[7] Milnes, A.G. (1977). *Deep Impurities in Semiconductors.* New York: John Wiley & Sons.

[8]   Mott, N.F., and Davis, E.A. (1979). *Electron Processes in Non-Crystalline Materials*. Oxford: Clarendon press.

[9]   Zhydkov, V., and Chemersky, G. Lava-like fuel-containing materials dynamic elastic constants determination in the workbench, (in Ukrainian). *Problems of Chornobyl*, 2004, **14**.

[10] Mott, N.F. (1974). *Metal-Insulator Transitions*. London: Taylor and Francis LTD.

[11] Shklovskii, B.I., and Efros, A.L. (1983). *Electronic Properties of Doped Semiconductors*. Springer.

[12] Zhydkov, O. (supervisor) et al. (2003). *Fuel-Containing Materials Electrodynamics Properties Study when Creating the Technology for its Separation from other "Shelter" Object Radioactive Waste*, (in Ukrainian). Annual Research Report of Interdisciplinary Scientific and Technical Centre "Shelter". Chornobyl, Government issue 0101U002570.

[13] Bonch-Bruevich, V.L. et al. (1981). *Electron Theory for Disordered Semiconductors*, (in Russian). Moscow: Nauka.

[14] Baryakhtar, V.G., ed. (1998). *Solid State Physics (Encyclopaedia dictionary)*, (in Russian). Kiev.

[15] Zhydkov, V. Coulomb explosion and steadiness of high-radioactive silicate glasses. *Condens. Matter Phys.*, 2004, **7**, No. 4, p. 845.

PARTICIPANTS

ARW on Ionic Soft Matter
Lviv, April 14-17, 2004

Jose ALEJANDRE
Departamento de Quimica
Universidad Autonoma
Metropolitana/Iztapalapa
Mexico D. F., Mexico
jra@xanum.uam.mx

Jean-Pierre BADIALI
Laboratorie D'Electrochimie
Analytique
Universite P. et M. Curie
4 Place Jussieu 75230
Paris Cedex 05, France
jpbadiali@wanadoo.fr

Mathias BALLAUFF
Physikalische Chemie I
Universität Bayreuth
95440 Bayreuth, F.R. Germany
matthias.ballauff@uni-bayreuth.de

Josef BARTHEL
Institut für Physikalische
und Theoretische Chemie
Universität Regensburg
93040 Regensburg, F.R. Germany
josef.barthel@chemie.uni-regensburg.de

Dezshő BODA
Department of Physical
Chemistry
University of Veszprém
H-8201 Veszprém, Hungary
boda@almos.vein.hu

Taras BRYK
Institute for Condensed
Matter Physics
National Academy of Science
of Ukraine
Lviv 79011, Ukraine
bryk@icmp.lviv.ua

Jean-Michel CAILLOL
LPT - CNRS (UMR 8627)
Bâtiment 210
Université de Paris Sud
F-91405 Paris, France
Jean-Michel.Caillol@th.u-psud.fr

Kwong-Yu CHAN
Department of Chemistry
The University of Hong Kong
Pokfulam Road, Hong Kong
hrsccky@hkucc.hku.hk

Alina CIACH
Institute of Physical Chemistry
Polish Academy of Sciences
ul. Kasprzaka 44/52
01-224 Warsaw, Poland
aciach@ichf.edu.pl

Maksym DRUCHOK
Institute for Condensed
Matter Physics
National Academy of Science
of Ukraine
Lviv 79011, Ukraine
maksym@icmp.lviv.ua

Bob EISENBERG
Department of Molecular
Biophysic and Physiology
Rush Medical College
Chicago IL 60612, USA
beisenbe@rush.edu

Christian von FERBER
Freiburg University
79104 Freiburg, F. R. Germany
ferber@uni-freiburg.de

Ron FAWCETT
Department of Chemistry
University of California
Davis CA 95616, USA
fawcett@indigo.ucdavis.edu

Anthony HAYMET
Department of Chemistry
University of Houston
Houston TX 77204, USA
haymet@uh.edu

Barbara HRIBAR
Faculty of Chemistry
and Chemical Technology
University of Ljubljana
Snezniska 5
1000 Ljubljana, Slovenia
barbara.hribar@uni-lj.si

Douglas HENDERSON
Department of Chemistry
and Biochemistry
Brigham Young University
Provo UT 84604, USA
doug@chem.byu.edu

Myroslav HOLOVKO
Institute for Condensed
Matter Physics
National Academy of Science
of Ukraine
Lviv 79011, Ukraine
holovko@icmp.lviv.ua

Fumio HIRATA
Institute for Molecular Science
Myodaiji, Okazaki 444-8585
Japan
hirata@ims.ac.jp

Yurij HOLOVATCH
Institute for Condensed
Matter Physics
National Academy of Science
of Ukraine
Lviv 79011, Ukraine
hol@icmp.lviv.ua

Julia KHALACK
Division of Physical Chemistry
Arrhenius Laboratory
Stockholm University
S-106 91 Stockholm, Sweden
julia@physc.su.se

Karl HEINZINGER
Menzelstrasse 2
D-55127 Mainz, F.R. Germany
khh@mpch-mainz.mpg.de

Olexandre IVANKIV
Institute for Condensed
Matter Physics
National Academy of Science
of Ukraine
Lviv 79011, Ukraine
oiva@icmp.lviv.ua

Volodymyr JAKIMCHUK
Institute for Postgraduate
Education
Odessa National University
Odessa 65026, Ukraine
jakimchuk@ukr.net

Svyatoslav KONDRAT
Institute of Physical Chemistry
Polish Academy of Sciences
ul. Kasprzaka 44/52
01-224 Warsaw, Poland
valiska@leon.ichf.edu.pl

Yurij KALYUZHNIY
Institute for Condensed
Matter Physics
National Academy of Science
of Ukraine
Lviv 79011, Ukraine
yukal@icmp.lviv.ua

Alexander KOBRYN
Institute for Molecular Science
Myodaiji, Okazaki 444-8585
Japan
alex@ims.ac.jp

Jiri KOLAFA
Department of Physical
Chemistry
Institute of Chemical
Technology
16628 Prague, Czech Republic
kolafa@cesnet.cz

Hartmut KRIENKE
Institut für Physikalische
und Theoretische Chemie
Universität Regensburg
93040 Regensburg, F.R. Germany
hartmut.krienke@chemie.uni-regensburg.de

Viktor LYKAH
National Technical University
'Kharkiv Polytechnic'
Kharkiv 61002, Ukraine
lykah@kpi.kharkov.ua

Volodymyr MARENKOV
Department of Thermophysics
Odessa National University
Odessa 65026, Ukraine
maren0@ukr.net

Miguel MOLERO
Departamento de Química
Física
Universidad de Sevilla
41071 Sevilla, Spain
molero@us.es

Yurij MEDVEDEVSKIKH
Pisarzhevsky Institute
of Physical Chemistry
National Academy of Sciences
of Ukraine
Lviv 79053, Ukraine
vfh@org.lviv.net

Roman MELNYK
Institute for Condensed
Matter Physics
National Academy of Sciences
of Ukraine
Lviv 79011, Ukraine
romanr@icmp.lviv.ua

Ihor MRYGLOD
Institute for Condensed
Matter Physics
National Academy of Science
of Ukraine
Lviv 79011, Ukraine
mryglod@icmp.lviv.ua

Timea NAGY
Department of Physical
Chemistry
University of Veszprem
H-8201 Veszprem, Hungary
Hungary
nagyt@almos.vein.hu

Ivo NEZBEDA
E. Hala Laboratory
of Thermodynamics
Institute of Chemical Process
Fundamentals
Czech Academy of Sciences
16502 Prague, Czech Republik
ivonez@icpf.cas.cz

Ihor OMELYAN
Institute for Condensed
Matter Physics
National Academy of Science
of Ukraine
Lviv 79011, Ukraine
omelyan@icmp.lviv.ua

Said ODINAEV
Tajik Technical University
Dushanbe 734025, Tajikistan
ods@ttu.tajik.net

Chris OUTHWAITE
Department of Applied
Mathematics
The University of Sheffield
Sheffield S3 7RH
United Kingdom
c.w.outhwaite@sheffield.ac.uk

Oksana PATSAHAN
Institute for Condensed
Matter Physics
National Academy of Science
of Ukraine
Lviv 79011, Ukraine
oksana@icmp.lviv.ua

Taras PATSAHAN
Institute for Condensed
Matter Physics
National Academy of Science
of Ukraine
Lviv 79011, Ukraine
tarpa@icmp.lviv.ua

Montgomery PETTITT
Department of Chemistry
University of Houston
Houston TX 77204, USA
pettitt@uh.edu

Stefan SOKOLOWSKI
Department for the Modelling
Physico-Chemical Processes
MCS University
Lublin 20031, Poland
stefan@zool.umcs.lublin.pl

Xueyu SONG
Department of Chemistry
Iowa State University
Ames IA 50011, USA
xsong@iastate.edu

Orest PIZIO
Instituto de Quimica
Univeridad National
Autonoma Mexico
Coyoacan 04510
Mexico D.F., Mexico
pizio@servidor.unam.mx

Roland ROTH
Max-Planck-Institut
für Metallforschung
Institut für Theoretische
und Angewandte Physik
Universität Stuttgart
70569 Stuttgart, F.R. Germany
roland.roth@mf.mpg.de

Wolffram SCHRÖER
Institut für Anorganische
und Physikalische Chemie
Universität Bremen
28359 Bremen, F.R. Germany
schroer@uni-bremen.de

Eckhard SPOHR
IWV-3
Forschungszentrum Juelich
52425 Juelich, F.R. Germany
e.spohr@fz-juelich.de

Janusz STAFIEJ
Department of Electrode
Processes
Institute of Physical Chemistry
Polish Academy of Sciences
01224 Warsaw, Poland
accjst@ichf.edu.pl

Andrij TROKHYMCHUK
Institute for Condensed
Matter Physics
National Academy of Science
of Ukraine
Lviv 79011, Ukraine
Brigham Young University
Provo UT 84602, USA
adt@icmp.lviv.ua

Mykhaylo TOKARCHUK
Institute for Condensed
Matter Physics
National Academy of Science
of Ukraine
Lviv 79011, Ukraine
mtok@icmp.lviv.ua

Ihor STASYUK
Institute for Condensed
Matter Physics
National Academy of Science
of Ukraine
Lviv 79011, Ukraine
ista@icmp.lviv.ua

Leonid SUKHODUB
Institute of Applied Physics
National Academy of Science
of Ukraine
Sumy 40030, Ukraine
isukhodub@yahoo.com

Tomaz URBIČ
Faculty of Chemistry
and Chemical Technology
University of Ljubljana
Ljubljana 1000, Slovenia
tomaz.urbic@uni-lj.si

Zoriana USATENKO
Institute for Condensed
Matter Physics
National Academy of Sciences
of Ukraine
Lviv 79011, Ukraine
pylyp@icmp.lviv.ua

Eduard VAKARIN
Laboratorie D'Electrochimie
Analytique
Universite P. et M. Curie
4 Place Jussieu 75230
Paris Cedex 05, France
vakarin@ccr.jussieu.fr

Mónika VALISKÓ
Department of Physical
Chemistry
University of Veszprém
H-8201 Veszprém, Hungary
valisko@almos.vein.hu

Vojko VLACHY
Faculty of Chemistry
and Chemical Technology
University of Ljubljana
Ljubljana 1000, Slovenia
vojko.vlachy@uni-lj.si

Ihor YUKHNOVSKII
Institute for Condensed
Matter Physics
National Academy of Sciences
of Ukraine
Lviv 79011, Ukraine
yukhn@icmp.lviv.ua

Anatoly ZAGORODNY
Bogolyubov Institute
for Theoretical Physics
National Academy of Science
of Ukraine
Kyiv 03143, Ukraine
azagorodny@bitp.kiev.ua

Oleksandr ZHYDKOV
Interdisciplinary Scientific
and Technical Centre "Shelter"
National Academy of Sciences
of Ukraine
Chornobyl 07270, Ukraine
ortmeiaj@slavutich.kiev.ua

Volodymyr ZHYDKOV
Interdisciplinary Scientific
and Technical Centre "Shelter"
National Academy of Sciences
of Ukraine
Chornobyl 07270, Ukraine
zhidkov@ukrpack.net

Index